# Acknowledgment

*Author, editors, publisher and the Carl Friedrich von Weizsäcker Foundation distinctly express their gratitude to the*

UDO KELLER STIFTUNG FORUM HUMANUM

*for its generous financial support, without which the present English translation of C. F. von Weizsäcker's* Aufbau der Physik *would not have been possible.*

*The* Udo Keller Stiftung Forum Humanum, *which is based in Neversdorf, Schleswig-Holstein, Germany, is concerned with crucial questions of human life putting an emphasis on religiousness and spirituality. The Foundation does not follow one particular outlook or confession, but rather tries to gain a deeper understanding of the limits, contradictions and possibilities of human knowledge. The aim of the organization is to support a dialog between a spiritual outlook, the natural sciences and the views of formal religion.*

*In line with this goal the* Udo Keller Stiftung Forum Humanum *supports a variety of projects and cooperates in particular with the Foundation Weltethos led by Professor Hans Küng, the Carl Friedrich von Weizsäcker Foundation and the Carl Friedrich von Weizsäcker Society.*

# The Structure of Physics

# Fundamental Theories of Physics

*An International Book Series on The Fundamental Theories of Physics:
Their Clarification, Development and Application*

**Editor:**
ALWYN VAN DER MERWE, *University of Denver, U.S.A.*

**Editorial Advisory Board:**
GIANCARLO GHIRARDI, *University of Trieste, Italy*
LAWRENCE P. HORWITZ, *Tel-Aviv University, Israel*
BRIAN D. JOSEPHSON, *University of Cambridge, U.K.*
CLIVE KILMISTER, *University of London, U.K.*
PEKKA J. LAHTI, *University of Turku, Finland*
FRANCO SELLERI, *Università di Bara, Italy*
TONY SUDBERY, *University of York, U.K.*
HANS-JÜRGEN TREDER, *Zentralinstitut für Astrophysik der Akademie der Wissenschaften, Germany*

# The Structure of Physics

*by*

Carl Friedrich von Weizsäcker

*edited, revised and enlarged by*

Thomas Görnitz
University of Frankfurt, Germany

and

Holger Lyre
University of Bonn, Germany

A C.I.P. Catalogue record for this book is available from the Library of Congress.

ISBN-10  1-4020-5234-0 (HB)
ISBN-13  978-1-4020-5234-7 (HB)
ISBN-10  1-4020-5235-9 (e-book)
ISBN-13  978-1-4020-5235-4 (e-book)

Published by Springer,
P.O. Box 17, 3300 AA Dordrecht, The Netherlands.

*www.springer.com*

*Printed on acid-free paper*

Original version: *Aufbau der Physik,* Hanser Verlag, Munich, 1985.
Translated into English by Helmut Biritz, Georgia Institute of Technology, School of Physics, Atlanta, USA.

Figures 6.7, 6.8, 6.9 and 6.10 (pp. 165–166) from Tonry, J.L. et al., *Astrophysical Journal* **594** (2003), 1-24, have been used with the kind permission of Dr. B. Leibundgut and the Astrophysical Journal.

All Rights Reserved
© 2006 Springer
No part of this work may be reproduced, stored in a retrieval system, or transmitted
in any form or by any means, electronic, mechanical, photocopying, microfilming, recording
or otherwise, without written permission from the Publisher, with the exception
of any material supplied specifically for the purpose of being entered
and executed on a computer system, for exclusive use by the purchaser of the work.

Albert Einstein

Niels Bohr

Werner Heisenberg

# Contents

Editors' Preface ................................................. xi

Preface (1985) ................................................. xiii

On Weizsäcker's philosophy of physics (by H. Lyre) ............ xix

1   **Introduction** ................................................. 1
    1.1   The question ................................................. 1
    1.2   Outline ..................................................... 2

## Part I The unity of physics

2   **The system of theories** ....................................... 13
    2.1   Preliminary ................................................. 13
    2.2   Classical point mechanics ................................... 16
    2.3   Mathematical forms of the laws of nature .................... 28
    2.4   Chemistry ................................................... 31
    2.5   Thermodynamics .............................................. 33
    2.6   Field theories .............................................. 35
    2.7   Non-Euclidean geometry and semantic consistency ............. 35
    2.8   The relativity problem ...................................... 37
    2.9   Special theory of relativity ................................ 41
    2.10  General theory of relativity ................................ 44
    2.11  Quantum theory, historical ................................. 51
    2.12  Quantum theory, plan of reconstruction ..................... 54

## 3  Probability and abstract quantum theory .................. 59
3.1  Probability and experience .................................. 59
3.2  The classical concept of probability ....................... 62
3.3  Empirical determination of probabilities .................. 66
3.4  Second quantization ....................................... 68
3.5  Methodological: Reconstruction of abstract quantum theory .. 71
3.6  Reconstruction via probabilities and the lattice of propositions  73

## 4  Quantum theory and spacetime ........................... 81
4.1  Concrete quantum theory ................................. 81
4.2  Reconstruction of quantum theory via variable alternatives ... 85
4.3  Space and time ............................................ 93

## 5  Models of particles and interaction ....................... 105
5.1  Open questions ............................................ 105
5.2  Representations in tensor space ........................... 109
5.3  Quasiparticles in rigid coordinate spaces ................. 117
5.4  Model of quantum electrodynamics ....................... 123
5.5  Elementary particles ...................................... 131
5.6  General theory of relativity .............................. 140

## 6  Cosmology and particle physics (by Th. Görnitz) .......... 149
6.1  Quantum theory of abstract binary alternatives and cosmology 149
6.2  Ur-theoretic vacuum and particle states .................. 169
6.3  Relativistic particles ..................................... 175
6.4  Outlook .................................................. 177

---

## Part II  Time and information

## 7  Irreversibility and entropy ............................... 181
7.1  Irreversibility as problem ................................ 181
7.2  A model of irreversible processes ........................ 187
7.3  Documents ............................................... 194
7.4  Cosmology and the theory of relativity ................... 201

## 8  Information and evolution ................................ 211
8.1  The systematic place of the chapter ...................... 211
8.2  What is information? .................................... 212
8.3  What is evolution? ....................................... 215
8.4  Information and probability .............................. 215
8.5  Evolution as growth of potential information ............. 218
8.6  Pragmatic information: Novelty and confirmation .......... 229
8.7  Biological preliminaries to logic ......................... 234

## Part III On the interpretation of physics

**9  The problem of the interpretation of quantum theory** ...... 243
    9.1  About the history of the interpretation .................... 243
    9.2  The semantic consistency of quantum theory ............... 260
    9.3  Paradoxes and alternatives ............................. 276

**10  The stream of information** ................................ 297
    10.1  The quest for substance ................................ 297
    10.2  The stream of information in quantum theory .............. 300
    10.3  Mind and form ........................................ 306

**11  Beyond quantum theory** .................................. 311
    11.1  Crossing the frontier ................................... 311
    11.2  Facticity of the future .................................. 316
    11.3  Possibility of the past .................................. 321
    11.4  Comprehensive present ................................. 327
    11.5  Beyond physics ........................................ 330

**12  In the language of philosophers** ........................... 333
    12.1  Exposition ............................................ 333
    12.2  Philosophy of science .................................. 334
    12.3  Physics .............................................. 337
    12.4  Metaphysics .......................................... 342

**References** .................................................. 347

**Index** ...................................................... 353

# Editors' Preface

Carl Friedrich von Weizsäcker is certainly one of the most distinguished German physicists and philosophers of the 20th century—equally renowned for his early contributions to nuclear physics and his life-long research on the foundations of quantum theory. At the same time, Weizsäcker is highly esteemed by a much broader audience for his sociocultural, political, and religious thought. His writings comprise more than 20 books, many of which have been translated into several languages.

But throughout his life, Weizsäcker's main concern was an understanding of the unity of physics. For decades he and his collaborators have been pursuing the idea of a quantum theory of binary alternatives (so-called *ur theory*), a unified quantum theoretical framework in which spinorial symmetry groups are considered to give rise to the structure of space and time. *Aufbau der Physik*, first published exactly 20 years ago, in 1985, and followed by numerous reprints, was primarily intended to give an overview and update of this enterprise. But the book was only published in German, and thus could scarcely have contained the subsequent insights and results of ur-theoretic research of the late 1980s and the 1990s, due mainly to the work of Thomas Görnitz. These circumstances were the main incentive for producing the present edition, which is a newly arranged and revised version of the *Aufbau*, translated into English, in which some original chapters and sections have been skipped, and a new chapter on ur theory and a general introduction to Weizsäcker's philosophy of physics have been added. A comparison of the present book's structure to that of the original book can be found on page XIV, footnote 2).

*The Structure of Physics* should be of value to anybody with interests in physics, its history, or its philosophy, since it contains far more than the particular focus on ur theory in the central Chaps. 4, 5, and 6 of the first part. As a prominent eyewitness to the historical development of quantum mechanics, Weizsäcker's presentation of the system of physical theories in the second chapter and his way of presenting the general interpretive issues of quantum mechanics in Chap. 9 are both of special importance. Furthermore, Weizsäcker's discussion of time and information in the second part, along with his analyses in the last three chapters of the third part, reveal him to be an original and outstanding philosophical thinker.

We are very grateful to the many people and institutions without whom the present edition would have been impossible: the Kluwer and Springer publishing houses for adopting the project; the Carl Friedrich von Weizsäcker-Stiftung—in particular Bruno Redeker—for administrative support; and the Udo Keller Stiftung Forum Humanum for a generous donation. Special thanks are due to Helmut Biritz, who provided a careful translation of the *Aufbau* and who was both a pleasant and patient collaborator. It is our hope that this edition will help to make Weizsäcker's unique ideas in the philosophy of physics more accessible to the English-speaking world.

*Pentecost 2005*

*Thomas Görnitz*
*Holger Lyre*

# Preface (1985)

The book reports on an attempt to understand the unity of physics. This unity began to manifest itself in rather unexpected form in this century. The most important step in that direction was the development of quantum theory; the emphasis of this book is therefore on the endeavor to understand quantum theory. Here, *understand* refers not merely to practical application of the theory—in that sense it has been understood for a long time. It means being able to say what one does when applying the theory. This endeavor has led me, on the one hand, to reflect upon the foundations of probability theory and the logic of temporal propositions, and on the other to progress to what appears to me a promising attempt to generalize the theory in such a way that relativity and the basic ideas of elementary particle theory could be derived from it. If this attempt were successful, we would come one step closer to the actual unity of physics as an understood theory. The understanding of the unity of physics is on the other hand no doubt a prerequisite for insight into its philosophical meaning and its role in our endeavor to perceive the oneness of reality. This might finally be necessary if we wish to comprehend the significance of natural science in the cultural development of our times, as a key to deep, effective, and perilous insights.

I have placed the three names Albert Einstein, Niels Bohr, Werner Heisenberg at the head of the book. Einstein was the genius of the century. The theory of relativity is his work, and it was on his account that quantum theory got under way. All younger workers remain under the spell cast by his insights. Bohr was the inquiring master of atomic theory. He pressed onward into realms from which Einstein shut himself off; the completion of quantum theory was the handiwork of his followers. Heisenberg, with matrix mechanics, took the first steps on solid ground. Among the generation of the creators of quantum theory he was *primus inter pares*. As his equals one might perhaps mention Dirac, Pauli, and Fermi. The creation of the new physics was a collective undertaking. Indispensable work was carried out by Planck, who opened the door to quantum theory; by Rutherford, who in the experimental investigation of atoms was the master and teacher that his student Bohr became in theory; by Sommerfeld, de Broglie, and Schrödinger; by Born and Jordan; and by a great many more experimentalists.

For me, the mention of these three names also carries the personal significance of admiring and affectionate remembrance. I unfortunately never met Einstein, but his name was familiar to me by the time I was a schoolboy, and from decade to decade I learned to better understand his greatness. When I was nineteen years old, Bohr revealed to me the philosophical dimension of physics. He gave me what I had been looking for in physics. From him I learned to understand the influence that Socrates must have exerted over his followers. I had the good fortune to meet Heisenberg when I was fifteen. He brought me into physics, taught me its craft and its beauty, and became a lifelong friend.[1]

One might perhaps mention here an amusing play on round numbers: without being pre-planned as such, the present book will be published, almost to the day, on Bohr's one-hundredth birthday, October 7, 1985. Sixty years ago (Pentecost 1925) Heisenberg, while in Helgoland, discovered the foundations of quantum mechanics. Fifty years ago (1935) Einstein published his quantum mechanics thought experiment with co-authors Podolsky and Rosen.

As for the genesis of this book, when the investigations reported here began, the work of the pioneers had long since come to a close. Heisenberg told me as early as April 1927, two months after we first met, about his yet-unpublished uncertainty relations. From that time onward I wanted to study physics to understand quantum theory. But the longer I was a physicist the clearer it became to me that I still did not understand the theory. In 1954 I came to the conclusion that the classical horizon of thought must be transcended even in the realm of logic; about 1963 I realized that this had to do with the logic of time. Both steps were prepared. The central role of time became clear to me in a study of the second law of thermodynamics (1939), described in this book in Chap. 4.[2]

---

[1] I might very well mention here more elaborate accounts of the three: *Einstein* (1979), *Bohr und Heisenberg: Eine Erinnerung aus dem Jahr 1932* (1982), *Werner Heisenberg* (1977, 1985). References can be found in the bibliography.

[2] Editors' note: Weizsäcker refers to the original *Aufbau*, the present book has the following, different arrangement:
Chapter 1: Aufbau 1.1, 1.3,
Chapter 2: Aufbau 6
Chapter 3: Aufbau 3.1–3.3, 7.4, 8.1–8.2,
Chapter 4: Aufbau 9,
Chapter 5: Aufbau 10.1–10.2, 10.4–10.7,
Chapter 6: new (by Th. Görnitz),
Chapter 7: Aufbau 4,
Chapter 8: Aufbau 5.1–5.5, 5.7–5.8,
Chapter 9: Aufbau 11,
Chapter 10: Aufbau 12,

# Preface (1985)

I have written philosophical essays on quantum theory since 1931, with the more tenable ones being published in the book *Zum Weltbild der Physik* (1943, finished 1957, 7th edition). The path to the logical interpretation is now described in 7.7. Only after I had found this interpretation could I—that was my feeling—make firm progress. But the road was very long. In 1971 I published an interim report in the book *Die Einheit der Natur*, still only a collection of essays. Since then I have continued working steadily.

The length of the path was due in part to the difficulty of the subject matter, and in part to the limitations of my mathematical ability. Had more colleagues been interested in this research the mathematical problems could have been solved much sooner, but I could not arouse their curiosity. The path of this reflection lay beyond the successful line of approach of the topical research in physics. Even Heisenberg, who always wanted to stay informed on the progress and problems of my work, told me: "You are on a good track, but I cannot help you. I cannot think so abstractly." Success alone rouses the productive curiosity of scientists, and I needed the help of that curiosity before success could follow. On the other hand, the apparent distractions in my life due to politics and philosophy only slightly slowed the pace of this work. Philosophy was indispensable for a philosophically oriented analysis of physics; attempting to understand Plato, Aristotle, Descartes, Kant, Frege or Heidegger was no distraction at all from the main topic, and hence entailed no loss of time. Politics was a different matter. But for me it would have been morally impossible to do physics while ignoring political, probably catastrophic consequences of physical research. Politics cost me perhaps a total of ten working years, perhaps more. Yet alongside politics the work continued steadily; subconscious contemplation does not stop when other matters temporarily occupy the conscious mind. Worse, though, was the inevitability of political failure, given the prevailing denial of inherent risks.

The work is not finished. I am writing this account with the feeling that there is probably not much time left to me, partly on account of my age, and partly in view of the uncertain times. In contrast to *Einheit der Natur*, this book is designed as a single continuous train of thought. One shortcoming is its bulk. Apparently I had needed to portray many details and to follow many and varied alternative paths to attain a clear view of the entire subject, which might ultimately have enabled me to say everything in a fraction of the present scope. But, with novel thoughts, a more elaborate presentation might help the reader's comprehension. At any rate, I have never striven for that hermetical terseness so prevalent in mathematics.

The amount of material has led this report being divided into two books. The present book, appearing first, portrays in one direct progression the reconstruction of physics that I aspire to. I have also chosen *Aufbau der Physik* as its title. *Einheit der Physik* (The Unity of Physics) would have been factu-

---

Chapter 11: Aufbau 13,
Chapter 12: Aufbau 14.

ally more accurate, but I avoided that title solely to preclude confusion with *Einheit der Natur* (The Unity of Nature). A second book, under the title *Zeit und Wissen* (Time and Knowledge), will contain philosophical reflections. At present I am undecided as to whether that latter book will also be subdivided.

This book is a research report and not a textbook. It therefore requires of the reader certain prior knowledge of the topics under consideration. But I have taken pains to develop the physical and philosophical ideas broadly, and to avoid mathematical details as much as possible. An expert will be able to fill in mathematical details; they would remain incomprehensible to the layman. I do not deny, however, that in the verbal presentation, the only one I was capable of, there might be hidden unresolved mathematical problems that I myself have not sufficiently recognized. Chapters 1 to 6, 12, and 14 should be immediately readable by a natural scientist or philosopher reasonably familiar with physics. Chapters 7–11 and 13 assume a knowledge of quantum theory.

Material spanning about twenty years was available for this book. I have not attempted to write everything anew but used some of those materials verbatim. Hence there remains a certain unevenness, and repetitions of the same ideas in different contexts. Some of the texts are more pedagogically formulated, others are more like technical reports or programmatic. The reader will more easily orient himself by being able to keep them apart. For this I have identified each of the old texts according to their date of origin and first usage. In brief: Chaps. 2 and 4 are from a first draft of the book written in 1965, in the form of a lecture. In Chap. 3 the older formulation has been replaced by texts from around 1970. A few texts from the 1970s or reports of such are contained in Chaps. 5–7 and 12. Chapters 1, 8–10, 13, and 14 have been written anew. The texts are now incorporated into a continuous train of thought, with the exception of Chaps. 2–4, which were already coherent.

The investigations described here would not have been possible without decades of collaboration. The first more elaborate publication, in 1958, was coauthored by E. Scheibe and G. Süßmann. R. Ebert participated in the daily discussions at that time. The thesis of H. Kunsemüller contributed to the understanding of quantum logic. K. M. Meyer-Abich clarified the genesis and meaning of the basic concepts of N. Bohr. From 1965 through 1978 M. Drieschner carried out a significant part of the work on probability, irreversibility, and the axiomatic foundations of quantum theory. F. J. Zucker, during his stay in Germany, contributed substantially—along with philosophical ideas—to an understanding of the concept of information,, as did E. and C. v. Weizsäcker in the Heidelberg "Offene Systeme" discussion group. In America F. J. Zucker then established contacts, in part through an exemplary translation of *Einheit der Natur*. L. Castell provided an essential stimulus in 1968 and for all further investigations by introducing group-theoretical ways of thinking. From 1970 through 1984 he led the Starnberg group; essential parts of Chaps. 9–10 are reports on his work and that of his students. Among external contacts, discussions with H.-P. Dürr spanning decades were essential. In 1971 I encountered in D. Finkelstein the only physicist who, independently of us, had developed the same ideas about the relationship between quantum theory and spacetime continuum. Periodic contact for discussions followed. Several times, P. Roman was our guest in Starnberg for months, and he made the first and continuing contributions to the cosmological applications of ur theory. In recent years, I owe significant ideas on the problem of evolution to a discussion with H. Haken and B.O. Küppers; Regrettably, it was not possible to take into account a new book by K. Kornwachs. In Starnberg, the work was carried by K. Drühl, J. Becker, P. Jacob, F. Berdjis, P. Tataru-Mihaj, W. Heidenreich, Th. Künemund. In 1979, Th. Görnitz joined our working group; the present form of Chaps. 9 and 10 owes much to his significant new ideas, especially on the problem of space and the general theory of relativity. In exemplary fashion, Käte Hügel, Erika Heyn, Ruth Grosse, Traudl Lehmeier performed the thankless secretarial duties of a group that moved solely in abstract, unintelligible spheres. Without the dedicated efforts of Ruth Grosse, this book would not exist today.

*Pentecost 1985*                                                    *C. F. v. Weizsäcker*

# On Weizsäcker's philosophy of physics

*by Holger Lyre*

*Aufbau der Physik* appeared exactly twenty years ago in its first edition.[3] Weizsäcker considers it his physical–philosophical magnum opus—the fruit and quintessence of especially those of his papers that deal with a philosophically motivated program that bases the fundamental structures of physics based on a rigorous and consistent quantum theory of binary alternatives. The title of the program is "ur theory," and the *Aufbau* deals with it extensively. This introduction attempts to explain the basic ideas of ur theory, its rank in Weizsäcker's thinking, and why the present publication of the *Aufbau* in English is justified.

The *Aufbau* is the last in a series of physical–philosophical books Weizsäcker wrote during his lifetime:[4] *Die Atomkerne* 1937, *Zum Weltbild der Physik* 1943, *Die Geschichte der Natur* 1948, *Physik der Gegenwart* (with J. Juilfs) 1952, *Die Tragweite der Wissenschaft* 1964, and *Die Einheit der Natur* 1971. These books, however, are only some of his publications, as the full range of Weizsäcker's œuvre encompasses altogether four great subject areas: physics, philosophy, politics, and religion. Weizsäcker's publications in each of these areas alone would suffice to form the highly visible work of an outstanding scientist. In concert, however, they represent a life's work unmatched in its universality in the twentieth century. Nevertheless, physics always stood at the center of Weizsäcker's thinking. With physics he started out (as pupil of Heisenberg and Bohr), and to it he fully returned early in the 1980s, especially after the closing of his Max Planck Institute "Zur Erforschung der Lebensbedingungen der wissenschaftlich-technischen Welt" (Research into Conditions of Life in a Scientific and Technological World) in Starnberg. In between, there were important way stations of a scientist and *homo politicus*, beginning in 1942 as professor of nuclear physics in Strasbourg, and his indisputedly controversial participation in the "Uranverein" (the German atomic research project

---

[3] C. F. von Weizsäcker. *Aufbau der Physik*. Hanser, Munich, 1985.
[4] Cf. the list of main book publications of C. F. von Weizsäcker at page XXXII.

under pressure of the Nazis); rebuilding and group leader at the Max Planck Institute for Physics in Göttingen (where he conducted research on cosmogony and the theory of turbulence); the sensational Göttingen declaration of well-known German scientists late in the 1950s, opposing the atomic armament of the German army; the transition to a chair of philosophy in Hamburg ("an incomparable stroke of luck"); founding and directing the aforementioned institute at Starnberg in 1970; and finally, after his retirement in the early 1980s, returning full-time to the philosophy of physics, as witnessed by the publication of the *Aufbau*, and of his last and largest philosophical work *Zeit und Wissen*.[5] Weizsäcker received numerous international distinctions and honorary degrees; twice he declined when approached for the candidacy of Federal President of Germany. In physics textbooks one can find his name under headings such as Bethe–Weizsäcker mass formula, Bethe–Weizsäcker cycle, origin of the planetary system, and Weizsäcker–Williams approximation.

## Quantum information theory of urs

The *locus classicus* of ur theory,[6] Weizsäcker's basic framework of a philosophically motivated reconstruction of physics, is the essay on *complementarity and logic* (KL I) dated 1955.[7] It was followed in 1958 by the *quantum theory of the simple alternative* (KL II),[8] and the "three-men" paper on *multiple quantization* (KL III) co-authored by with Erhard Scheibe, and Georg Süßmann.[9] As early as KL I (p. 552) Weizsäcker had formulated the basic idea of his later theory:

> *The quantum logic of simple alternatives leads to a manifold of states, which can be assigned to the totality of directions in three-dimensional real space... This is the well-known mathematics of spinors. Neglecting normalization, one then obtains a manifold of states which can be assigned to that of points in three-dimensional space. I would suspect that the mathematical properties of actual physical space follow in this way from the logic of complementarity. The argument, which thus far I have not been able to formulate rigorously, uses the consistency postulate of logic for multiple quantization: If physics admits of*

---

[5] C. F. von Weizsäcker. *Zeit und Wissen*. Hanser, Munich, 1992.

[6] The German prefix *Ur* means *original, elementary*, or *pre-*.

[7] C. F. von Weizsäcker. Komplementarität und Logik. *Die Naturwissenschaften*, 42: 521–529, 545–555, 1955. (Reprinted in: *Zum Weltbild der Physik*. 7th edition, 1958).

[8] C. F. von Weizsäcker. Die Quantentheorie der einfachen Alternative (Komplementarität und Logik II). *Zeitschrift für Naturforschung*, 13 a: 245–253, 1958.

[9] C. F. von Weizsäcker, E. Scheibe, and G. Süßmann. Komplementarität und Logik, III. Mehrfache Quantelung. *Zeitschrift für Naturforschung*, 13 a: 705–721, 1958.

*simple alternatives at all, they always define, initially abstractly, three-dimensional spaces. Thus one must expect that there is a representation of physics in which it describes processes in three-dimensional real spaces, or perhaps in one such space.*

As Weizsäcker writes in a later autobiographical essay, the crucial idea occurred to him at a spa in Bad Wildungen in the autumn of 1954, "upon waking one morning at six o'clock."[10] An interesting previous hint, however, is to be found in an earlier short note from 1952.[11] There Weizsäcker points out the remarkable fact that the metrics of Hilbert space as well as position space are quadratic forms, and that this may indicate that the latter is a consequence of the former.

All in all, ur theory is based on two central assumptions:

1. The predictions of empirical science can be reduced to smallest units, binary alternatives, and permit a decomposition of state spaces into atoms of information (information-theoretical atomism).
2. The smallest possible nontrivial state space of quantum theory, a two-dimensional Hilbert space, permits a symmetry group which itself represents a three-dimensional space. Mathematically this is the well-known connection between spinors and tensors (spinorism).

In the 1950s, both assumptions were anything but self-evident, and were quite revolutionary. Even more remarkable is the fact that both themes play a central role in present-day fundamental physics. The first assumption, before the background of quantum theory, is nothing but an anticipation of the concept of qubits of present quantum information theory. Nevertheless, Weizsäcker goes in a decisive manner beyond the usual (quantum) information theory: he wants to consider the "abstract structure" of quantum theory as fundamental to the reconstruction of empirical science. Physics, in the sense of the general dynamics of objects in space and time, is therefore preceded by abstract quantum theory methodologically, epistemically, and as we will see, even ontologically. Philosophically speaking, abstract quantum theory consists of a catalog of the most general conditions for the possibility of empirical science. Here we see, taken over from Kant, the transcendental–philosophical character trait of Weizsäcker's thinking—abstract quantum theory comprises, so to speak, the *Metaphysical Foundation of Natural Science*[12] in the twentieth and twenty-first centuries.

What exactly is to be understood with abstract quantum theory will become apparent in Chap. 3, where Weizsäcker discusses various paths of reconstruction. In particular, the first path contains a recapitulation of the logical

---

[10] C. F. von Weizsäcker. *Der Garten des Menschlichen*, p. 562. Hanser, Munich, 1977.
[11] C. F. von Weizsäcker. Eine Frage über die Rolle der quadratischen Metrik in der Physik. *Zeitschrift für Naturforschung*, 7 a: 141, 1952.
[12] I. Kant. *Metaphysische Anfangsgründe der Naturwissenschaft*. Riga, 1786.

structure of quantum theory. It is well known that the set of subspaces of a Hilbert space form a nondistributive lattice, generally referred to as quantum logic. If one interprets quantum theory abstractly as the (meta-)theory of empirical theories, as Weizsäcker does, then the most general form of an empirical theory of predictions can be expressed in quantum logic—specifically, the structure of the lattice of empirically verifiable predictions or, in general, empirically decidable alternatives. The fact that abstract quantum theory can be interpreted as logic thus lends support to aprioristic intuition, the axioms of logic always being good candidates for synthetic judgments a priori.

We can indeed consider the aprioristic interpretation and justification of the structure of abstract quantum theory to be an additional assumption—one which methodologically comes before the two assumptions mentioned above. There are certain problems associated with this, which can merely be touched upon here. It is unfortunately not immediately evident whether the axioms of abstract quantum theory, like the ones presented in 3.2 and based on investigations by Michael Drieschner into the postulates of quantum logic, are immediately obvious a priori.[13] The very special structure of Hilbert space has yet to be exhaustively justified in this fashion. Secondly, Weizsäcker does not pursue a strict Kantianism: his method of the so-called *Kreisgang*[14] mixes a naturalistic strategy—the "semicircle" of man and his apparatus of perception being part of nature—with a reflection on the conditions which make naturalism possible—the "semicircle" of transcendental philosophy.[15] The details of this philosophical methodology cannot, however, be elaborated here; for present purposes we simply wish to start with the a priori character of abstract quantum theory in a heuristic sense.

At this point, the transition from abstract to "concrete" quantum theory is of interest. For in the abstract reconstruction most of what are usually considered central concepts of physics like "energy," "matter," and "interaction," along with "space" or "spacetime," have yet to be mentioned. Abstract quantum theory merely requires concepts like "system," "state," "state space," "transitions between states" (dynamical or due to an apparently discontin-

---

[13] M. Drieschner. *Voraussage–Wahrscheinlichkeit–Objekt. Über die begrifflichen Grundlagen der Quantenmechanik.* Springer, Berlin, 1979.
M. Drieschner, Th. Görnitz, and C. F. von Weizsäcker. Reconstruction of Abstract Quantum Theory. *International Journal of Theoretical Physics*, 27 (3): 289–306, 1988.

[14] Weizsäcker chose the word "Kreisgang" to characterize his overall philosophical method. The term is difficult to translate (and is not a common German notion, either), and will be used as a *terminus technicus* throughout the book. In its literal meaning it refers to a "circular movement" of knowledge and cognition. The largest circle possible is captured by Weizsäcker's often used phrase: *Nature is older than humankind, humankind is older than natural science*, which should indicate the inextricable intertwining of a naturalistic and a transcendental attitude.

[15] C. F. von Weizsäcker. *Zeit und Wissen.* Hanser, Munich, p. 29f, 543f, 1992.

uous "measurement,") and "observable." Ur theory represents just such a transition to concrete physics. The first assumption serves again as the point of departure: *all* alternatives which can empirically be decided at all are obtained in the context of abstract quantum theory. This also includes empirical decisions about positions in space and time. Thus the structure of space or spacetime itself ought to follow from abstract quantum theory.

Here a digression is in order. The structure of time, meaning the sequence of its modes of past, present, and future, can according to Weizsäcker's interpretation decidedly not be derived. Rather, it is one of the essential prerequisites of any empirical science whatsoever. If one does physics, an empirical science, then in Weizsäcker's opinion one tacitly already knows about the structure of time, for experience entails applying lessons learned from the facts of the past to the open questions of the future. The use of time as parameter-time—i.e., within the concept "spacetime"—is therefore to be distinguished from the asymmetric directedness of time. This basically corresponds to McTaggart's distinction between B- and A-series of time.[16] The two essential a priori assumptions of Weizsäcker's philosophy of physics may therefore be characterized as *temporality*—the distinction between factual past and open future, and *distinguishability*—the possibility of making distinctions within the empirically accessible domain, which is inherent in the concept of an alternative.[17]

To return to the derivation of space and spacetime from the quantum theory of binary alternatives—those atomic alternatives into which every complex alternative can in principle be decomposed—it is precisely this fact that led Weizsäcker to the idea that the quantum theory of binary alternatives (in modern terms, the theory of qubits) assumes a special role, as every version of physics had ultimately to be reducible to this abstract foundation, and thus ultimately to quantum information theory. Long before the introduction of the term *qubit*, Weizsäcker denoted the smallest possible building blocks of empirical sciences by the German word *Ur alternative* (*urs*, for short, and correspondingly *ur theory*). If the ur hypothesis is correct, the symmetry of urs must play a distinguished role in physics. At this point the second pillar of the ur theoretic structure comes in: the quantum theory of alternatives, ur theory, *is* the theory of three-dimensional space.

Mathematically, Weizsäcker had come across the known fact that $SU(2)$, the basic symmetry group of urs, is locally isomorphic to $SO(3)$, the group of rotations in space, as mentioned in the introductory quote. This idea was then subsequently developed in various directions. In the papers KL I–II Weizsäcker essentially attempts to justify $SL(2, \mathbb{C})$, the unimodular group in the space of two-spinors, in terms of quantum logic, and then to interpret its mathematical relationship to the homogeneous Lorentz group $SO(1,3)$ as a physical

---

[16] J. M. E. McTaggart. The unreality of time. *Mind*, 17 (68): 457–474, 1908.

[17] H. Lyre. *Quantentheorie der Information*. Springer, Wien, 1998 (2nd ed. Mentis, Paderborn, 2004).

derivation of special relativity from the quantum theory of binary alternatives. In the ur theoretic path of the reconstruction, detailed in the present Chap. 4, a slightly different strategy is employed, but with the same basic motive of justifying space in terms of quantum theory. Following general custom, one now starts with normalized vectors in Hilbert space, and the largest possible symmetry group of urs then encompasses the groups

$$SU(2),\ U(1)\ \text{and}\ \mathbb{K}, \qquad (0.1)$$

where $\mathbb{K}$ represents complex conjugation. Weizsäcker uses the fact that $SU(2) = \mathbb{S}^3$, i.e., that the basic symmetry group of the ur itself is a three-dimensional manifold. The basic assumption of ur theory means then that $\mathbb{S}^3$ represents the simplest position-space model of the universe. Thomas Görnitz, having analyzed the regular representations of $SU(2)$ in more fully developed mathematical form (Sect. 6.1), was able to combine this with equally central ur theoretic discussions of the physics of large numbers.[18] This will be addressed in more detail in the next section.

Besides establishing the global model of space, the investigations of Lutz Castell and coworkers in the 1970s were important for the representation of the local spacetime structure based on ur theory.[19] Castell was interested in the conformal group $SO(4,2)$, from which its spinorial representation $SU(2,2)$ follows naturally if one doubles the space of urs, going from two- to four-spinors. In this way, complex conjugation in (0.1) is naturally taken into account and one is led to urs and anti-urs, as described in Sect. 4.1.

In discussions Weizsäcker sometimes joked that his book *Aufbau der Physik* was written "around page 407" (the present page 100), the page where one can find the generators of $SU(2,2)$ and also, as a subgroup, of the Poincaré group, which is important for the representation of massive particles. Dirk Graudenz succeeded, on the basis of this representation, in deriving a general Poincaré-invariant vacuum state of urs.[20] Görnitz demonstrates in Sect. 6.2 how to obtain particle states from it by means of ur creation and annihilation in Minkowski space. One would hope that one day ur theory will enable at this point a connection with the quantum field theory of particles and their interactions. This too will be discussed in the next section.

By this point the basic theme of ur theory should have become apparent, namely, the derivation of the structure of spacetime in an abstract and strictly quantum theoretical manner. Recently this theme has also been mentioned by workers in modern quantum information theory:

---

[18] T. Görnitz. Abstract Quantum Theory and Space-Time Structure. I. Ur Theory and Bekenstein–Hawking Entropy. *International Journal of Theoretical Physics*, 27 (5): 527–542, 1988.

[19] L. Castell, M. Drieschner, and C. F. von Weizsäcker (eds.). *Quantum Theory and the Structures of Time and Space*, 6 vols. Hanser, Munich, 1975–1986.

[20] T. Görnitz, D. Graudenz, and C. F. von Weizsäcker. Quantum Field Theory of Binary alternatives. *International Journal of Theoretical Physics*, 31 (11): 1929–1959, 1992.

*It turns out that the lowest symmetry common for all elementary systems is the invariance of their total information content with respect to a rotation in a three-dimensional space. The three-dimensionality of the information space is a consequence of the minimal number (3) of mutually exclusive experimental questions we can pose to an elementary system. This seems to justify the use of three-dimensional space as the space of the inferred universe.*[21]

Time will tell whether such a promising contact with quantum information theory—a deep-seated possible realization of Wheeler's motto[22] "It from Bit"—can actually be worked out. In this sense Weizsäcker might be considered the godfather of quantum information theory.

## Spinorism, quantum gravity, interaction, and large numbers

The second basic assumption of ur theory means that Weizsäcker's program can be interpreted as a form of "spinorism." David Finkelstein expresses this as:

*Spinorism [is] the doctrine and program of describing all the fundamental entities of nature solely by spinors... By 1957 Penrose was already deep into his theory of spin networks, and Weizsäcker's spinorial theory of fundamental binary quantum alternatives, or urs, was several years old. Their work provides the house of spinorism with two wings. Spinorists like Penrose develop the classical geometric meaning of spinors and seek such meaning for other $\psi$ functions as well, shaping a quantum theory that partakes more of the classical. Spinorists like Weizsäcker regard spinors as describing a fundamental quantum two-valuedness and seek to leave the present quantum theory by the exit facing away from the classical.*[23]

Finkelstein himself "inhabits" the same wing as Weizsäcker insofar as both share the opinion that "a fundamental two-valuedness" is at the heart of a reconstruction of physics. But in contrast to Weizsäcker, Finkelstein emphasizes even in his early papers on *Space-time code* the discrete network character and

---

[21] C. Brukner and A. Zeilinger. Information and fundamental elements of the structure of quantum theory. In L. Castell and O. Ischebeck, (eds.). *Time, Quantum, and Information.* Springer, Berlin, 2003.

[22] J. A. Wheeler. Information, physics, quantum: the search for links. In S. Kobayashi, H. Ezawa, Y. Murayama, and S. Nomura (eds.). *Proceedings of the 3rd International Symposium on the Foundations of Quantum Mechanics,* pages 354–368. Physical Society Japan, Tokyo, 1989.

[23] D. Finkelstein. Finite Physics. In R. Herken (ed.), *The Universal Turing Machine—A Half-Century Survey,* pages 349–376. Springer, Wien, 1994.

the process orientation of quantum models of spacetime, with close connection to cellular automata.[24] Yet despite all differences in execution, in all three of the great one-man programs of Penrose, Finkelstein, and Weizsäcker, one can nevertheless discern a family resemblance among certain basic assumptions.

Comparing the Penrose–Finkelstein–Weizsäcker trio with programs that have come up in the meantime, it is perhaps Finkelstein's approach that most easily permits connections to Alain Connes' *Noncommutative Geometry*,[25] while the spinoristic element, as also emphasized by Finkelstein in his paper, can be recognized in the way quantum gravity is treated by the school of Ashtekar. Ashtekar recognized that spin variables permit important progress in the canonical quantization of gravity.[26] The transition to the loop representation of Rovelli and Smolin, and the geometric interpretation of models of canonical quantum gravity, underscore the significance of spinorism for these programs.[27]

In contrast to the "heavy machinery" of string theories, all of the aforementioned programs clearly emphasize the *background independence* of their models from the very outset. Weizsäcker's ur theory can claim for itself to have been one of the first programs of this kind. However, compared to other programs, one must clearly concede that ur theory is considerably lacking in its mathematical exposition. It is more of a programmatic blueprint whose attraction lies perhaps mostly in its conceptual integration of fundamental philosophical reflections. *The Structure of Physics* should thus also be of interest to present-day physicists working in the aforementioned programs, as Weizsäcker's deep epistemological and methodological reflections might also stimulate neighboring programs.

It is instructive to examine in more detail both a persistent weakness of ur theory—its almost complete lack thus far of a description of interaction—and its single empirically suggestive strong point, namely its new perspective and potential strength in explaining the physics of large numbers. Let us consider first the question of interaction. In KL II and III, as well as the present Sect. 4.9, one finds an attempt at an ur theoretic model of quantum electrodynamics. The starting point is the representation of a light-like four-vector in the form of Pauli matrices according to

$$k_\mu = \sigma^\mu_{\dot{A}B} u^{\dot{A}} u^B. \tag{0.2}$$

---

[24] D. Finkelstein. *Quantum Relativity: A Synthesis of the Ideas of Einstein and Heisenberg*. Springer, New York, 1996. (See references on the *Spacetime code* papers I–V, Phys. Rev. D 1969–1974, therein.)

[25] A. Connes. *Noncommutative Geometry*. Academic Press, New York, 1994.

[26] A. Ashtekar. *Lectures on Nonperturbative Canonical Gravity*. World Scientific, Singapore, 1991.

[27] C. Rovelli. *Quantum Gravity*. Cambridge University Press, Cambridge, 2004.
L. Smolin. *Three Roads to Quantum Gravity*. Weidenfeld & Nicolson, London, 2000.

There $u^A$ denotes an ur spinor, dotted indices represent complex-conjugate components. Weizsäcker is now interested in a procedure he calls "multiple quantization." By quantization one usually means taking two steps: first a transition from a discrete number of degrees of freedom to a continuum, and then a transition to operator-valued quantities with corresponding commutation relations. Consider first a simple classical yes/no alternative $a_A$. Then the first step involves constructing a wave function $\phi(a_A)$, i.e., a spinor $u_A \equiv \phi(a_A)$. According to (0.2) we can obtain from this a four-vector $k_\mu$, which as the second step we then write as the operator $\hat{k}_\mu$. In this way one obtains the first quantization of a binary alternative.

Following the same scheme, one obtains wave functions like $\varphi(k_\mu)$ at the level of second quantization. The previously introduced operators $\hat{k}_\mu$ act on these wave functions. If as usual we now interpret $k_\mu$ as an energy–momentum vector, then the functions $\varphi(k_\mu)$ can be considered, after a Fourier transform, to be ordinary quantum mechanical wave functions $\psi(x_\mu)$. Through second quantization of a binary alternative one thus obtains relativistic quantum mechanics. A second iteration of this procedure, i.e., the third quantization of urs, would then correspond to the quantum field theory of free fields.

But what about the dynamics of fields? As the relation $k_\mu k^\mu = 0$ holds for (0.2), one obtains from the Fourier transform of $\hat{k}_\mu \hat{k}^\mu \varphi(k_\mu) = 0$ the wave equation $\Box \psi(x_\mu) = 0$ as a purely algebraic identity. Weizsäcker, Scheibe and Süßmann discovered in KL III that in a similar way one can obtain the Weyl, Dirac, Klein–Gordon, and Maxwell equations. For the latter three cases, however, it is again necessary to first make the transition from ur spinors to bispinors.

In a certain way one has thus reconstructed the free dynamics, but not yet a coupling of fields. This is still a basic deficiency of ur theory. Yet another point is striking: why is the multiple quantization procedure apparently only suitable for an "ur theoretic derivation" of free Maxwell equations? How could one obtain the additional interacting fields? Here it is particularly remarkable that a theory that aims at a justification of spacetime does not lead in an equally natural manner to a description of gravity.

A first step in this direction might perhaps be taken in the following way. It is well known that a spinor dyad is equivalent to a system of tetrads of light-like four vectors (null tetrad). As functions on $SU(2)$, urs form in a natural way a spinor dyad (with spinors $u^A$, $v^A$ satisfying $u_A v^A = -v_A u^A = 1$). The tetrad vectors have the form (0.2), but consisting in general of mixed combinations of $u_A$ and $v_A$. By appropriately manipulation, a null tetrad can always be brought into the real-valued form $\theta_\mu^\alpha = (t_\mu, x_\mu, y_\mu, z_\mu)$, where the spacelike vectors $x^\mu, y^\mu, z^\mu$ form a tangent-triad on $\mathbb{S}^3$ with an orthogonal timelike vector $t^\mu$. Insofar as such a tetrad is built from ur spinors, a quantization of urs induces a quantization of the tetrad. Such a quantized ur tetrad could be interpreted, under the assumption of $SU(2) = \mathbb{S}^3$, as a global model of position space, a quantization of spacetime coordinates. In the manner of multiple quantization it is equally possible to derive a wave equation for the four massless spin-1 bosons $\theta_\mu^\alpha$, which might perhaps be interpreted as gravitons.[28]

Yet again these considerations do not lead to a derivation of the full dynamics, i.e., the coupling of the interaction fields with matter. The usual procedure in physics is by means of gauge theories, i.e., by postulating certain local symmetries. At least one can show that the algebra of those ur creation and annihilation operators, which form an ur tetrad, is close to a 12-dimensional Lie algebra, the corresponding Lie group being isomorphic to $SL(2,\mathbb{C}) \otimes SL(2,\mathbb{C})$. But it is completely unclear whether this group could furnish a suitable ur-theoretic candidate for a gauge theory of gravity. This obscurity results above all from the completely different understanding of space which characterizes ur theory, as compared to standard physics. In particular, the concept of gauge-theoretic locality and the assumption of a spacetime continuum associated with it can at best be understood as a limiting case from the standpoint of ur theory.

This becomes even more apparent when we examine the ur-theoretic discussion of the physics of large numbers. Here it suffices to illustrate the main idea; the detailed calculations can be found in Sect. 5.1 and 6.2. How many binary alternatives (urs) are possible in the universe, and how many make up our world? Or to phrase it differently, how large is the physical information content of the universe? As already mentioned, an ur can be represented as a function on its own symmetry group $SU(2)$. There the essential idea is that an ur is not a small particle, but constitutes space itself. As a function on $SU(2)$ it can be visualized as a possible binary decomposition of space, i.e., perhaps telling us whether a thing is to be found in the "upper" or "lower" half of the universe. Weizsäcker now asks how many binary decompositions one needs to perform to find, e.g., a proton in the universe. For this it is sufficient to localize the proton within its Compton wavelength $\lambda_p$. As empirical input he uses the ratio of the radius of the universe $R$ to $\lambda_p$, about $10^{40}$ (the so-called first Eddington number). Hence one must perform $10^{40}$ subdivisions of space to physically localize a proton. Of course one has must do this three times, once for each dimension of space, but this scarcely affects the order of magnitude of the number $10^{40}$ as additional information content (like the specification of the charge of the proton or its spin). The assumption is then that $10^{40}$ urs or quantum bits constitute a proton.

Now $\lambda_p$ is not an arbitrary unit of length, as every length measurement is related to an energy. If one were to choose the total energy content of the universe for the simultaneous decomposition of space into equal intervals, then

---

[28] H. Lyre. Quantum Space-Time and Tetrads. *International Journal of Theoretical Physics*, 37 (1): 393–400, 1998.

H. Lyre. C. F. von Weizsäcker's Reconstruction of Physics: Yesterday, Today, Tomorrow. In L. Castell and O. Ischebeck (eds.), *Time, Quantum, and Information*. Springer, Berlin, 2003.

a length of order $\lambda_p$ again follows. In this sense the number

$$N = \left(\frac{R}{\lambda_p}\right)^3 = 10^{120} \qquad (0.3)$$

is at the same time the maximum number of elementary cells in space and the total number of urs—in quantum bits it is the number of decisions needed to specify whether a certain cell is occupied by a nucleon or not. The state space of urs is therefore of dimension $2^{10^{120}}$. It follows as an *empirically verified result* that the total number of nucleons in the universe is $10^{120}/10^{40} = 10^{80}$. This is the second Eddington number; its quadratic relationship to the first, considered a deep riddle by many physicists, acquires in this way an ur-theoretic explanation.

In Sect. 6.1 Görnitz carries out more detailed calculations of large numbers, and shows in particular how the ur-theoretic results match the large numbers which follow from Bekenstein and Hawking's entropy calculations within the context of the thermodynamics of black holes. There one relies upon the conceptual assumption, proposed by Weizsäcker in Sect. 8.1, that entropy can be interpreted as *potential information*. In this sense an ur represents an elementary physical unit of (quantum theoretical) potential information.

Let us add one further observation: Bekenstein's work is known to lead to the remarkable connection $S = \frac{1}{4}A$ between entropy and the area of the event horizon of a black hole.[29] As Gerard 't Hooft has emphasized, this can be interpreted to mean that physical objects are characterized by the amount of information derivable by "projecting" all their degrees of freedom onto a surface.[30] In other words, rather than a volume, a surface suffices for the representation of the information that completely characterizes an object physically. This is reminiscent of a holographic representation and is called the *holographic principle*. In Planck units (with Planck length $l_0$), one obtains from Bekenstein's formula the ur-theoretic result $S_u = \frac{1}{4}(R/l_o)^2 \approx (10^{60})^2 = 10^{120}$ bits for the total information content of the universe. Similarly, one obtains the first Eddington number $S_p = \frac{1}{4}(\lambda_p/l_o)^2 \approx 10^{40}$ for the information content of a proton. For other objects, however, there are systematic discrepancies between the results of ur theory and the holographic principle. For example, for an electron one finds $S_e \approx R/\lambda_e \approx 10^{37}$ urs. Bekenstein's formula, however, leads to $S_e = \frac{1}{4}(\lambda_e/l_o)^2 \approx 10^{46}$. Ur theory and the holographic principle thus lead to different conclusions about the nature of space. In the absence of an experimental test, it would appear worthwhile to continue to pursue both avenues.

---

[29] Cf. J. Bekenstein. The limits of information. *Studies in History and Philosophy of Modern Physics*, 32 (4): 511–524, 2001.

[30] Cf. G. 't Hooft. Obstacles on the way toward the quantization of space, time, and matter—and possible resolutions. *Studies in History and Philosophy of Modern Physics*, 32: 157–180, 2001.

## Ontology of form

In conclusion, we again consider the ontological context in which ur theory operates. Traditional, Aristotelian-flavored ontology is an ontology of substance. Aristotle imagined a physical thing, an object, to be composed of matter with the object's form impressed on it. In this view matter functions as the carrier of form. If one asks about the essence of a thing, that which makes a thing a thing, the answer of Aristotle is form and not matter. It is the form of a statue that constitutes its essence, not the material it is made of. Nevertheless, the idea of pure, i.e., carrier-less form is erroneous to Aristotle in the case of concrete things. One last point of abstraction to him is the idea of primordial matter. It is devoid of any form, and therefore unrecognizable, yet indispensable as the ultimate carrier of all existing forms.

The concept of substance might nowadays be considered the generic term for concepts like matter and energy—it is a central assumption of substance-ontology that no thing is ontologically conceivable without a substance as carrier. In contrast, however, the world view of ur theory appears to lead to the anti-substantialistic, and in that respect radical, idea of pure form—or information, as its modern concept of quantification. The question would be whether such a radical ontology of form is consistently defensible at all.

The Aristotelian view draws its plausibility from the fact that Aristotle distinguishes between essential and accidental properties, where only the former contribute to the essence of a thing. It is then almost imperative to assume that essential properties are those due to the thing in and of itself, and not due to the existence of other things. Such properties are called intrinsic properties, in contrast to relational properties. Thus the mass of a particle, also according to the notions of modern physics, is an intrinsic property, whereas Carl Friedrich's being the brother of Richard is a relational property (as it does not pertain to Carl Friedrich per se, but only to Carl Friedrich in relation to the existence of Richard).

It is now quite natural to assume that intrinsic properties must be affixed to "something," and that something is just the substantial carrier of the thing. In the case of quantum mechanics, the basic fact of correlations between states in Hilbert space leads to the conclusion that all quantum mechanical properties, i.e., those that are tied to eigenvalues of states, are not intrinsic but relational properties, as correlations always involve several distinct quantum systems. In this sense, quantum theory itself appears to retreat from traditional substance-ontology and lead instead to an ontology of relations. In present-day philosophy of science, this is actually discussed in this way.[31]

Nevertheless, properties still remain which according to the modern standard model of elementary particles also represent candidates for intrinsic properties. These are generally properties that are not described by operators in Hilbert space, i.e., not as quantum properties—in particular, masses

---

[31] Compare e.g. M. Esfeld. *Holism in Philosophy of Mind and Philosophy of Physics.* Synthese Library No. 298. Kluwer, Dordrecht, 2001.

and charges. The common vision of all unification programs is some day to describe these properties too in genuine quantum theoretic terms. At that point, only spacetime would come into question as the ultimate carrier of all quantum fields. Programs like ur theory, which also attempt to reconstruct spacetime quantum theoretically, are therefore faced with the ontologically radical consequence of leaving no carrier of the universe at all. In that respect, Weizsäcker's program is one of the earliest and perhaps also most radical attempts to apply a strict ontology of form to a physical research project, as witness the most recent debate in the philosophy of science on so-called structural realism (and in particular its radical ontic variant[32]). Here as well, Weizsäcker also anticipated topical ideas quite early.

*The Structure of Physics* is therefore in many respects still an important book: as an eyewitness to the physics of the twentieth century, due to his personal acquaintance with physicists like Heisenberg and Bohr, Weizsäcker's observations, especially in the third part of the book, are of extraordinary value for the history of science. The physical core of the book, the presentation of ur theory in the first part, offers an enormous and still not completely exhausted wellspring of ideas—from our present perspective, especially in its anticipation of the essential themes of modern programs in quantum gravity. Finally, Weizsäcker outlines a philosophically challenging world view—as regards a radical ontology of form as information, as well as in the sense of a methodological and epistemological emphasis on the structure of time—which especially affects his understanding of thermodynamics (in the second part of the book) and his own original interpretation of quantum theory, and in this way pervades the entire book.

May the present English edition find many readers and offer many stimulating insights.

---

[32] S. French and J. Ladyman. Remodeling structural realism: Quantum physics and the metaphysics of structure. *Synthese*, 136 (1): 31–56, 2003.

# Major published books by C. F. von Weizsäcker

- Die Atomkerne. Hirzel, Leipzig, 1937.
- Zum Weltbild der Physik. Hirzel, Leipzig, 1943.
  (7th enlarged edition, Hirzel, Stuttgart, 1958; *The World View of Physics*, University of Chicago Press, Chicago, 1952).
- Die Geschichte der Natur. Hirzel, Zürich, 1948.
  (*The History of Nature*, University of Chicago Press, Chicago, 1949).
- Der begriffliche Aufbau der theoretischen Physik, 1948.
  (Lectures at the University of Göttingen, edited by H. Lyre, Hirzel, Stuttgart, 2004).
- With J. Juilfs: *Physik der Gegenwart*. Athenäum, Bonn, 1952.
  (*The Rise of Modern Physics*, Braziller, New York, 1957, and *Contemporary Physics*, completely rev. ed., 1962).
- Die Tragweite der Wissenschaft. Hirzel, Stuttgart, 1964.
  (6th edition 1990 with a new 2nd part; *The Relevance of Science: Creation and Cosmogony*, Collins, London, 1964).
- Die Einheit der Natur. Hanser, Munich, 1971.
  (*The Unity of Nature*, Farrar, Straus, and Giroux, New York, 1980).
- Fragen zur Weltpolitik. Hanser, Munich, 1975.
- Wege in der Gefahr. Hanser, Munich, 1976.
- Der Garten des Menschlichen. Hanser, Munich, 1977.
  (*The Ambivalence of Progress: Essays on Historical Anthropology*, Paragon House, New York, 1988).
- Deutlichkeit. Hanser, Munich, 1978.
- Diagnosen zur Aktualität. Hanser, Munich, 1979.
- Der bedrohte Friede. Hanser, Munich, 1981.
- Ein Blick auf Platon. Reclam, Stuttgart, 1981.
- Wahrnehmung der Neuzeit. Hanser, Munich, 1983.
- Aufbau der Physik. Hanser, Munich, 1985.
- Die Zeit drängt. Hanser, Munich, 1986.
- Bewußtseinswandel. Hanser, Munich, 1988.
- Bedingungen der Freiheit. Hanser, Munich, 1990.
- Der Mensch in seiner Geschichte. Hanser, Munich, 1991.
- Zeit und Wissen. Hanser, Munich, 1992.
- Wohin gehen wir? Hanser, Munich, 1997.

# Books on Weizsäcker's philosophy of physics

- P. Ackermann, W. Eisenberg, H. Herwig, and K. Kannegießer (eds.) (1989). *Erfahrung des Denkens—Wahrnehmung des Ganzen: Carl Friedrich von Weizsäcker als Physiker und Philosoph*, Akademie, Berlin.
- L. Castell, M. Drieschner, and C. F. v. Weizsäcker, (eds.) (1975, 1977, 1979, 1981, 1983, 1986). Quantum Theory and the Structures of Time and Space, Vol. 1–6. Hanser, Munich.
- L. Castell and O. Ischebeck (eds.) (2003). Time, Quantum, and Information. (Festschrift on the occasion of Weizsäcker's 90th birthday). Springer, Berlin.
- M. Drieschner (1979). Voraussage—Wahrscheinlichkeit—Objekt. Über die begrifflichen Grundlagen der Quantenmechanik. Springer, Berlin.
- M. Drieschner (1992). Carl Friedrich von Weizsäcker zur Einführung. Junius, Hamburg.
- Th. Görnitz (1992). Carl Friedrich von Weizsäcker. Ein Denker an der Schwelle zum neuen Jahrtausend. Herder, Freiburg i. Br.
- Th. Görnitz (1999). Quanten sind anders. Spektrum Akademischer Verlag, Heidelberg.
- D. Hattrup (2004). Carl Friedrich von Weizsäcker—Physiker und Philosoph. Primus, Darmstadt.
- W. Köhler (ed.) (1992). Carl Friedrich von Weizsäckers Reden in der Leopoldina: Zum 80. Geburtstag des Physikers, Philosophen und Leopoldina-Mitglieds. Barth, Leipzig. (Nova Acta Leopoldina, Abhandlungen der Deutschen Akademie der Naturforscher Leopoldina, Neue Folge, Nr. 282, Band 68).
- W. Krohn and K. M. Meyer-Abich (eds.) (1997). Einheit der Natur—Entwurf der Geschichte. Begegnungen mit C. F. v. Weizsäcker. Hanser, Munich.
- H. Lyre (1998). Quantentheorie der Information. Springer, Vienna, New York.
- K. M. Meyer-Abich (ed.) (1982). Physik, Philosophie und Politik. (Festschrift on the occasion of Weizsäcker's 70th birthday). Hanser, Munich.
- E. Scheibe and G. Süßmann (eds.) (1973). Einheit und Vielheit. (Festschrift on the occasion of Weizsäcker's 60th birthday). Hanser, Göttingen.
- M. Schüz (1986). Die Einheit des Wirklichen. Carl Friedrich von Weizsäckers Denkweg. Neske, Pfullingen.

# 1
# Introduction

## 1.1 The question

*Sapere aude*

What is the truth of physics?

Physics is based on experience. Theories formulate laws that apply to experience. The system of physical theories developed over the past few centuries is converging to a unified, comprehensive theory. Quantum theory is presently the closest approximation to such a general theory of physics known. This theory appears to be valid for all of nature, and nowadays is also believed by the majority of scientists to be valid for the realm of organic life.

It is useful to learn to be surprised about the right things. We often fail to wonder about what is most astonishing because it has been familiar to us for a long time, and therefore taken for granted. How can there be comprehensive theories at all? The basic assumptions of quantum theory can be formulated, for a mathematically versed reader, on one printed page. About a billion presently known experimental facts are consistent with quantum theory, and not a single experiment is known to have convincingly given the impression of contradicting quantum theory. How can we understand this success?

This is what is known as a philosophical question, which one then shoves aside in favor of the everyday tasks of science. Normal science, which solves its problems according to fixed "paradigms," (Kuhn 1962) functions on a "level" at which one can dispense with the mountaineering expertise of philosophy. But scientific revolutions (Kuhn), transitions to new closed theories (Heisenberg 1948), *do* require philosophical questions. The present book studies the structure of physics, starting with the philosophical question of how comprehensive theories are possible at all, in the expectation of achieving a new level of theoretical investigation with respect to such questions, and with respect to physics itself.

How is theory possible? It never follows with logical necessity from experience. What will happen in the future never follows with logical necessity from laws that have proved themselves in the past. Yet thus far the predictions of theories we still believe in have proved themselves. How were these predictions justified while the predicted outcomes were still in the future? to this question of Hume, Kant answers that the basic, general insights of physics always prove themselves *in* experience because they express necessary conditions *for* experience. We will adopt this idea of Kant, not as a certainty, but as a heuristic conjecture. We will find out how far it will take us.

Experience unfolds in time. The logical forms in which we speak of events in time are therefore our first topic of study. From there we proceed to the concept of probability, which we understand prognostically. We interpret quantum theory as a general theory of probabilistic predictions relating to individual, empirically decidable alternatives. We claim to derive from this interpretation of quantum theory both the three-dimensionality of space and the theory of relativity.

Physics is thus as generally valid as the separability of alternatives, i.e., the divisibility of our knowledge into individually decidable yes/no questions. This basis for its success—its empowering form—at the same time defines the limits of its truth.

## 1.2 Outline

This section follows in detail the line of argument of the entire book. Originally it was planned as a final summarizing chapter, but perhaps it serves better as an initial overview. It remains for the reader either to read it immediately as an introduction to the entire book, to use it as a "road map" while perusing the book, or treat it as a review at the end.

### 1.2.1 Methodology

The theme of this book is the unity of nature as manifested to us by the unity of physics. The historical form of the unity of physics (Chap. 2) is a sequence or system of closed theories. Following Heisenberg (Sect. 2.1 and 12.2) we call a theory closed if it cannot be further improved by small changes. A later theory usually differs radically from its predecessor in certain basic concepts, but explains the success of the predecessor within a range of applicability. The most comprehensive closed theory nowadays is quantum theory. This book adopts the working hypothesis that all of present-day physics can be reduced to quantum theory.

We seek to describe the unity of physics, and to justify it as far as possible.

A theory of modern physics is presented in mathematical form (Sect. 2.2a). The mathematical concepts employed acquire physical meaning (semantics) according the way in which colloquial speech describes our relationship to nature. Colloquial speech for newer theories is mostly the language available from older theories. Certain fundamental statements are declared to be laws of nature. The mathematical form of the laws of nature developed historically. We distinguish four such forms (Sect. 2.3): morphology, differential equations, extremum principles, symmetry groups. In a certain sense, each of these forms justifies the previous one. We will tentatively trace the newest form, that of symmetry groups, back to the separability of alternatives.

This description of the laws of nature demands an explanation. We say that physics is based on experience. A law of nature, considered logically, is a general statement. In the generality thus implied, it cannot be verified by experience. It should hold for an essentially unlimited number of individual cases, including all those which still lie in the future. According to Kant a statement will in general hold in experience if it enunciates prerequisites for any possible experience. We will have explained the laws of nature if we have reduced them to the prerequisites of experience.

Experience means learning from the past for the sake of the future. The past, present, and future tenses are thus prerequisites for experience. We will attempt to reconstruct all of physics by starting with the modalities of time.

### 1.2.2 Temporal logic

Logic is the science that formulates prerequisites for any science, including physics. Empiricism too, understood as scientifically collected and interpreted experience, ought to obey the laws of logic. We find, however, that traditional logic does not adequately describe those propositions relating to the modalities of time, in particular to the present and future. Specifically, we propose not to assign in principle the values "true" and "false" to future statements, but modalities like "possible, necessary, impossible." The relationship between this logic of temporal propositions and the general science of logic will be discussed in *Zeit und Wissen* Chap. I 6.

In the classical theory of probability and its quantum mechanical generalization (Chap. 3), we refer to catalogs of formally possible temporal statements. In classical theory such statements ought to satisfy the three conditions of decidability, repeatability, and compatibility of decisions. In quantum theory the third constraint is dropped. The catalogs have the mathematical structure of lattices.

### 1.2.3 Probability

We define the probability of a formally possible temporal statement, or of the formally possible event described by that statement, as a quantified future modality: it is the predicted relative frequency of an event of the given type. From this one can derive the classical laws of probability according to Kolmogorov's axioms. The relationship of this definition of probability to the traditional logical, empirical, and subjective definitions will be the topic of Chap. 4 in *Zeit und Wissen*.

Our definition of probability is "recursive" (Sect. 3.2). Mathematically speaking, one must describe the prediction of a relative frequency as its expectation value. The expectation value of a relative frequency in an ensemble of possible cases is defined in terms of the probability for the occurrence of that relative frequency, and thus in terms of the expectation value of the relative frequency of that relative frequency in a "meta-ensemble" of ensembles. It will be shown that this recursive definition is not a weakness of the definition, but in fact the only way in which the prediction of an empirical quantity (and thus also of an empirically interpreted probability) can be rigorously interpreted at all.

Abstract quantum theory in Hilbert space can be reconstructed as a generalized probability theory (Sect. 3.6). This might be the reason for its universal validity.

### 1.2.4 Irreversibility, evolution, stream of information

The starting point of the interpretation of time presented here, and of the entire subsequent reconstruction of physics, was an analysis of Boltzmann's derivation of the second law of thermodynamics by means of statistical mechanics (Chap. 7). The derivation is consistent only if the concept of probability is applied there exclusively to future events. As a consistency check one can then show afterwards that the facticity of the past and the openness of the future (in the form of the existence of documents of the past but not the future) then follows from the irreversibility of events according to the Second Law. The difference that exists between Now and past and future points in time cannot, however, be reconstructed from laws of nature that are formally valid at every point in time; it is an assumption, but not a consequence of the general laws of nature. Strangely enough, almost all physicists recoil emotionally from this conclusion (see Sect. 9.3d$\delta$; *Zeit und Wissen* Chap. I.3.6).

Shannon's definition of information as (positive) entropy is correct if information and entropy are understood as potential knowledge (Sect. 8.4). One can then show that evolution and thermodynamic irreversibility are necessary statistical consequences of the same structure of time—the difference between past facticity and future possibility. In the case of evolution, increasing entropy means an increase in the multiplicity of forms, and thus of potential information (Sect. 8.5).

Because perception can also be interpreted as enhancing information, evolution is similar in form to perception (Sect. 8.7b). The structures of animal behavior turn out to be biological precursors to logic (Sect. 8.7). This justifies equating the "subjective" concept of utility with the "objective" concept of information. In a non-hierarchical reconstruction of science, it is legitimate to recover the structures of logic with which we started the reconstruction as attributes of human behavior as living beings. This is the idea of the *Kreisgang*.[1]

In philosophical tradition one calls that which persists in time a substance. The foregoing discussion, as well as the consistent interpretation of quantum theory, suggest (Sect. 9.2e) renouncing the Cartesian distinction between "extended" and "thinking" substance ("matter" and "mind"). According to classical Greek philosophy, that which persists is "Eidos" (form). Now one can define information as a multiplicity of forms. Events in time can then be interpreted as stream of information (Chap. 10).

These abstract deliberations, however, become topics of discussion only in terms of the actual structure of physical theories.

### 1.2.5 The system of theories

Classical mechanics presents us with a foursome of entities: matter, forces, space, time (Sect. 2.2). In the mechanistic world view of the seventeenth century, one attempted to reduce forces to a defining property of bodies, their impenetrability. The historical development of physics took another path. The details of this path were determined mostly by new discoveries, sometimes also by changing modes of thought. Yet in retrospect one can discern an inner logic to this path, determined by the structure of the concepts themselves.

The decisive conceptual problem, at the end of a long theoretical development, turned out to be the dynamics of the continuum. The volume of space occupied by an extended object is mathematically infinitely divisible into smaller volume elements. What forces hold the parts of the body that occupy those volume elements together? Chemistry led to the picture of identical, stable space-filling atoms for each element (Sect. 2.4). Physics could not offer a consistent mechanical model of such atoms. The success of celestial mechanics and the problems of the dynamics of continua led instead to the model of mass points subject to action at a distance. The forces, thus interpreted as separate entities, turned out to be fields, i.e., dynamical continua themselves (Sect. 2.6). The inescapable severity of the problem manifested itself in the most abstract and thus most unshakable of the classical disciplines of physics, statistically based thermodynamics (Sect. 2.5). From the development which led to quantum theory we deduce in retrospect the impossibility of a fundamental classical physics, namely a classical dynamics of continua of bodies and fields. Classically, the infinite number of degrees of freedom of a continuum does not permit thermodynamic equilibrium.

With awareness of the conceptual problems of classical physics, quantum theory then enters physics not as a conceptual embarrassment, forced upon us by new discoveries, but on the contrary as the resolution of a conceptual dilemma that is unsolvable without it. It makes possible the thermodynamic

---

[1] Cf. p. XXII, fn. 14.

equilibrium of a continuum, explains the stability and identity of the atoms of an element, and offers a universal framework for physics.

The physics of the past century also began to fuse the other two foundations of classical mechanics, space and time, into a new union. The old problem of the relativity of motion (Sect. 2.8) found a group theoretical solution in the special theory of relativity (Sect. 2.9). The heart of the problem was the law of inertia, inexplicable in terms of the classical concept of causality (Sect. 2.2c). The special theory of relativity makes measurements of space and time dependent on the state of motion of the objects but—contrary to a widely used figure of speech—does not abolish the difference between space and time; the distinction between spacelike and timelike separations is Lorentz invariant. Nor does the theory of relativity abolish our description of the modes of time; the distinction between past and future is also Lorentz invariant. The mathematical discovery of non-Euclidean geometries and Einstein's idea of the local equivalence of a gravitational field and an accelerated reference frame led, in the general theory of relativity (Sect. 2.10), to a description of the spacetime metric patterned on field theories. Contrary to Einstein's original intentions, the theory remained dualistic: matter and metric field could not be reduced to one another. These two inherently complex entities are what remains of the fourfold foundations of classical mechanics. Understanding their interrelationship would be part of a program to unify physics.

### 1.2.6 Abstract quantum theory

By abstract quantum theory we mean the general laws of quantum theory in more or less the mathematical form given by J. v. Neumann (Sect. 3.5b). The states of an arbitrary object are described by the linear subspaces of a Hilbert space. The metric of this Hilbert space determines the conditional probability $p(x, y)$ of finding a state $y$, given that the state $x$ is present. The states of a composite object reside in the tensor product of the Hilbert spaces of its parts. The dynamics of an object is given by a unitary group of mappings of its Hilbert space onto itself, depending on a time parameter $t$.

We call this theory abstract because it is universally valid for arbitrary objects. It says nothing about the existence of an (empirically three-dimensional) position space, about bodies or point masses, or about the specific forces acting between objects (i.e., about the choice of the Hamiltonian operator which generates the dynamics). Because of this general validity we interpret it as a theory of probability which differs from classical probability theory only in the choice of the underlying lattice of propositions. This lattice defines so-called quantum logic (Sect. 3.6). To the recursive definition of probability corresponds the procedure of second or multiple quantization (Sect. 3.4). Following Dirac, Feynman interpreted Hamilton's principle of classical mechanics as Huygens' principle of wave mechanics; analogously we read the extremum principle of wave mechanics as Huygens' principle at the next higher level of quantization.

Historically quantum theory arose out of concrete physical problems. Its abstract generality, however, suggests attempting to reconstruct the theory from postulates which only embody plausible prerequisites of possible experience (Chap. 3). The logical starting point is the concept of an $n$-fold alternative, i.e., an empirically decidable question which admits exactly $n$ mutually exclusive answers. Independently of it, in Sect. 3.6 the concept of an object is used, which might be described as the mathematical stylization of a material thing. An alternative then belongs to an object; its answers denote possible properties (states) of the object. The concept of an object is probably employed in all axiomatic formulations of quantum theory. However, quantum theory itself shows that this only describes an approximation (Sect. 3.6e): every object can be combined with objects in its environment to form a composite object. In the Hilbert space of the composite object, however, the states in which the sub-objects also have well-defined states are only a set of measure zero. The success of quantum theory (and all the more so of its limiting case, classical physics) must stem from the factually good separability of objects in terms of their corresponding alternatives.

The appropriate use of "finitism" (Sects. 3.6d, 4.2a$\alpha$2) will be useful in the reconstruction. Empirically, only finite alternatives can be decided; the quantum theory that developed historically, on the other hand, uses a Hilbert space of denumerably infinite dimensions. The problem is dealt with broadly in Sect. 4.2a$\alpha$2 under the title "open finitism." There only finite alternatives, but for arbitrarily large $n$, are considered and treated in a common state space, which is consequently denumerably infinite-dimensional. "Objects" belonging to alternatives of fixed finite dimensions are called subobjects; the state space of an object is then the vector sum of the spaces of infinitely many subobjects.

The decisive assumption of quantum theory is referred to by the term *expansion*, or alternatively indeterminism (Sect. 3.6h). It says that for any two mutually exclusive states $x$ and $y$ of some alternative, there is at least one state $z$ that does not rule out either of the two. We define $z$ in terms of the conditional probabilities $p(z,x)$ and $p(z,y)$. We reconstruct from these first the quantum logical lattice of propositions, prove that it is a projective geometry, and introduce a Hilbert space as the vector space in which this projective geometry can be defined (Sect. 3.6). Dynamics is introduced at the end as an invariance group of the probability metric.

### 1.2.7 Concrete quantum theory

By *concrete quantum theory* we mean the theory of objects that actually exist. In the form presented in this book it is unfinished, but intended to be a comprehensive program. For details we refer the reader to Chap. 4 and 5. Here we merely discuss the basic formulation of the question.

The distinction between general and special laws of nature is an old one. One can, however, ask whether it is of a fundamental nature. Special laws describe special areas of experience. To the extent to which physics approaches a unified whole, the general laws assume a form such that they themselves determine their corresponding special areas, e.g., as special solutions of general equations. Bohr's quantum theory of atomic structure thus explained the previously empirically determined periodic system of the elements. A similar hope exists nowadays for systematizing the elementary particles. Thus, it is conceivable that the general theory itself determines what special solutions, and in particular what elementary particles, are possible.

By and large, one assumes nowadays that for this to happen, one must supplement abstract quantum theory at least with special dynamical laws. Chapters 4 and 5 explore the conjecture that this may not be necessary, with the exception of one single hypothesis whose complete triviality we have not been able to prove: namely that all actual alternatives can be constructed from binary ur alternatives (the "ur hypothesis").

Present-day elementary particle physics seeks to explain the system of elementary particles in terms of symmetry groups. The fundamental group is the Poincaré group, defined by the special theory of relativity; in addition there are compact groups of "internal" symmetries. Now the ur hypothesis implies the existence of a real three-dimensional position space and the validity of special relativity (Sect. 4.1). In this way, coordinate space and special relativity are derived purely quantum theoretically; apart from the ur hypothesis they do not require any additional assumptions of abstract quantum theory. The existence of particles follows immediately from the special theory of relativity; they are the irreducible representations of the Poincaré group.

Beyond that, the theory is at present merely a program whose implementation depends on overcoming mathematical difficulties. The formal elaboration of the theory, leading to its empirical verification, is an as-yet unsolved mathematical problem. We present a model of quantum electrodynamics (Sect. 5.4) and a proposal for the justification of gauge groups in the systematics of particles (Sect. 5.5). The explanation of sharp rest masses is likely to depend on the solution of a statistical problem (Sect. 5.5d).

The general theory of relativity expresses in this context precisely the interrelationship among local Minkowski spaces that was left unresolved in the quantum theoretical reconstruction of the spacetime continuum (Sect. 5.6).

### 1.2.8 Questions of interpretation

The protracted debate over the interpretation of quantum theory can only be understood in terms of its historical roots. Quantum theory originated from classical physics. Despite its overwhelming empirical success, its deviation from the classical world view was felt to be a sacrifice. Between Bohr and Einstein the issue was whether this success justified the sacrifice or not (Sect. 9.1, 9.3a–e). Both held on to the importance of classical physics: Bohr for the description of empirical phenomena (Sect. 9.1g), Einstein for its concept of reality (Sects. 9.1i, 9.3d).

From our point of view, this debate appears more to conceal the true unsolved problems of quantum theory. None of the concepts in the debate would be intelligible without a pre-existing understanding of events and processes unfolding in time—now, between the factual past and possible future. Bohr's thesis that experiments must always be described in terms of classical concepts is based on the requirement of factual, irreversible results; in that respect it can be explained in a temporal theory and is legitimate if explained. Einstein's concept of reality transfers the attributes of facticity pertaining to the past to future, i.e., possible events as well.

In our opinion this debate is, for historical reasons, too heavily oriented toward "concrete" instead of "abstract" physics; it favors concrete pictures of events which historically one could naturally work out before explaining them on the basis of general laws. To us, classical physics represents a limiting case of concrete quantum theory; concrete quantum theory is presumably a consequence of abstract quantum theory; and abstract quantum theory is a general theory of probabilistic predictions. Each of these three steps involves unresolved questions which, however, in the debate over the interpretation, have not even come up for discussion, and which we mention here in conclusion.

Classical physics as a limiting case of concrete quantum theory: a limiting case is much poorer in information than the sequence from which it is derived. We have emphasized this as "quantum theoretical extra knowledge" (Sect. 9.3f). This is the main point of Heisenberg's uncertainty relation: the classical path must not exist merely in order for the immeasurably richer information of the Schrödinger equation to exist.

Concrete quantum theory as a consequence of the abstract: I cannot but suspect that the ur hypothesis is trivial, i.e., a necessary consequence of abstract quantum theory, if the latter is reconstructed according to the postulate of interaction (Sect. 4.2b3). Be that as it may, the startling derivation of position space as the representation space in the quantum theory of a binary alternative is in any event a beautiful example of quantum theoretical extra knowledge. Knowing quantum theory, one immediately obtains a three-dimensional metric space of possibilities for every yes/no decision.

Abstract quantum theory as generalized probability theory: this formulation indeed correctly reveals the degree of abstraction, and thus the presumptive reason for the general validity of quantum theory. It then follows, among other things, that we have no reason to rule out the applicability of quantum theory to psychic phenomena (Sect. 9.2e). We have already made use of this in the concept of the stream of information (Chap. 10). However, at this level the concept of probability is perhaps an inadequate means of expression, as is the concept of indeterminism. Here we encounter one of the remaining frontiers of quantum theory (Chap. 11). Indeed, with the logical concept of an empirically decidable alternative, we assume the facticity of the measured result after the decision, and the consequent loss of information that is inherent in irreversibility (Sect. 9.2c$\beta$). Scrutinizing quantum theory thus leads us to a critique of the premises without which we would not have been able to reconstruct it. How is it possible that such a successful theory can be built upon such a shaky foundation?

We formulate the self-critique and the limited self-justification of quantum theory in Chap. 11.

Classical logic, as the "mathematics of what is true and false" (*Zeit und Wissen* Chap. I.6), is a theory of yes/no decisions. We have introduced in it the quantum theoretical extra knowledge through the expansion postulate which, however, utilizes the concept of probability, i.e., the counting of favorable and unfavorable cases, thus again yes/no decisions. However, to obtain a mathematically precise theory we must postulate symmetry requirements which allow the application of group theory. We can justify these requirements only by virtue of the approximate separability of alternatives, thus again on the basis of ignorance. Concrete quantum theory provides us a quantitative estimate of the validity of this approximation (Sect. 11.4). The universe indeed turns out to be "almost empty"; particles can be very far apart from one another. The self-justification of quantum theory therefore says that in most cases we can describe this "extra knowledge" to a very good approximation by means of representations of the symmetry group, and thus in terms of Hilbert vectors (Schrödinger functions). The self-criticism, however, persists: behind this good approximation there may lurk far-reaching connections that are missed by our present theories. It cannot be ruled out that these even transcend the methodological starting point of our reconstruction of all physics, i.e., the distinction between past, present, and future (Sects. 11.4–5).

In the language of the philosophers (Chap. 12), we formulate the interpretation based on the Aristotelian triad of logic, physics, and metaphysics. For logic we draw upon the methodology of the modern theory of science; in a *Kreisgang*[2] it leads us back to the beginning. Metaphysics teaches us to see the limits of the truth of physics.

---

[2] Cf. p. XXII, fn. 14.

# Part I

# The unity of physics

# 2
# The system of theories

## 2.1 Preliminary

The title *The Unity of Physics* expresses the conjecture that it might be feasible to summarize all of physics, as far as fundamental laws are concerned, in one single theory. We have divided the outline of such a reconstruction of physics into two parts. The second part will deal with the significance of time, or more to the point, the mutual relevance of *past*, *present*, and *future* in physics. There only time will be at issue, but not the unity of physics.

Here, in the first part, we begin in the present chapter with the historical development of the great theories of physics, as they were understood by their creators. We interpret this historical progression factually as a "system," i.e., that the theories relate to one another and, in the way they developed historically, that the older ones have in each case been put into a new perspective by the newer ones.

This historical development was assessed by Kuhn (1962), and previously by Heisenberg (1948) as regards its internal dynamics, its factual exigency. Kuhn speaks of normal science governed by a paradigm prevailing at that time, and about paradigm shifts he calls revolutions. Each paradigm has only a limited scope. A revolution begins with the appearance of difficulties that ultimately prove to be intractable within the prevailing paradigm. Most of the time this intractability will only be recognized after a new paradigm has resolved it; one can then see why the old paradigm was bound to fail. The historical process of the emergence of the great theories of physics, which interests us here, had been more specifically and thus more accurately described by Heisenberg as a sequence of "closed theories." Every closed theory, i.e., one that cannot be improved by small modifications, has a range of applicability, the limits of which, however, can only be delineated by the subsequent theory.

Here, however, we do not wish to start with a methodological concept but rather be instructed by retracing the actual history of physics, as Heisenberg and Kuhn naturally had done originally. Of course this is not a book about the history of physics. The author can emphasize only very briefly, for each of the

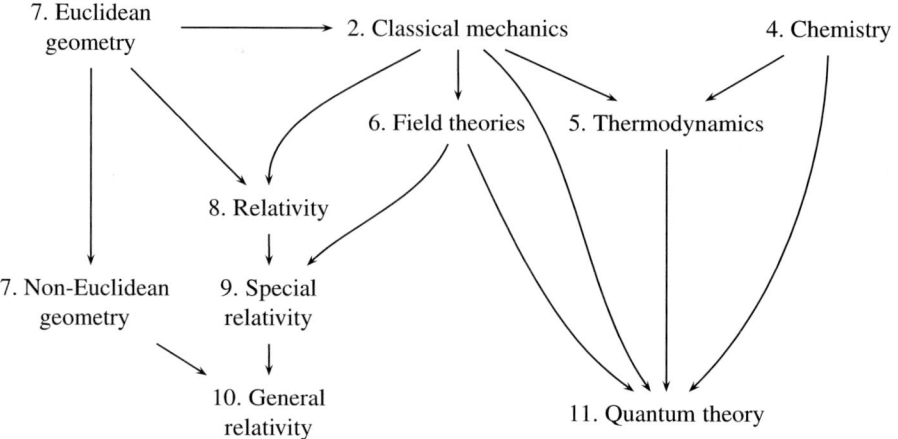

**Fig. 2.1.** The system of theories

theories, those points that appear to him to be most important from the points of view mentioned. The "closed theories" do not comprise a linear progression, but contain instead various branch points. In addition, the predicate *closed*, and even the characterization as a theory, do not pertain to all of them to the same degree. The historical/factual interest that gave rise to this account was always related, within my limited powers, to the corresponding technical issues that propelled the theories forward. I strove to learn from the particular about the general. The chapter progresses through the theories, and pauses twice to reflect on methodology, first in Sect. 3 on the mathematical form of the laws of nature, and then in Sect. 7 on semantic consistency. The exposition in the twelve sections is not quite uniform. This is due partly to their relationship to other parts of the book, and partly to having been written at different times.

Let us diagram the relationships among the theories, with numbers indicating the relevant sections of the book (Fig. 2.1). As an introduction to this chapter we comment on the diagram. In the center of the first row is *classical mechanics*. It is the starting point of modern physics. If we also wanted to identify its historical origin, we would have to draw, besides the arrow leading to it from Euclidean geometry, two additional arrows from the disciplines of astronomy and technology, which are not included in the diagram. To geometry it owes its methods, and to astronomy and technology its concrete issues. Taking it as a model, Sect. 2 also serves as an introduction to the problems we will meet in the system of theories.

Four arrows go from classical mechanics, to the problem of relativity, the field theories, thermodynamics, and directly to quantum theory. Two independent sciences stand nearby, both with their origins in antiquity: geometry and chemistry. From Euclidean geometry two arrows lead to the problem of relativity, of which there was awareness in earlier modern times, and to the non-Euclidean geometry of the nineteenth century. Two arrows extend outward from chemistry, to thermodynamics in the nineteenth century and to quantum theory in the twentieth century.

We consider first of all the center and left half of the diagram. The *field theories* are in part a direct consequence of classical mechanics as the mechanics of continua, and in part they derive from the empirical sciences of light, electricity, and magnetism. In the late nineteenth century they established the local-interaction approach to physics that has prevailed ever since.

Two arrows lead to the *relativity problem*, one from geometry and one from mechanics. Section 8 reviews inevitable questions concerning kinematic reference frames engendered by the emergence of mechanics in the seventeenth century.

*Special relativity* provides the solution to this problem by incorporating the lessons learned from field theories.

*Non-Euclidean geometry* is one result of modern inquiries into problems inherent in Greek geometry, which were recognized even by their creators. One can now choose among several geometrical theories, raising questions about the empirical significance of geometry.

Three arrows actually ought to point to *general relativity*—from special relativity, non-Euclidean geometry, and the theory of gravitation, which historically was a component of mechanics. Section 10 deals in some detail with its unresolved problems.

Next we turn to the right side of the diagram.

*Thermodynamics* is an independent fundamental discipline. It originated from the empirical phenomena of heat and assimilated the experience gained about chemical reactions. Its First Law extends the conservation of energy, first defined in mechanics, to all of Nature. Its Second Law establishes the universal phenomenon of irreversibility. Its statistical interpretation introduces probability into physics as a basic concept. Under the heading *thermodynamics of the continuum* we describe the inevitable failure of the thermodynamics of classical field theories, which inevitably put us on the path to quantum theory.

Four arrows point to *quantum mechanics*, which evolved from the thermodynamics of the continuum. In Bohr's model of the atom it accomplishes a unification of mechanics and chemistry. In wave mechanics it assimilates the field theories. Nowadays it can be considered the fundamental theory of physics. How the system of theories will appear after quantum theory has assimilated the other theories, in particular the theories of relativity, we describe programmatically in the twelfth section of this chapter.

## 2.2 Classical point mechanics[1]

### 2.2.1 First analysis of the meaning of the basic equations

$$m_i \frac{\mathrm{d}^2 x_{ik}}{\mathrm{d}t^2} = f_{ik} \quad (i = 1 \ldots n,\ k = 1, 2, 3) \tag{2.1}$$

If, without any further explanation, one asks a physicist nowadays for the meaning of the above formula, one can expect an answer something like: "Well, those are perhaps the equations of motion of classical mechanics." Here we begin an analysis of the understanding which is implied by this answer.

We have put the correct answer into the mouth of the physicist in question, i.e., the one we had meant, but with expressions of hesitation. Indeed, there are widespread conventions in science about the meaning of symbols in formulas which often make it easier for the reader, e.g., in articles written in a foreign language, to understand the formulas rather than the accompanying prose. But these conventions are not unambiguous, in fact in both possible directions: one and the same symbol does not always represent the same concept, and a certain concept is not always represented by the same symbol. The first of these ambiguities is the fundamental one. The multiplicity of concepts in science is so much greater than the set of symbols utilized (and reasonably remembered) that one symbol must necessarily represent very many different concepts. Which one is meant follows from the context or is decided through explicit definition. Precisely because the meaning depends on the context or definition, divergent traditions emerge concerning the usage of symbols, which at times might again be reconciled later on. What results is the theoretically avoidable but historically inevitable description of one and the same concept using different symbols.

The study of the meaning of symbols, of the combinations and operations performed on them, is known as *semantics*. The meaning of a formula like the one written above can be investigated from the standpoint of the three sciences of logic, mathematics, and physics. For *logical semantics* it is the expression of a *proposition* that claims to be true. For *mathematical semantics* it is an equation among quantities which may assume real numbers as values. For *physical semantics* it expresses a *law of nature*, here the law of motion of a system of point masses. In this case, its "actual" meaning, the formula is usually written down; we investigate it first. There, however, we must clearly rely upon a preconception of its logical and mathematical meaning. What a true proposition and a mathematical equation is we can assume to be known pragmatically. On the other hand, we must explain what mathematically is not self-evident, namely how the quantities denoted by letters are to be interpreted mathematically.

---

[1] Subsections a–c are notes from 1971 which were meant to be the beginning of a chapter on the analysis of physics, in a first draft of the present book. Subsections d and e are from the present; d contains a lengthy quote from the book *Die Einheit der Natur*.

## 2.2 Classical point mechanics

The formula is an equation, in fact a differential equation of second order; the symbols

$$= \text{ and } \frac{d^2}{dt^2}$$

have the usual meaning, with $t$ denoting the variable of the argument. Of the other symbols there are the letters $i$ and $k$ for the indices, representing natural numbers from a range indicated in parentheses. $k$ may assume one of the three values 1, 2, or 3; $i$ is one of the values from 1 to $n$; and $n$ is not fixed in general, but for every concrete usage of the equation assumes a definite natural value $\geq 1$. For every choice of $i$ and $k$ we obtain a differential equation. Our formula is thus a concise symbolic representation of $3n$ equations. This representation—to include here this anticipation of logic—is possible because all $3n$ equations have the same "form."

This representation of different entities that take the same form is accomplished by using indices as *variables* in the sense which Frege has stated more precisely: they are not "changing" quantities, which would not make any sense in logic, but they "indicate generality." The remaining letters $m, x, f$, and the already mentioned $t$ are also variables in this logical sense, for which, however, real numbers (for $m$ actually only positive ones) must be substituted. In the mathematics of differential equations one distinguishes, in a linguistic usage different from that in logic, $x$ and $t$ as the *variables*, whose functional dependence is studied, from $m$ and $f$ as *parameters*, i.e., quantities that can be chosen arbitrarily once, but must then be kept fixed for the form of the equation determined by this choice and the resulting nature of its solutions. There one usually calls $t$ the independent and $x$ the dependent variable, thus indicating that $x$ is sought as a function of $t$. For every choice of the indices $i, k$ there is a separate dependent variable $x_{ik}$, whereas $t$ is assumed always to be the same quantity. We are thus looking for $3n$ functions $x_{ik}(t)$. $m_i$ is there meant as a constant parameter, i.e., independent of $x_{ik}$ and $t$. On the other hand, $f_{ik}$ (which is not expressed in compact notation) will be a function of all $3n$ quantities $x_{ik}$, and may furthermore depend explicitly on $t$.

One can set up and solve these equations—i.e., find the functions $x_{ik}(t)$ for given values of the parameters $m_i$ and the parameter functions $f_{ik}(x_{\ell m})$, with $\ell = 1 \ldots n$, $m = 1, 2, 3$—without any knowledge of the physical meaning attributed to them. This is reflected in the semantic usage by the physicist. When asked about the detailed meaning of Eq. (1), he will in general not speak of the physical meaning of the *symbols* $x, t, \ldots$, but about the physical meaning of the mathematical quantities denoted by these symbols. That which mathematically appears to be the meaning of the symbols, the mathematical interpretation itself, is to the physicist a mathematical "formalism" that still needs to be interpreted with regard to "reality," "nature," or "experience." Just as mathematics, historically, had existed before the mathematical symbolism, many concepts of physics had also existed historically before their mathematical specification. As for the problem of how the concepts of "specification," "interpretation," "experience," and "semantics" are related, see also Sect. 2.7 on "semantic consistency" and *Zeit und Wissen* I 5.2.7. But first we present the physical interpretation of the equation as it is usually understood by physicists.

The equation determines the behavior of certain physical *objects* called *point masses*. Empirically, however, we do not know any objects that behave exactly like point masses. The physicist calls a concept that is related to experience, such as that of a point mass, an *idealization*. We then immediately face a new semantic problem. The physical interpretation of the formalism again requires two steps. First, we assign to the mathematical concepts new *names* derived from a linguistic usage in which mathematical quantities describe the properties of certain idealized objects—in the present case, the properties of point masses. We then consider what *objects* in reality, or what elements of our experience, are perhaps described by such idealized objects, how such a description is possible, how accurate it is, and so on. Naturally, here as well, the historical sequence was just the opposite. The concepts describing idealized objects are the result of a long historical process. Nowadays we can more easily understand the result than the process that led to it; we therefore begin with an account of the result. But the actual task of interpretation, which we are not spared, remains to be addressed in our second step. It will eventually compel us to explicitly examine the structure of the historical development. We now take the first step and consider the physical naming of the various mathematical quantities in the equation.

The number $n$ specifies how many point masses there are, each labeled by the index $i$. The equation, as it stands, thus holds for an arbitrary but finite number of point masses, which must be chosen for each application. For a given $i$, the three quantities $x_{ik}$ ($k = 1, 2, 3$) denote the three *spatial coordinates* of the corresponding $i$-th point mass, relative to an arbitrarily chosen rectangular coordinate system. The quantity $t$ is called the *time*. The constant parameter $m_i$ is called the mass of the $i$-th point mass. The function $f_{ik}(x_{\ell m}, t)$ is called the $k$-th component of the force acting on the $i$-th point mass. If the masses are given and the force functions are such that the system of $3n$ equations can be solved,[2] then every solution consists of $3n$ functions $x_{ik}(t)$. For a given $i$ one calls the three functions $x_{ik}(t)$ a coordinate representation of the *path* of the $i$-th point mass. One also calls the $3n$ quantities $x_{ik}$ the spatial coordinates of the *phase point* of the entire system of $n$ point masses and the $3n$ functions a coordinate representation of the *path* of this system in *phase space*. A solution of this system of $3n$ second-order differential equations contains $6n$ constants of integration, also known as parameters of the solution. The solutions form a $6n$-dimensional manifold. As parameters of the solution one usually chooses

---

[2] Here we do not discuss the mathematical conditions for the existence of solutions that a physicist would consider reasonable.

## 2.2 Classical point mechanics

in physics the values of the $x_{ik}$ and their first time derivatives[3]

$$v_{ik} \lessgtr \frac{\mathrm{d} x_{ik}}{\mathrm{d} t} \tag{2.2}$$

at a fixed point in time $t_0$, and calls them the *initial conditions*.

Behind this terminology there are notions whose abstract structure one might describe as follows. A system of $n$ point masses is a physical object whose overall behavior in time can be determined theoretically if three things are known:

1. general laws
2. parameters of the system
3. parameters of the state

The general laws can be subdivided into

1a. specification of the formally possible properties of the system
1b. general laws that describe how these properties vary in time

We call those properties *Formally possible* which can be attributed to a system if it is a possible object of the particular theory at all. All quantities $t, x_{ik}, m_i, f_{ik}$ denote formally possible properties, such that what a physicist calls a *quantity* (e.g., a coordinate $x_{ik}$, for a fixed pair of indices $i, k$) represents a *class of properties* for which one and only one can apply to the particular system in question. This is then called the *value* of that quantity for the system. That the system *might* have a value of a certain quantity as its property is expressed by the word *possible*. That this possibility is determined by the general theory is expressed by the word *formal*. In special cases, it can turn out that a certain value of a quantity, although formally possible, may not be *actually possible*. The specifications 1b, 2, and 3 successively restrict the range of what is actually possible. Formally possible properties we also call *conditional in a broad sense*, thereby giving the philosophically ambiguous word "conditional" a special meaning.

There is an essential difference between the quantity $t$, the *time*, and the other quantities, which is in fact manifest in our manner of speaking. Let $t'$ be a special value of $t$, such as a point in time signified by the hands of a watch. Then one does not say that the system has the property $t'$, but that it has the properties $x_{ik} \ldots$ at time $t'$. The value of time is not a property of any special system. This corresponds mathematically to $t$ in (2.1) being the only independent variable. For this reason, in 1b we have called the only general law restricting the range of formally possible properties, namely (2.1), the law of temporal development of those quantities.

Some formally possible properties occur in 2 and 3 under the designation *parameters*. These are the values of quantities that can still be freely chosen

---

[3] In this book we use the symbol "$\lessgtr$" to mean "equal by definition"; this is also often written as "$=_{\mathrm{Def}}$"

from the standpoint of the general laws. Mathematically these are the specifications necessary to determine the solution of the equation, i.e., the trajectory in phase space. As parameters of the system we have designated those that are required to fix the precise form of the equation itself; they already occurred in the mathematical description as "parameters." These are the $m_i$ and $f_{ik}$. We call parameters of the state those that select the solution, such as the initial conditions. Changing the parameters of the system, one passes to the description of another system; the fact that according to the general laws systems with different parameters are possible characterizes these parameters as conditional in the broader sense. Changing only the parameters of the state, such as $x_{ik}$ and $v_{ik}$ at a certain time, one passes according to theory to a different possible state of the same system, another solution of the same equation. For the time being, we call the parameters of the state *conditional in a narrower sense*.

Physical semantics therefore faces the task of interpreting the four types of quantities $m_i, f_{ik}, x_{ik}, t$. In a casual manner we assign to them the four basic concepts *point mass, force, space*, and *time*, to be discussed in the following four sections. We will thus not get much farther in this section than to formulate certain main problems.

### 2.2.2 Bodies, point masses, systems of point masses

Sometimes one designates a single point mass, sometimes a system of mass points, as an *object*. This manner of speaking assumes that an object can be part of a more comprehensive object, or, as one also puts it, that an object may *consist* or be *composed* of parts which themselves are objects. This manner of speaking, which we will criticize in quantum theory, corresponds to the mindset of classical physics. It arises naturally, considering that the two concepts, "point masses" and "systems of point masses," are idealizations of the concept of a *body*.

Historically, the mechanics of point masses developed out of *celestial mechanics*. The latter provides the empirical application of the concepts developed in Sect. 2.2a, their "setting" or "place in life."[4] For example, the planets are bodies with diameters so small compared to their distances from one another and from the sun that for most calculations they can be treated as points with a single constant parameter, namely mass.[5] They interact according to a known force law, the law of gravitation:

---

[4] The term *setting in life* (in German "Sitz im Leben" which means *setting, place*, or *situation in life* from which it evolved, or setting in the life of a community) became established in the philology of the New Testament to describe the real historical use of concepts recorded in abstract and dogmatized form.

[5] We should add that for a perfectly spherically symmetric body its gravitational influence on another body outside its radius is exactly the same as if its mass were concentrated at its center.

$$f_{ik} = \sum_{j \neq i} f_k^{ij}, \tag{2.3}$$

$$f_k^{ij} = G\frac{x_{jk} - x_{ik}}{r_{ij}}\frac{m_i m_j}{r_{ij}^2}, \tag{2.4}$$

with

$$r_{ij}^2 = \sum_k (x_{ik} - x_{jk})^2, \tag{2.5}$$

and a universal constant $G$. Their positions at a certain time $t_0$ can be determined empirically.[6] These positions are the initial conditions from which the future positions can then be predicted. It was, incidentally, in Babylonian astrology that the prediction of planetary positions as a means of forecasting the influence of the heavens on human affairs found its first "place in life."

In this description of the planets, the concept of a *body* is simple. A body might be described as a thing that is extended, cohesive, and moves through its environment (see *Zeit und Wissen* II.6.2). Attempting a more precise definition of a body leads to technical problems of physics, but we can at least get started. "Extended" can be taken to mean "space-filling." This immediately poses the problem of how one envisions the space to be filled by the body; about this see *d* below. Assuming the expression to be understood, we furthermore must consider that the space filled by the body can be subdivided into smaller spaces, e.g., a "right" and a "left" half, etc. These subspaces are filled by parts of the body. Hence the body seems to consist of adjoining parts which are bodies themselves. To be sure, they are mostly not moving relative to one another inside the body. But internal oscillations, elastic and plastic deformations, and actual divisions of bodies show their parts as being movable in principle.

Now a body appears to us as a *continuum*, at least in thought being infinitely divisible. Two questions arise: Does a body actually consist of infinitely many parts, as we can imagine it to be? What keeps the parts of a body together? One attempt at an answer is given by *classical atomic theory*: Every body consists of a finite number of indivisible bodies of finite size, the atoms. Evidently, however, this doctrine only defers the theoretical problem. If the atoms themselves are bodies, what about *their* parts? Here is now a second, purely theoretical "place in life" for point mechanics; since Boscovich, especially in the late nineteenth century, it serves as a radicalization of the atomic hypothesis, taking the form: Bodies are in reality systems of point masses.

This thesis of *point-like atoms* interprets the parts of a body as ultimate, irreducible facts—mobile points—that maintain their identity over time. The mass of a body is related to the idea of the amount of matter in a way that is never completely clarified, and is now merely a number attributed to the point-like atoms. How a point mass itself holds together is a problem that is

---

[6] For methodological reasons, we should emphasize that an accurate determination of the position of a planet itself uses celestial mechanics from the outset.

no longer considered. The mass points making up a body are considered to be held together by forces acting at a distance between them. The concept of the force is necessarily correlated with the concept of the point mass. Historically we should at once add that explaining the existence of almost rigid bodies in terms of forces acting at a distance between point masses never succeeded, and that we explain the approximate incompressibility of rigid bodies essentially in non-classical terms, namely in terms of quantum theory. The thesis of point-like atoms obeying a classical point mechanics was an admittedly important mistake for the clarification of the concept, or more cautiously, a one-sidedness.

We close this section with some thoughts about the empirical determination of the masses of bodies. As a matter of fact, this happens mostly through weighing, that is, using the law of gravitation in which the mass occurs again. Each such measurement presumes a concept of the force beforehand, a special form of the force law, and thus the basic laws of mechanics. Now it is also one aim of physical semantics to permit an empirical examination and verification of the claim that the formalism in question is the "correct" physical theory, or at least a good one. Can one determine the values of the physical quantities that enter into the equations without first taking the validity of these equations for granted? Here for the first time we encounter the *foundational problem of physics*.

Let us first assume that positions and times (and thus the velocities) can be measured independently of the basic laws of mechanics. One could then envision an empirical check of the hypothetically assumed basic equation, such that one measures the initial conditions for a system of point masses (e.g., planets), predicts from these the paths at later times, and compares them with the subsequent actual motion. Although an agreement of both for individual cases would not constitute a logically rigorous empirical verification of the basic equation as a *general* law, a disagreement would be considered an empirical *refutation*. But the prediction is only possible for already known masses $m_i$ and force functions $f_{ik}$. In celestial mechanics we posses the additional hypothetical equations (3)–(5) for the forces. Now there remains nothing else but to use different values of the masses $m_i$ and to try to find an assumption that leads to agreement with experience. In fact this happens in successive approximations, which permit a gradual improvement of the assumptions. Historically this procedure led to complete success; over centuries, celestial mechanics proved itself in terms of millions of empirical data.

This procedure depends on investigators coming up with promising hypotheses. Much thought has been devoted to whether the successive empirical determination of all relevant quantities is even possible, such that eventually the laws can be read off as a simple description the empirical data. It is one thesis of the present book that this is not possible. Now in empirical science, an alleged impossibility is also incapable of rigorous logical proof. We can only give factual examples of failure and plausibility arguments for the inevitable failure of this "strictly empirical" procedure. We can try to learn from these examples what minimum set of theoretical ideas must be taken for granted, at least hypothetically, for the determination of empirical quantities.

In our case we may attempt next to determine empirically the masses of (terrestrial) bodies, still assuming that positions and momenta are measurable, independent of the theory under scrutiny; here we mean the *inertial mass* used in (2.1). It can be measured empirically in terms of the inertia of a body: two bodies subject to the same force suffer accelerations directly proportional to their masses. One can then also show empirically that mass is an extensive quantity to a very good approximation, i.e., if one joins two bodies into one, then the mass of the composite body is equal to the sum of the masses of its parts. This procedure, however, differs only in detail from the historically older one of celestial mechanics. It again hypothetically presumes the general theoretical framework which it is ultimately meant to prove. As we have no direct sensory perception of inertial mass, it is a hypothesis that there is actually one and only one parameter that determines the inertia of every body. Inertial masses can only be compared if it is already understood that a force is measured in terms of the *acceleration* it produces (see Subsection c) In addition, there must be assumptions about when the forces acting on two bodies with different masses are equal. All these assumptions stand the test of experience, but the precise formulation of this experience was at least historically only possible in terms of just these assumptions, and the author is not aware of any subsequent precise account of this experience that does not make these assumptions.

### 2.2.3 Force, inertia, interaction

The conceptual meaning of force is that it is the *cause of change*. Formally we see this from (2.1), where the function $f_{ik}$, whose form enters into the solution as a system parameter, is proportional to the time derivative of a state variable. In words: it is already assumed how the force will depend on the state, and this force then determines the change of the state in time. The concepts of time and change, like other concepts, we assume as given.

Equation (2.1) will determine the change of the state completely. One then understands $f_{ik}$ not as *one* cause among others but as *the* cause of the change of state. This idea, however, immediately confronts us with a *causal paradox* of classical mechanics, which we still do not find resolved in modern physics. As this paradox is scarcely recognized nowadays, as compared to the time of Galileo, we should like to emphasize it here specifically. Mathematically it is due to the differential equation (2.1) being of second order. The state variable $x_{ik}$ alone does not determine its own subsequent development, but does so only in conjunction with its time derivative $v_{ik}$ as an arbitrary initial condition. In particular, setting the force, which is assumed to be the cause of the change of state, equal to zero, there are still solutions with constant velocity:

$$x_{ik} = a_{ik}t + b_{ik}$$
$$v_{ik} = a_{ik} \tag{2.6}$$

A body with no motive force acting upon it moves with constant velocity.

One avoids the linguistic appearance of a paradox by appropriately calling force not the cause of motion but the cause of acceleration, and letting the state be described by position *and* velocity (or momentum). But even when one interprets the $x_{ik}$ and $v_{ik}$ as the state parameters, the solution (2.6) represents a state that varies in time, without Eq. (2.1), where one has put $f_{ik} = 0$, providing any indication of an external influence that might be regarded as the cause of this persistent change of state. In contrast to the sensitive causal conscience of Aristotle and the scholastics, who searched for an explanation of the continued motion of a freely thrown body, in modern times we have simply renounced such an explanation of inertial motion. This renunciation is not a resignation in principle regarding causal explanations; it is none other than surrender in the face of an unsolved problem. We will seek the solution in a radical formulation of the concept of time (Sect. 4.3c).

We set aside this problem for the time being, then, and continue to speak in a sloppy manner of the force being the cause of the change of state.

Continuing with the end of the previous section, when asking about the *empirical* justification of the general expression for the force according to (2.1), we first must ask about the empirical foundation of the law of inertia, i.e., the absence of terms of first order in the differential equation. There was no completely force-free motion, i.e., true inertial motion, observable by humans within the range of experience of early classical physics, and strictly speaking, it certainly does not exist at all. Hence once again one depended of the formulation of a hypothesis, inspired by the observations of free fall, objects rolling down inclined planes, etc. Again we content ourselves with ascertaining the success of this hypothesis. The proof that is impossible to explain all presently known facts by means of a mechanics based on an equation of motion of first order (or higher than second order) would probably be very difficult, and rigorously impossible. Taking (2.1) for granted, one can, after determining the the masses $m_i$, empirically find forces step by step, in many cases most conveniently by means of static measurements, i.e., by adding several forces up to zero. The vectorial addition of forces follows from the vectorial addition of acceleration if one assumes (2.1).

We take up the question of the *theoretical* justification of Eq. (2.1) itself later. Here we inquire, for the given general form (2.1), into the theoretical justification for the special force laws $f_{ik}$ to be used. The attempt of the seventeenth century to deduce the forces between bodies from the concept of a body, namely its impenetrability, thus explaining forces in terms of pressure and collision, was unsuccessful. It contained, however, the causal notion that the state at a particular location can only be influenced by the state in its immediate vicinity, the principle of local interaction, which was taken up again in physics at the the end of the nineteenth century and formulated as a law of nature in the special theory of relativity. The mechanics of mass points abandons this idea completely. Instead, it must make the force a separate physical reality in addition to the mass compounded of the point masses. The classical adoption of action at a distance (Coulomb's, Weber's law of electrostatics and electrodynamics) enabled the force, following the example of Newton's law of gravitation, to depend solely on the relative coordinates of the bodies (or on their time derivatives), and thus retained the idea of force being the influence of one body upon another. In the field theories, however, forces develop into independent realities in addition to the bodies.

**2.2.4 Space**

The three coordinates $x_{i1}, x_{i2}, x_{i3}$ of the $i$-th point mass denote its *position* in *space*. Just as the mechanics of point masses found itself compelled to introduce, besides point masses, forces as realities of a different kind, classical mechanics ever since Newton considered space to be an independent reality, of a different kind from bodies (and forces). This was by no means self-evident; rather it was the conclusion of an effort of abstraction that had been accomplished to this precision perhaps first by Newton.

We illustrate this abstraction by means of the expression *absolute space* first introduced by Newton. He distinguishes it from *relative* space as the embodiment of the *relative positions*, i.e., the positions relative to other bodies, where a body can be found. "In my room" or "on top of the Mt. Blanc" are relative places; "my room," bound by its walls in my house, or "the Alps," localized relative to our planet, the Earth, are relative spaces in the sense of Newton. To understand the importance of this distinction a review of the philosophical development of the concept of space is perhaps indispensable.

The first forerunner of the Newtonian abstraction was probably the concept of the *void* of the Greek Atomists (Leucippus, Democritus). This itself has a philosophical prehistory. Parmenides of Elea had introduced the concept of the *Being* (eon) as a fundamental concept. The Being cannot arise or dissolve, as it would have to arise from non-Being and dissolve into non-Being; there is, however, no non-Being. The changes we experience in the world are then a mere illusion (doxa). Here we do not attempt an adequate interpretation of this doctrine (see *Die Einheit der Natur* section IV.6.3 and Picht, *Die Epiphanie der ewigen Gegenwart*, 1960). The atomists were at any rate followers of Parmenides insofar as they admitted the unchangeability of the Being; they, however, wanted to save the reality of change. They did this by assuming a multiplicity of Beings, namely the atoms. The change in the world is then nothing but a change of the relative positions of the atoms. For this to be possible atoms must be movable relative to one another without changing their shape or size. To that end they must be in "the void." Atomists must therefore introduce two principles: the plenum (pleron) and the void (kenon), which they also called the Something (to den) and the Nothing (to meden). The indivisibility of atoms was not an additional postulate; in this way of thinking it followed directly from the atoms as Being, thus unchangeable.

The classical tradition of philosophy rejected atomism. The ultimate reason for this was its "materialistic" denial of a highest spiritual principle. But the paradoxical assertion of the existence of a Nothing was one of the theoretical points of criticism. Plato uses a word (chora), which we must translate as "space," to denote the principle of a pure possibility that Aristotle called "matter" (hyle). Aristotle repudiated the existence of a completely empty space with precise arguments. For a body in such a space there is no reason why it should move rather faster than slower, rather here than there. For a completely empty, structureless space this argument is irrefutable; what physics nowadays calls a vacuum is filled with fields, thus not empty in the sense of Aristotle and the Greek atomists. Aristotle avoided the need to speak of space at all by defining the location (topos) of an object as the surface of the bodies surrounding it. Indeed spatial correlations of bodies with no empty intervals between them can be defined as mere relations. Newton distances himself from this notion.

The void of the atomists was postulated as being infinite. The world of Aristotle was finite; therefore the position of a body relative to the Earth, at rest at the center of the universe, could be understood as "absolute" position. Nicholas of Cusa again introduced the idea of an infinite universe, and immediately drew the conclusion that position and motion were not absolute but defined only relative to other bodies.[7] The emerging natural science of modern times soon turned again to the convenient method of explanation by means of atoms and the vacuum. Galileo needed the vacuum, as only in vacuum does the law of falling bodies assume a simple form.

Newton draws his conclusions from this development. We can read off the power of his argument from Eq. (2.1). The equation is invariant under a "Galilean transformation," i.e., under transformation to a coordinate system that moves uniformly with respect to the original coordinate system, but not under transformation to one which is accelerated with respect to the original system. In the latter case, "fictitious forces" appear. Acceleration relative to absolute space is thus objectively defined.

This[8] dualism of space and matter left many thinkers dissatisfied. Leibniz denied the existence of an independent object "space," ascribing it instead to the domain of the properties of bodies. A body has size, shape, distance from another body; but it would be meaningless if one wanted to distinguish the entire world, as it is, from an imaginary identical one which had been moved ten miles to the right "in space." This criticism, which Mach took up again, paved the way to the theory of relativity. But initially it was not accepted. Newton adopted the strong physical argument that at least an "absolute" acceleration of a body (like a rotation in the experiment with the pail) can be measured in the body itself, without comparison with other bodies. The entire

---

[7] See *Zum Weltbild der Physik* on the history of this development, in particular the essay *Die Unendlichkeit der Welt. Eine Studie über das Symbolische in der Naturwissenschaft.*

[8] The following four paragraphs are taken from *Die Einheit der Natur* II,1d, p. 143–144.

subsequent development of physics can only be understood if one accepts the dualism of space and matter from the outset.

Space is thus a proper subject of physics, but a unique one. The mere assertion of its existence carries a palpably different sense than asserting the existence of matter. *There is matter* means *There are objects somewhere.* *Somewhere* means *somewhere in space.* But it would be meaningless to say *Somewhere there is space.* Space is in fact precisely that which gives meaning to the concept *somewhere*.

Likewise, the relationship between space and matter from the standpoint of causality exhibits the same lack of symmetry. That a rotation "relative to space" gives rise to centrifugal and Coriolis forces appears to be an influence of space on matter. But in classical physics there is no influence of matter on space; its structure is fixed a priori. We can regard all these facts, which are difficult to formulate correctly in words, as evidence of past, and to some extent present, unresolved problems in the unity of physics.

Alongside the problem of the nature of space looms the problem of how we can know anything about it. Newton, to be sure, invoked evidence for his postulate of absolute space. From the fact that we already know with certainty, before any individual experience of "external" things, that it will be experience in space, Kant concluded that space is a subjective prerequisite for all sensory cognition, a form of our perception. Geometry, interpreted as the science of space, appears in both Newton and Kant not as a branch of physics but a prerequisite.

We revisit the subsequent development of the concept of space in the section on the general theory of relativity. Regarding the development of geometry, see *Zeit und Wissen*, I.5.2.

### 2.2.5 Time

Newtonian mechanics is also the starting point for a formally parallel treatment of space and time. In Eq. (1), however, the time $t$ is clearly distinguished from the spatial coordinates $x_{ik}$ as the sole independent variable; the equation is a differential equation in time, but not in space. In other words, mechanics describes the motion of bodies, and conceives of motion to be change in position. Change, however, is change " over time." Yet here Newton treats time in the same way as space by postulating, by analogy with absolute space, and for analogous reasons, *absolute time*.

Up to the present century, this blurring of the qualitative difference between space and time has continued unabated. Even in point mechanics, time can be formally introduced as an additional variable. Field theories have then used partial differential equations based not on the $3n$ spatial coordinates of $n$ point masses, but four coordinates instead, three for space, one for time, followed by differentiation. The differential equations, however, are always hyperbolic and do not admit transformations such as Euclidean rotation, which would transform a spatial coordinate $x_k$ into the time coordinate $t$ (and $t$

into $-x_k$). The appearance of hyperbolic "rotations" of space and time in the special theory of relativity changed nothing here; it gave rise, however, to the popular saying that time is "nothing but" a fourth spatial coordinate.

The present book has chosen a radically different starting point. It treats time in isolation, and from the outset completely avoids defining it as a real variable $t$, treating it instead in terms of the structure of present, past, and future, or of factuality and possibility. The reason for this was the realization, as described in Chap. 6, that thermodynamic irreversibility can only be reconciled with mechanics by explicitly assuming this structure of time. As we now begin, in the second part of the present book, with a chapter describing the traditional structure of theoretical physics, we must also accommodate this traditional structure in our line of reasoning. We must show that all empirically established propositions of physics can be effortlessly formulated in our language, if not more precisely. Ultimately, we will see that this language alone provides the means to formulate quantum theory without apparent paradoxes.

## 2.3 Mathematical forms of the laws of nature[9]

In contemporary physics one can formulate a general law of nature in at least four forms. One can specify

a) a family of functions
b) a differential equation
c) an extremum principle
d) a symmetry group

Mathematically these forms are closely related. The solutions of a differential equation comprise a family of functions. For a family of functions one can construct a differential equation with its manifold of solutions being this given family (see Courant–Hilbert, Vol. II, Chap. I). An extremum principle implies differential equations for its Euler equations. The converse is in general not possible; only certain classes of differential equations are associated with extremum principles. A differential equation (and, if it exists, the corresponding extremum principle) is in general invariant under a symmetry group (typically a Lie group), which transforms the solutions of the differential equation amongst themselves. Conversely, a group generates on its homogeneous spaces a family of functions which permit representations of the group.

In probing the foundations of physics it is of interest, in terms of content, whether one can ascribe higher priority to one of these forms of the laws than

---

[9] This note was written in 1982 as a preliminary study of a discourse on symmetries in quantum theory and the theory of ur alternatives, and in some respects therefore anticipates topics treated in subsequent chapters. I present it here because it contributes to the language in which the classical theories are described in the present chapter.

the others. The above arrangement of the four forms corresponds approximately to their historical order of appearance. We can in the first instance vaguely designate as *law* a general rule that applies to many individual cases. This concept has been sharpened over the course of history.

The concept of a *family of functions* is a modern mathematization of the *morphological* type of law going back to antiquity; one describes a multiplicity of forms by their similarities and differences. For the history of physics the spacetime morphology of planetary motion became important. Its mature form comprised Kepler's laws. Here a fundamental problem arises: Kepler's three laws characterize *possible* planetary orbits; they do not determine which ones actually occur. Kepler, persisting in morphology, searched for a comprehensive law for the structure of the entire planetary system. That initially did not turn out to be a fruitful approach.

What prevailed was rather the formulation of the laws of nature as *differential equations with respect to time*. This represents a *causal* concept of the laws: the forces present at a given time govern the evolution of the state of the system. Wigner[10] spoke in Tutzing 1982 about "Newton's greatest discovery," the distinction between law and initial conditions. The law determines all *possible* motions, the totality of all solutions of the differential equation. The initial conditions dictate what motion actually occurs. The overall structural form of the planetary system, for example, can be ascribed (since Buffon and Kant) to the evolutionary history of the system.

The *variational principles*, from Fermat to Hamilton, were in the eighteenth century likely to be understood as an expression of a *final* cause. Ultimately, however, the development of the variational calculus showed that the differential and integral mode of description can be equivalent for many types of laws.[11] One must instead ask why precisely such types of laws exist at all. Hamilton's principle, for example, shows that Newton's equations of motion can be the Euler equations of a variational principle, as they are of *second order*.

I emphasize here a special problem, of importance for the following, which was previously mentioned in 2.c. The *law of inertia*, which empirically enforces the occurrence of second derivatives in the equation of motion, is fundamental for classical mechanics. It, however, represents a *causal paradox*. Aristotle understood motion as a change of state, and thus force as the cause of motion. In classical mechanics, however, inertial motion is just the motion without any forces acting. In the seventeenth century one still felt the paradox therein; Descartes and, following him, Newton defined the state of a body in terms of its velocity such that only acceleration was seen as a change of state. But this is inconsistent, as two bodies with the same velocity but at different locations are in different states, as correctly put by the modern description in phase space; and during inertial motion the point in phase space varies. If one wants

---

[10] See also Wigner (1983).
[11] See *Naturgesetz und Theodizee*, in: *Zum Weltbild der Physik*, 1943, 1957.

to think causally in a consistent way, one must radicalize Mach's ideas and interpret the inertial motion as being caused by the universe (the "distant masses"). This I have attempted in the ur theory, but now I doubt whether this is an adequate formulation. The other possibility is to regard causality, as expressed by a differential equation with respect to time ("relativistic causality" in the language of present-day physics), merely as an aspect of the foreground, some sort of classical limit of a different temporal connection. Both possibilities are discussed in Chap. 3 and 4.

A model for such a discussion can be found in the description of inertial motion according to Hamilton's principle. The time integral of a state function $L$ over all kinematicly possible paths from a point $(x_1, t_1)$ fixed in space and time to another fixed point $(x_2, t_2)$ is assumed to be an extremum for the actual path. Note that only the position $x$ is prescribed at the initial and final time, not the state $(x, p)$. In any case, for $x_1 \neq x_2$ these boundary conditions already anticipate that in the meantime the body moves: the form of $L$ when external forces are absent is then chosen so that the motion is rectilinear and uniform. Hamilton's equations for $\dot{x}$ and $\dot{p}$ stipulate, as the Euler equations of the variational principle, the form of causality sufficient for this requirement: the state defines its own changes, but according to this causality almost all states (those for which $p \neq 0$ initially) are compelled to change. The intuitively plausible image of "acting causes" has here been lost unless one could conceive the variational principle itself as the effect of such causes. This claim is honored only in wave mechanics, where Hamilton's principle of corpuscular mechanics appears as the Huygens principle of waves (Dirac 1933, Feynman 1948). But now there arises the question of the causal explanation of the equations of motion for waves. The matter wave obeys the Schrödinger equation, which itself can be derived from a field theoretical Hamilton principle. Apparently, despite the great plausibility of Dirac's idea, one still has not gained anything in principle, unless one attempted to understand the Hamilton principle of the Schrödinger equation for the one-particle problem as a Huygens principle for second quantization. This I have attempted (Weizsäcker 1973c); we return to it in Chap. 4.

Thus, attempting a serious causal understanding of force-free motion leads directly to obscurity, even in contemporary physics. Under the influence of an empiristic philosophy, we have merely become accustomed just to assert the incomprehensible.

*Symmetry groups* denote a type of law that cannot be integrated in the three-way alternative of morphology, causality, and finality, but instead points to a possible common origin of these three forms. Historically the concept of symmetry stems from morphology. Heisenberg saw in Plato's use of the regular solids as "atomic models" a forerunner of his own group theoretical way of thinking; Kepler used the same solids in an imaginative construction of his comprehensive model of the solar system. The group theoretical treatment of geometry and physics, introduced about 1870 by Sophus Lie and Felix Klein (the "wise" and the "lucky"), achieved great success in Einstein's derivation of the Lorentz transformation based on simple group theoretic postulates. Nowadays one derives variational principles and differential equations as far as possible from invariance postulates. The question then becomes how best to justify the groups employed. Heisenberg, in fact, saw in them an empirical, aesthetic factor that is perhaps not further reducible. It has been my longstanding goal to derive them within the framework of axiomatic quantum theory, based upon requirements of indistinguishability.

I might be permitted to cite here a preliminary stage of these considerations. For a Diploma thesis (Franz 1952) I had proposed to investigate possible dynamics of differential equations of $n$-th order in time, in particular to compare the "Aristotelian" ($n = 1$) and the Newtonian dynamics ($n = 2$) with possible higher orders ($n > 2$). All appeared to be internally consistent. I then sought to identify Newtonian dynamics as the classical limit of wave mechanics, for which I postulated a rotationally invariant wave equation which appeared to require the Laplace operator, i.e., a differential operator of second order. The proposal was not followed up mathematically, but in fact it pointed in the direction adopted since then.

## 2.4 Chemistry

Chemistry is the empirically based science of the qualities and transformations of material substances. From the wording of this casual definition one can immediately infer a number of empirical facts which were always known in everyday life.

The subject at hand is *substances*. Examples are water, silver, benzene. This use of the word differs from the somewhat shaky usage in philosophy. "Sub-stance" philosophically is what underlies the changes. Here the chemical usage of the word follows the philosophical one. "A substance" in the sense of chemistry, however, is neither a universal principle nor a single phenomenon. It is a specific material characterized by certain *qualities* (hardness or softness, taste, smell, color, etc.), some *stuff* of which many different things might consist (a creek, a raindrop, the ocean; a silver spoon, a coin, etc.). The qualities may in general pass through a continuous scale of "values" (from hard to soft, from hot to cold, from violet to red, etc.). The substances, however, can be classified in a discrete manner, and the continuous transitions between them prove to be mere mixtures. There is pure water in which all sorts of substances can be "dissolved"; pure silver or all kinds of alloys; pure benzene, etc. The *discreteness* of the pure substances is a basic fact of chemistry which a priori is not self-evident at all.

The *transformations* of substances are partly changes of the state of one and the same substance—especially under the influence of *heat* (solid, liquid, gaseous)—and partly *chemical reactions*. The latter concept owes its origin again to the empirical, altogether nontrivial discovery that there is a difference between a *mixture* and a *compound* of two substances. The qualities of a mixture pass through a continuum between the qualities of the pure substances; the qualities of a compound are novel and characterize a new substance.

This observation supports the very early chemical theory that attempts to explain all substances as compounds of a few basic materials, the *elements*. The Latin word "elementum" is a translation of the Greek "stoicheion," which means letter. As the entire *Iliad* can be written with the twenty-four letters of the alphabet, so does this rich and colorful world consist of only a few elements. This comparison is due to the atomists, but could equally well be applied to Empedocles' doctrine of the four elements earth, water, air, and fire, the latter being explained by Aristotle as the four combinations of the two pairs of basic qualities, wet/dry and warm/cold. These are philosophical attempts at a comprehensive abstract theory. Historically, from our present point of view, this theory came too early. The great advance in chemistry, fostered by the intermediate stage of alchemy, occurred at the end of the eighteenth century when, on empirical grounds, a limited number of "chemical elements" such as hydrogen, oxygen, etc. were decided upon as a basis.

We have recalled these universally known facts to point out that even in everyday life the discreteness of substances indicates a basic structure of reality whose scientific explanation, as we shall see, can only be expected from quantum theory. It is precisely the familiarity of this phenomenon that makes it seem so mundane. If not for the phenomena of discreteness in nature—straightforward discrimination among a finite number of substances being one example—we would likely find it impossible to formulate concepts. But how can this discreteness persist in light of the continuous variability of forces and qualities?

Before we continue let us mention one price the modern doctrine of chemical elements has paid. The four elements of Greek philosophy represented at the same time sensory, psychological, and intellectual qualities. For example, the four elements were linked, in terms of character traits, to the four "temperaments." In the basic principles of alchemy the chemical meaning is inseparable from the psychological/symbolic one (which was emphasized, among others, by C.G. Jung). The modern theory of chemical elements, however, is meant to be strictly "materialistic." Oxygen is completely characterized by its physical properties. The link between *material* substances and psychological, intellectual, and even sensory qualities is no longer the subject matter of chemistry. In the interpretation of quantum theory (Chap. 9 and 12), we will again be reminded of these displaced questions.

The discreteness of the substances obtained in reactions was quantitatively formulated at the beginning of the nineteenth century in the law of constant and multiple proportions (Dalton 1808). A theoretical interpretation was immediately offered by the atoms of chemistry, according to which an element is characterized by a certain kind of atom, and a chemical substance by a molecule consisting of atoms in a certain way. The discreteness of the possible sorts of molecules then explains the discreteness of the substances defined not as mixtures but as compounds, and allows the first definitive distinction between mixtures and compounds. At the same time, the old "four elements" are explained as being different phases. Only in a gas do the molecules have an independent existence; therefore only in a gas is the distinction between a mixture and a compound always precisely defined.

Many scientists had already leaned toward atomic theory, but it was chemistry that first conferred empirically based legitimacy on the discipline. Yet here as well there was a price to be paid.[12] Among the criticisms of the atomic hypothesis by classical philosophy (see Sect. 2c) was also that this doctrine by no means resolves the problems associated with the concept of continuously extended matter. Instead, it merely shifts them to the interior of the atom. Kant argued that an extended atom fills space, which can be thought of geometrically as being divided into subspaces. These subspaces are obviously filled by parts of the atom. Therefore the atom consists of parts, and it is only a question of the forces acting between them whether the so-called atom will hold together under any circumstances. One cannot dispose of the problem of the nature of forces by means of the atomic hypothesis. One may say that we owe the chemical theory of the atom to the serendipitous philosophical naïveté of scientists, in this case chemists. For just under a century the description of the empirically required building blocks of the elements by means of extended atoms was sufficient. At the end of the nineteenth century, physics began to realize the unresolved problems of this concept, and only the revolution of quantum theory was for the time being able to separate right from wrong.

The multiplicity of chemical elements demanded further explanation, even in the context of chemical atoms. The hypothesis of Prout (1815) that all elements consisted of hydrogen offered a unity of substance, but failed in that many atomic weights turned out not to be integral multiples of the atomic weight of hydrogen. This was only explained by the atomic theory of the twentieth century in terms of mixtures of isotopes and mass defects. The periodic system of the elements (Mendeleev 1869), on the other hand, pointed to a definite relationship which, however, could only be explained by quantum theory (Bohr 1919).

Chemistry represents decisive progress in the system of theories which, however, by the end of the nineteenth century, proved to be more an empirical setting of the problem.

## 2.5 Thermodynamics

Classical thermodynamics can be considered the greatest achievement of abstraction in physics; perhaps of physics altogether, certainly of the physics before Einstein. Although the theory of heat began as a description of the phenomena of certain sensory experiences—a quality, according to our earlier

---

[12] See *Die Atomlehre der modernen Physik* in: *Zum Weltbild der Physik*, 1943, $1957^7$.

definition of chemistry—its two major laws span the entirety of physics. In hindsight, the feat of abstraction lay primarily in the renunciation of deriving of these laws from specific models. As we know now, the atomism the founders of these theories actually believed in was, in the form it then had (chemical theory of atoms or, for Helmholtz, point mechanics), false or rather inadequate. But the thermodynamic argumentation is of such generality that it remained unaffected by later corrections of these model assumptions; it fact, it did not depend on the form of the model at all. Einstein expressed the singular status of thermodynamics in the following way: "A theory is the more impressive the greater the simplicity of its premises is, the more different kinds of things it relates, and the more extended is its area of applicability. Therefore the deep impression which classical thermodynamics made upon me. It is the only physical theory of universal content concerning which I am convinced that, within the applicability of its basic concepts, it will never be overthrown (for the special attention of those who are skeptics on principle)."(Einstein 1949, p. 32)

In fact, both laws of thermodynamics are based on two abstract and therefore universal concepts, *energy* and *temperature*, and, definable in terms of these two, the concept of entropy. Both concepts have a specific origin, energy from mechanics, and temperature from the theory of heat. But their characteristic abstract properties have a universal *temporal* meaning: "Energy remains conserved, temperatures become equalized." Thus the first law is a *conservation law*, the second a *law of irreversibility*.

Naturally such a general statement does not yet explain why each of these two laws refers to a *single* fundamental quantity: the one about conservation referring to energy remaining the same, the one about irreversibility, in an appropriate form, to entropy not decreasing.

Indeed, physics nowadays knows many conservation laws. The singular role of energy becomes understandable only in terms of the approach from the special theory of relativity, which makes it possible to equate the conservation of energy with the conservation of mass. We return to this problem in Chap. 10. But here we point out that one is then dealing with a property of *time*: conservation of energy, viewed relativistically, expresses the *homogeneity of time*.

On the other hand, physics knows but one quantity whose variations define irreversibility, namely *entropy*. This becomes clear in the statistical interpretation of thermodynamics in terms of the reduction of entropy to *probability*. This we discuss in detail in Chap. 6. Here, too, we see the distinctive role of *time* in physics.

The abstract and universal character of the Second Law, as Einstein characterized it, becomes particularly evident in the formulation found in Gibbs (1902). This formulation abjures any special model of atomic processes, employing instead only those completely general arguments which, as Gibbs knew, would outlast any change of model.

## 2.6 Field theories

Field theories, in a form we still find acceptable, were first derived from the mechanics of continua: acoustics, hydrodynamics, theory of elasticity. There, effectively, it was not important whether matter was in principle conceived as a continuum or having atomic structure. In the latter, the treatment as a continuum was empirically a good approximation, and that is how we interpret these theories today.

As soon as mechanics had made the distinction between matter and forces a central theme, a question arose as to whether the forces might not be effects of some space-filling matter, and must therefore themselves be described by some continuum mechanics. Newton, who himself had introduced gravity as an action at a distance, considered this only a successful description of a phenomenon still not fully understood. He says in the famous *Scholium generale* of his *Principia* (1687): "I have not as yet been able to deduce from phenomena the reason for these properties of gravity, and I do not 'feign' hypotheses."[13] Thus he postulates a cause which he presumes in his later "Opticks" (1706) to be a local interaction.

The belief in a point mechanics with actions at a distance remained a transitory phase of physics. Ever since Faraday's analysis of electrodynamics and its mathematical formulation by Maxwell, the forces themselves were considered a physical reality with internal dynamics, i.e., a *field*. The field variables were understood as functions of space at some instant of time. For the first time space and time were described as a four-dimensional continuum, the dynamics by means of a system of linear hyperbolic differential equations. Hypothetically one thought about this as the dynamics of a continuum, of a medium not directly observable mechanically, the ether. Einstein's special theory of relativity showed that no particular velocity (a rest velocity, for example) could be assigned to this medium relative to absolute space, and that the ether, as an explanation, had been a superfluous fiction. We will therefore subsequently describe the field theories only in relativistic terms.

## 2.7 Non-Euclidean geometry and semantic consistency

The transition to the great theories of the twentieth century was a twofold scientific revolution. We are interested in the "system of theories," i.e., the factual connection between old and new theories. This we discuss, here in a preliminary way, methodologically under the title of the *semantic consistency* of theories. What is meant by this we first illustrate with an example that

---

[13] In the phrase *hypotheses non fingo*, one ought perhaps to translate *hypotheses* in the old, already platonic sense of assumption, i.e., fiction, and *fingo*, which literally means conceive, as invent or devise. Should Newton have meant the newer meaning of *hypotheses* as conjecture, one ought to translate *fingo* as the sharper *feign*.

had emerged by the nineteenth century, concerning the possible validity of a non-Euclidean geometry.[14]

That the Euclidean parallel axiom cannot be deduced from Euclid's other postulates was in fact recognized by Saccheri and Lambert in the eighteenth century. Basically, this question was a direct consequence of skepticism about the direct evidence of the postulate. That Euclid had postulated it separately shows that even he did not believe in its possible derivation from the other postulates. Gauss conceived of a non-Euclidean geometry, which Bolyai and Lobachevsky worked out, independently of him and of one another. Naturally, the question arises as to which geometry is valid in physical space.

We leave aside the difficult epistemological questions that arise here.[15] We describe the circumstances as they would appear to a present-day physicist. If the physicist believes in a non-Euclidean geometry of physical space, he will initially give a mathematically precise formulation of that geometry; in the nineteenth century that would have been Lobachevsky's hyperbolic geometry. In the pedantic fashion of Sect. 2.2a above, he will introduce symbols which he describes with a *mathematical semantics*, e.g., $x_1, x_2, x_3$, with the meaning of coordinates in a hyperbolic space. Now there is the question of the *physical semantics*. For the question what actual, physically "measurable" quantities are meant by, say, $x_1, x_2, x_3$, he must appeal to a preconception, which is provided to him by some language to answer this question. The quantities are assumed to be, e.g., the projections of the distance vector of a point mass from a given point mass at rest, onto the axes of an arbitrarily chosen coordinate system. One can see, to put it succinctly, that I myself am using the largely mathematical language of a physicist. The preconception actually invoked is thus not a pre-mathematical version of colloquial speech, or even one of an early civilization still untouched by science and technology. The path from there to the problem faced by Gauss and Lobachevsky would be too long and cumbersome for a description in this book. In other words, the preconceptions required for a modern theory themselves constitute a modern theory, but an earlier one. In Heisenberg's formulation, the preconceptions for a closed theory are provided by the previous closed theories.

In our case, the new closed theory which is assumed to originate from hyperbolic geometry, supplied with a physical semantics, is obviously contained in the thesis of physical space being a hyperbolic space. The preconceptions we need in order to say what "physical space" is reside in the older closed theory, that of Newtonian mechanics, which teaches that physical space is Euclidean. But in that sense, in terms of the imposed physical semantics, we appear to introduce an immediate contradiction. We appear to assert $x_1, x_2, x_3$ to be coordinates of a hyperbolic space (per mathematical definition) and at the

---

[14] More details about this connection can be found in *Zeit und Wissen* I.5.2; the semantic consistency in particular is treated in *Zeit und Wissen* I.5.2.7.
[15] See *Zeit und Wissen*, I.5.2.1–6.

same time of a Euclidean space (per physical semantics). The physical theory created this way would therefore be "semantically inconsistent."

Any physicist knows how to resolve this apparent problem. The new theory is supposed to replace the old one. The old theory, strictly speaking, is then false. Presumably it is still useful, but only as a limiting case or local approximation of the new one. To make the new theory "semantically consistent" we must say something like "$x_1, x_2, x_3$ are coordinates which, as is generally known, we can only measure inaccurately. Thus far we have ascribed to them values consistent with the accuracy of the measured values, which were compatible with an interpretation of space being Euclidean. From now on we want to ascribe to them values that are compatible with an interpretation of space being hyperbolic." In other words, for the sake of semantic consistency of the new theory, we must adjust our preconceptions. As explained in 2.2$d$, because every theory, old as well as new, is based on idealizations, we may hope that this will work.

The methodological question of how to assure oneself of the correctness of a theory, briefly touched upon in 6.2, we will further discuss only in *Zeit und Wissen* I.1–4 and I.5.2. Here we already suspect that a rigorous methodology of empirical science is in fact impossible. In the present chapter we stay within the usual understanding of physicists. We accept the established theories and only analyze their interrelationships.

## 2.8 The relativity problem

Astronomy is an older science than physics. The first example of mathematical laws of nature we discussed in Sect. 2.3 above are from astronomy. The problem of the relativity of motion arises in astronomy.

The oldest, morphological version of the laws of planetary motion is inextricably intertwined with cosmology, which we might define as the specification of a model of the structure of the universe. Here Greek astronomy was compelled to reach a decision. The geocentric model was not self-evident. The Pythagoreans had already discussed other possibilities. Aristarchus offered the heliocentric system which Copernicus took up again later. If one follows the arguments of this ancient controversy,[16] one can see the decisive importance of convictions which nowadays we would assign to physics. One argument against the rotation of the Earth on its axis was that the air, in remaining behind as seen from the rotating Earth, would generate a permanent, enormous storm moving east to west. We are no longer impressed by this argument, as we long ago became accustomed to the notion that the air is rotating with the Earth. This, however, is a consequence of the *law of inertia*. In antiquity, i.e., in Aristotelian physics, a constant force was required to act

---

[16] Cf. Bartel Leendert van der Waerden: *Ontwakende Wetenschaft*, 1950 (Translated as *Science Awakening*, 1954).

to keep a sub-lunar body, thus also air, in motion. One sees here once more the key role of the law of inertia in modern physics.

*After* the Copernican model had been accepted in modern science, its recognition became the battle cry of the apostles of progress. This politicization of the conflict diverted attention from the actual and most interesting theoretical problems, which should have been resolved by the discussion. Even the debate among the experts, as in the early, still objective phases of the disputes between Galileo[17] and the church (Bellarmin 1615), reveals a lack of clarity in the arguments of both sides. From the point of view developed in the present chapter, one ought to look upon the dispute in perhaps the following manner.

Historically, there is an absolutistic and relativistic interpretation of motion. In the absolutistic tradition one can count among others Ptolemy, Copernicus, Kepler, Galileo, and Newton; in the relativistic are Cusanus, Bellarmin, Leibniz, and Mach. Einstein's intention was relativistic, but his results contained elements of the absolutistic point of view. The question, to begin with, has to do with the meaning of the vocabulary in which the dispute was carried out.

The dispute regarding the geocentric and heliocentric system naïvely presumes that one already knows the meaning of rest and motion, inasmuch as it is naïve/absolutistic. Nicolas of Cusa (Cusanus) had already overcome this naïveté in 1450. He assumes the universe to be infinite. Then there is no criterion of rest and motion in the shape of the universe itself (as there was before in the shape of the heavenly sphere and the motionless central body). Hence motion is by definition only a relative motion of bodies. Cardinal Bellarmin had understood this in 1615. He demanded of Galileo to present the Copernican system as a mathematical hypothesis but not as truth. "Hypothesis" here[18] means not "assumption" but "conjecture"; with the word "model" we come close to the manner of speech desired by Bellarmin. Bellarmin was capable long ago of thinking about the relativity of motion. One can describe, as one pleases, the same motion geocentrically or heliocentrically. Galileo at that time complied diplomatically. But he remained convinced of defending not a "hypothesis" but the truth.[19] Nowadays we can say in what sense he was right.

The astronomical argument for Copernicus was the greater geometrical and dynamical plausibility of his model. Aristarchus already knew that the Sun is bigger than the Earth. Why should not the bigger and luminous body be the center of the universe, with the others revolving around it? Even sharper is the geometrical argument: the fact that Mercury and Venus, averaged over

---

[17] See *Tragweite der Wissenschaft*, Lecture 6.
[18] See also Sect. 2.6 regarding the same meaning of the word in Newton.
[19] For that reason it came to the second trial 15 years later, with outwardly catastrophic consequences for him; he was incapable of successfully hiding his conviction.

the long term, have exactly the same period of revolution around the Earth (considered to be at rest) as does the Sun is a curiosity in the geocentric system. In the heliocentric system it is the natural consequence of moving closer and faster around the Sun than the Earth. Tycho Brahe therefore took over precisely *this* part of the Copernican system and made Mercury and Venus special satellites of the Sun. Lastly, Kepler's ellipses were heliocentrically simple, but geocentrically highly complicated curves.

But all these remarks do not strike at the core of the relativistic argument, which says that the very *question* of which of the two models is the correct one does not make any sense. As long as one visualized heaven as a finite sphere one still could hope to describe the motion relative to this sphere as "true motion." The victory of the belief in an infinite universe eliminated this argument as well. Astronomers and physicists nevertheless did not yield to the relativistic philosophy. The kinematic description might be conventional. The dynamics, however, only assumed in the Copernican model a "reasonable" form, i.e., one that was initially simple. According to present-day judgment, what is the substance of this way of thinking?

First we wish to clarify the ontological background of this dispute. It is about the relationship between matter and space. One can distinguish a monistic and a dualistic perception of this relationship.[20] One can call the absolutist tradition *dualistic*. According to Newton there is absolute space *and* matter. The tendency of the relativistic tradition is *monistic*: according to Leibniz and Mach there is, in physical language,[21] only substance; its spatial relations are then consequences of their defining attributes, their extent. Einstein's intention followed that of Mach; it was monistic.

We discussed in Sect. 2d the emergence and justification of the Newtonian dualistic point of view. We will see in Sect. 10 the extent to which Einstein could not maintain his monistic intention; the general theory of relativity, as Einstein managed to realize it, is dualistic. In Chap. 5.2, however, we return on a quantum theoretical basis to a monistic point of view. Here, in preparation, we examine Leibniz' mode of thought and contrast it with Newton's.

Leibniz, the philosophically more erudite of the two antagonists, thinks in terms of substance and attribute, logically speaking of subject and predicate. This is a dualism which physics and logic never could dispose of. Logically (in modern formulation), a class is a predicate and is to be distinguished from its elements as the subjects to which the predicate belongs; Russell rightly sharply distinguishes a class with only one element from the element itself (the class of the first emperors of France has as its only element Napoleon I). In physics one can further distinguish between essential and conditional predicates. The essential predicates characterize the physical object (the logical subject of the

---

[20] Here I am following Mittelstaedt (1979).
[21] Philosophically, both went farther in monism and also included consciousness, as being primary, Leibniz in his monads, Mach through perceptions as elements of all reality.

propositions). For example, in classical mechanics a point mass is an object which has mass, position, and momentum, and nothing else, as its attributes. The value of the mass is an essential predicate of the particular individual point mass; the concept "mass" is an essential predicate of the concept "point mass." The essential predicate of a concept is a class of conditional predicates, like the mass being the class of all possible values of mass. Similarly there is for every point mass the position as an essential predicate of the class of its possible positions; a point mass "is an object that has position." Likewise for momentum. Quantum theory further splits the attributes of objects into two fundamentally different classes: observables and states. By means of this distinction we formulate in 5.2 the relationship between quantum theory and the theory of relativity.

Leibniz now understands space to be the essential predicate of matter, namely as relation (in Russell's language: as binary predicate). Bodies have relative positions. The unavoidable dualism between matter and spatial data like distance and direction is thus already embodied in the dualism of substance and attribute. Leibniz finds therefore, philosophically, a dualism of the substances—space and matter—superfluous. Newton avoids the term "substance" for space but, logically, he treats space and time as substances ("entities").

If space, according to Leibniz, is to be understood as the epitome of *relationships* among bodies, then in fact the physical meaning of relativity must be decisive for this controversy. Leibniz had good arguments for the *relativity of space*.[22] In modern language: in terms of the translation group of Euclidean space, there is no *geometrical* difference between two locations. Also Galileo had already understood the *relativity of rectilinear uniform motion*. Mechanics, especially the law of free fall, has the same form on a uniformly sailing ship as on the Earth at rest. The decision for Newton was brought about not by philosophical but by mathematical and empirical arguments. Newton saw that his laws of motion did not admit *the relativity of accelerated motion* and he verified this experimentally in the experiment with the pail.

In our century, Einstein has brought the thesis of the relativity of uniform motion to perhaps its definitive form in the special theory of relativity, and has again raised the problem of accelerated motion in the general theory of relativity. We too now follow this course.

---

[22] Compare his argument, in which the actual universe and an exact duplicate shifted 10 miles to the right are identical. There he was referring to his postulate of the Identity of Indiscernibles. Clarke thereupon answered that as Sir Isaac Newton had proven the existence of absolute space, the two universes were objectively different. Leibniz then invoked the Principle of Sufficient Reason: Why should God have created the universe here rather than there? Clarke replied that there was a sufficient reason: God's will. Leibniz had to respond that God never acts arbitrarily, but according to rational reasons.

## 2.9 Special theory of relativity[23]

The preconceptions of the special theory of relativity comprise among others Euclidean geometry and Newtonian mechanics. Here both may be considered closed theories whose semantic consistency can be questioned.

A physical axiomatics of Euclidean geometry will appropriately be based upon the Helmholtz–Dingler operations on rigid bodies.[24] In this way it justifies the six-parameter Euclidean group of three-dimensional rotations and translations. The theory is then semantically consistent in the sense explained above. Its preconceptions, however, assume the existence of rigid bodies. Firstly, this is an idealization. Phenomena only approximately demonstrate the existence of rigid bodies. They do not justify, in and of themselves, the fiction of an arbitrarily close approximation to the ideal. One must therefore be prepared for the possibility that the theory has a limited range of applicability. Secondly, the existence of rigid bodies in geometry is only a preconception, and is not explained theoretically. Statistical physics of the late nineteenth century has gradually discovered that the application of classical mechanics to the interior of material bodies leads to difficulties which are presumably insurmountable in principle. It was in any event quantum mechanics that first solved this problem.

Classical mechanics adds, as was recognized from the group-theoretical point of view at the end of the nineteenth century (L. Lange[25]), a four-parameter extension to the Euclidean group, consisting of transformations that involve time. The one-parameter subgroup of time translations, which expresses the homogeneity of time, has been the easiest to accept, being a formulation of the assumption that the laws of nature are always valid. "Proper Galilean transformations," on the other hand, which transform between inertial systems, contain the principle of relativity which has evoked many philosophical discussions. Historically these discussions took place in two phases, which one may distinguish as the phase before Einstein and the phase after Einstein. Before Einstein the principle of relativity appeared to be a general principle of nature only if one regarded classical mechanics as the fundamental science of nature; one can describe this premise as the mechanistic world view. For example, under this premise the relativity of motion was asserted and discussed, e.g., by Leibniz (against Clarke, i.e., against Newton), by Kant (in the *Metaphysische Anfangsgründe der Naturwissenschaft*), and by Mach

---

[23] This section is, up to small changes and an addition at the end, Sect. 8 of the article *Geometrie und Physik* (1974), whose first seven sections will be incorporated into *Zeit und Wissen* I.5.3.
[24] Helmholtz: Torque-free motion of rigid bodies (Helmholtz 1868). Dingler: Construction of Euclidean planes by grinding three rigid bodies against one another (Dingler 1943).
[25] Editors' addition: Ludwig Lange, Ueber das Beharrungsgesetz, Berichte über die Verhandlungen der Königlich Sächsischen Gesellschaft der Wissenschaften zu Leipzig, mathematisch-physische Classe, 37, (1885), pp. 333–351.

(also against Newton). The physicists of the nineteenth century, however, mostly avoided the severity of the problem by assuming a special substance, the ether, at rest in space. Therefore only Michelson's experiment, along with Einstein's interpretation of this experiment, made the philosophical problem unavoidable, this time based on the Lorentz group. Only with Einstein did the principle of special relativity become, instead of a factual rule that holds for certain phenomena, an indispensable element in the chosen description of space and time. In this way, Einstein could correctly presume to have given, for the first time, precise physical form to the intent of the aforementioned philosophers (in particular Mach and perhaps Leibniz; Kant's views on the relativity of motion he evidently did not know).

This form, however, contains a philosophically disturbing problem, of which Einstein soon became aware after the formulation of the special theory of relativity. The principle of special relativity denies the existence of absolute space without, however, justifying the assumption of a general relativity of motion. It stands as an empirically justified, uncomfortable assumption between two that are seemingly more comfortable but unjustified. We delineate the two sides of this issue separately.

We can formulate the non-existence of absolute space in the following manner. The identity of a point in space over the course of time cannot be asserted in a verifiable way. If I look at a point twice in succession, then I cannot know whether I have looked at the same point both times. I could try to identify the point objectively by attaching a label to it, such as a recognizable spot on a body (in short, a body). But the invariance group of the equations of motion transforms a state where the body is at rest at that point into one where the body moves uniformly in a straight line, and passes that point only at one definite instant of time.

This argument could even be considered within classical mechanics. That had been done, e.g., by Lange in the form of mutually equivalent inertial systems. Einstein added that points in time also have no identity independent of the measurement instrument (the actual clock). Methodologically, his argument was effectively a sudden flash of light, as it was the first significant examination of what is here termed *semantic consistency*. We may therefore be permitted to paraphrase its methodological content by presenting first an incorrect and then a correct interpretation. The common starting point is that the wave theory of light leads to the postulate of a constant speed of light, and Michelson's experiment to the relativity postulate, also for light (and thus for all known natural phenomena). The two together yield the Lorentz invariance of the laws of nature. Then the wrong turn: "Therefore one cannot determine the absolute simultaneity of distant events with clocks. What cannot be determined does not exist. Hence absolute simultaneity does not exist." Correct is: "The concept of absolute simultaneity is not Lorentz invariant. *If* all laws of nature are Lorentz invariant, there *cannot* be absolute simultaneity. Our preconception must be corrected accordingly. Now someone might argue that one could even measure absolute simultaneity. On the other hand, the consistency of the theory dictates that clocks obeying Lorentz invariant laws of nature are also incapable of measuring absolute simultaneity." To put it logically: "What can be measured does exist" is assumed to be true. The erroneous version uses the logically unjustified negation, "What cannot be measured does not exist." The proper version merely correctly asserts the contrapositive: "What does not exist cannot be measured." It should be noted that critics of Heisenberg's uncertainty relations are subject to precisely the same misunderstanding.

Those physicists who developed a consistent interpretation of quantum theory, namely Bohr, Heisenberg and their successors, have always interpreted these arguments of Einstein as being the first introduction of the observer into a discussion of the meaning of physical concepts. Einstein always resisted this, and rightly so from his point of view, for two reasons. First, his measurement of lengths and time intervals can always be considered to be the mere reading of certain states of rulers and clocks that are present even if nobody observes them. Second, although neither space nor time per se are absolute in the sense defined here, the four-dimensional spacetime continuum, Minkowski's "world," is. A world point, or as Einstein liked to put it, an "event," is treated in the special and general theory of relativity as objectively identifiable. This, however, is not proved but assumed to be quasi evident. The manifold of all points in space and likewise the manifold of all points in time of a given, fixed inertial frame appear in this respect to be a conventional way to organize an unconventional manifold of events. (The semantic inconsistency of this assumption of an objectively given manifold of events first became an issue in quantum theory.)

However, for the foundation of physical geometry the objectivity of absolute acceleration in classical mechanics and the special theory of relativity is at least as important as the non-objectivity of absolute velocities. The Galilean transformation expresses, as clearly seen by Newton and demonstrated by means of the experiment with the pail, not a general relativity of motion but a consequence of a special dynamical law, the law of inertia. The fact that Newton's equation of motion is of second order in the time derivative implies that only velocities—not accelerations—can be considered to be relative. This scarcely nontrivial fact one prefers nowadays to relate to Newton's equation being the Euler equation of a Euclidean-invariant variational principle. This open problem was the incentive for Einstein to search for a general theory of relativity. Before we follow him in this, let us briefly characterize the relationship between the theories discussed here and actual experience.

All these theories start out with empirically proven principles which, in terms of their preconceptions, are not self-evident at all, but which have proved themselves in a convincing manner to the scientific community to be resistant to empirical falsification. One can call these principles the hard core of the respective theories. They are then hypothetically postulated as plainly universally valid principles. Anyone who accepts these principles is thus modifying his preconceptions. From the new preconceptions, the original nontrivial basic empirical facts are necessary phenomena, not in need of any further explanation.

The hard core of the "Galilean" principle of relativity is initially the empirical fact of the law of inertia. If one postulates the principle of relativity as a law of nature, then inertial trajectories, rather than points in space, have objective reality. This means that one cannot objectively label a point in space by means of a body and have that point remain identical over time, but one can very well label an inertial trajectory in this fashion. Then one can interpret the law of inertia as an obvious consequence of the previously postulated principle of relativity.

The hard core of the special theory of relativity is initially the empirical fact of the negative outcome of Michelson's experiment. If one stipulates Einstein's two postulates as laws of nature, then the absolute velocity of Michelson's apparatus has no objective reality. Then Michelson's result is the self-evident consequence of the postulated principles.

We emphasize that contrary to an oft-quoted but inaccurate remark of Minkowski, special relativity does not at all abolish the distinction between space and time. The objectively nonexistent points in space and likewise the objectively nonexistent points in time are combined (according to the special theory of relativity) into objectively existing "events" that fill the four-dimensional spacetime continuum. But due to the indefinite character of the Minkowski metric, it is impossible to transform timelike straight lines (thus possible inertial paths) into spacelike lines, and vice versa. Similarly, the positive light cone cannot be changed into the negative light cone by a continuous transformation, i.e., the distinction between past and future is also Lorentz invariant. Special relativity alters none of the considerations of the previous chapter (see Sect. 7.4).

## 2.10 General theory of relativity

### 2.10.1 Einstein's theory

Leibniz's idea of general relativity was taken up again by Mach, but now with the full physical understanding of the strength of Newton's mathematical-empirical argument (see the end of Sect. 2.8). He accepted the formal structure of Newton's mechanics but criticized Newton's interpretation of his pail experiment. Newton had not shown that inertial forces are not consequences of the relative motion of the pail with respect to the distant masses of the universe. Philosophically, this physical argument was a partial step in Mach's critique of the ontology of classical physics, one step on his way to an outline of a monistic world view. Within mechanics, only material bodies were assumed to have physical reality, but not points in space. Leibniz and Mach also criticized the narrow notion of physical reality due to Descartes dualism of matter and consciousness. They differed in the implementation of this further step. Leibniz adhered, with ontological use of logic, to the concept of substance; but his last substances, the monads, were assumed to be a medium of physical phenomena as well as consciousness. Mach discarded the concept of substance and considered material bodies (physical objects) as well as psychic subjects to be thought-economical bundling of elements he suggested to be called "sensations." Here we cannot yet follow this philosophical question. We will take it up again in Chaps. 10–12.

There was the physical question of how Newton's pail "knows" that it moves with respect to distant masses. One could imagine that for forces acting at a distance. Einstein, one generation younger than Mach, could only realize the same idea, under the influence of Maxwell's electrodynamics and his own theory of special relativity, in a theory of local action. For this he introduced Riemannian geometry. One can describe the relationship of this introduction to previous understanding by analogy with the introduction of the Galilean principle of relativity in terms of the concept of the "hard core" of the new theory. An initially merely empirical fact is postulated to be fundamental, and in light of this postulate the fact then appears necessary.

The hard core of the general theory of relativity is the initially empirical fact of gravitational and inertial mass being proportional. Einstein postulates this as a fundamental law and shows then that a homogeneous gravitational field is equivalent to a uniformly accelerated reference frame: the *equivalence principle*. This gives him hope to extend the principle of relativity to accelerated motions. Arbitrary gravitational fields and arbitrary accelerations, however, can only be matched locally. The mathematical tool for this is provided by the Riemannian geometry of the four-dimensional spacetime continuum. In this geometry there is no longer a special force called "gravitation". The proportionality of gravitational and inertial mass, hence the equivalence principle, has become as unavoidable as the law of inertia in consequence of the relativity of inertial systems.

Einstein was thus convinced of having established the general relativity of arbitrary motions. As expression of this relativity he called for a notation of the fundamental equations which must be invariant under arbitrary topological transformations: the *principle of general covariance*.

Initially, as a local action theory, the general theory of relativity was inevitably as *dualistic* as classical electrodynamics: it distinguished between the metric field and its sources. That was the new form of the Newtonian dualism of space and matter. The law of inertia then assumes the guise of motion along a timelike geodesic of the spacetime continuum. Weyl appropriately called the metric field the "guiding field."

Einstein, however, did not abandon the monistic intention, which yielded two approaches based on his theory of 1915: having the field reduced to the matter, or matter reduced to the field.

The back-influence of the metric field on matter could not be ontological (in the sense of both being identical), but only causal: the distribution of matter should determine the metric field completely. Einstein called this "Mach's principle." Einstein's 1916 model of the universe satisfied this principle. But in the present understanding Einstein's view on this issue did not prevail; we return to it in subsection c.

The second approach would be that in terms of "substance," there are not just bodies, as in Leibniz or Mach's mechanics, but there is, precisely the contrary, only the field. Einstein attempted this in his later years in the form of the unified field theory. By means of singularities of the nonlinear field equations, or additional rules to prevent the occurrence of singularities, he hoped to explain particles and quantization rules. At one time he also considered giving up continuity at small distances and replacing it by algebraic postulates. One must say that Einstein's attempt at a monistic theory has remained without success.

### 2.10.2 A note about the philosophical debate

It was the general theory of relativity which conveyed to a broader audience the impression of the revolutionary character of the newer physics. As regards geometry, Einstein, however, put up for debate only that philosophical stage which mathematics had already reached in the nineteenth century, from Gauss to Riemann. Decisive was the fusion of this geometry with physics in a theory which, in the bending of light at the sun's limb, had made an empirical prediction that was validated by 1919. Thereby the aprioristic philosophy was shaken to its foundations by means of a great counterexample.

For the contents of the debate we refer again to Sect. 7 of this chapter and to *Zeit und Wissen* I.5.2. To the objection that the deflection of light proved only that light rays are not straight lines Weyl (1918, p. 87) remarks "that only the entirety of geometry and physics is capable of an empirical verification." Einstein's comment that his theory would be incorrect if the deflection of light were not to occur empirically with the predicted value became for Popper the pivotal example of his thesis of falsification: science exists only where the researcher can specify what empirical fact would make him abandon a certain general proposition.

### 2.10.3 Deviating physical arguments

Here we are concerned with the physical development of the theory. In a somewhat perplexing way, the success of the theory, in the 70 years since then, was better corroborated than Einstein's original arguments. This becomes important for our own interpretation and further development of the theory (see Sect. 5.6). It appears that Einstein, with his own special genius, had correctly "guessed" a structure that was not justified from his initially available arguments. It would be important to understand what characterizes this structure. Hilbert, although stimulated by Einstein's question but independent of Einstein's result, also found in 1915 exactly the same structure (for this see, e.g., Mehra 1973a).

The question initially concerns the concept of general relativity. One can distinguish a *conventional* and *dynamical* interpretation of relativity. Conventionally one can call everything whatever one wishes: "A rose by any other name would smell as sweet." Conventionally admissible are, e.g., the following transformations:

a. arbitrary time-dependent Euclidean mappings of the infinite coordinate space onto itself (Cusanus),
b. arbitrary local Minkowski topological mappings of the spacetime continuum onto itself (Einstein),
c. arbitrary canonical mappings of the phase space onto itself (Hamilton).

Conventions, however, can only be communicated if the communicants understand one another beforehand; they cannot be more precise than the traditions they presume. One must already know what a Euclidean space, a spacetime continuum, a phase space is. This prior knowledge is tied to everyday speech.

Dynamical relativity assumes a system of laws in a traditionally given notation and permits only those transformations which leave the form of the laws invariant. This defines a rather restricted group, e.g., the Galilean or Poincaré group. The principle of special relativity is in this sense dynamical.

Einstein's postulate of *general covariance* means then that the initially conventional relativity in the sense of *b* ought to be at the same time a dynamical relativity, and thus determine the form of the fundamental laws of nature. It turns out, however, that by introducing the $g_{ik}$ into the formal expression of the law, any law can be written in a general covariant form. This is a special case of the rule that every law, given in traditional form, can be formulated to be invariant under changes in the wording, permissible within the convention, by explicit introduction of the defining characteristics of this convention. Another example is the fact that the equations of motion of classical mechanics, in canonical notation, are invariant under canonical transformations—although in that case the topology of the coordinate space is generally disrupted.

Einstein was mistaken to regard the equivalence principle as an extension of the dynamical invariance group. His freely falling elevator demonstrated something completely different from Galileo's uniformly sailing ship. The equivalence principle leads to Riemannian geometry. For a dynamical invariance group the existence of a Killing field in the Riemannian spacetime geometry would be necessary (see Sexl-Urbantke 1983, Chap. 2.9, p. 59). Lorentz invariance then holds only locally.

Einstein held on to the importance of the requirement of general covariance, even as it became clear to him that it can always be satisfied. He demanded now a field equation which should to be as *simple* as possible in covariant notation. Naturally the notion of simplicity is hard to specify. It presumably has less of a logical but more of an aesthetic character. If good scientific ideas can be considered to originally be perceptions of form, then "simplicity" is a conspicuous form to the perception-sensitive scientist.[26]

Einstein specifies more precisely the requirement of simplicity, namely that the interaction between matter and the metric field ought to be described by differential equations of lowest possible (i.e., second) order. From Einstein's later conjecture of a fundamental algebraic structure it is conceivable that the differential equations of the interaction are actually only an approximation and that the differential equation of second order appears as the first term of an approximation expansion. Simplicity is then manifested in other mathematical concepts.

A further topic of plausible criticism was "*Mach's principle.*" In a fundamental local interaction theory it remains a foreign object. Mittelstaedt (1979) analyzed this criticism in detail. Einstein's basic equation determines, for a given matter tensor $T_{ik}$, only the tensors $R$ and $R_{ik}$, not the complete Riemannian curvature tensor $R_{ik\ell m}$; to set the remaining conformal Weyl tensor $C_{ik\ell m}$ equal to zero is not necessary in this theory. If $T_{ik} = 0$ everywhere, there still remain many different "vacuum solutions" of the Einstein equations possible. In quantum field theory the "vacuum" turns out to be the origin of creation and annihilation processes of real matter. Thus the dualistic character of the theory appears to be unavoidable. We return to this in Sect. 5.6.

Through quantum theory a field theory of all matter has become possible. There, besides the dualistic interpretation of matter and the metric, also a formally *pluralistic* one is possible. It was devised by Gupta (1950, 1954) and Thirring (1961). In the special theory of relativity, thus in Minkowski space, a gravitational field is introduced by means of tensors $g_{ik}$, as one among several other fields. Then the invariance requirements for the interaction show that this field must couple to the total energy–momentum tensor $T_{ik}$ of all the fields present. This is synonymous with Einstein's equivalence principle. Therefore the *measurable* metric does not have to be the formally assumed Minkowski metric, but a Riemannian metric in which Einstein's field equation holds.

In its conceptual substance, however, this theory is still dualistic, only that in the first step it puts gravity not on the side of the metric field but of matter. It must assume the existence of a spacetime continuum in which a topology and a local Lorentz symmetry is defined. The assumption that this symmetry also holds globally, i.e., that the spacetime continuum is a Minkowski space, is, however, not necessary. It is enough to postulate that one can introduce in a sufficiently large neighborhood of any point a pseudo-Euclidean (thus Minkowski) coordinate system. The idea that the metric of this coordinate system could be measured with rulers and clocks turns out to be semantically inconsistent and is irrelevant for the argument. Rather matter itself, and precisely only its form directly coupled to the $T_{ik}$, namely the gravitational field, determines what lengths and time intervals are measurable.

---

[26] For this, see *Zeit und Wissen* I.5 and, for Heisenberg's aesthetic perception of form, see *Wahrnehmung der Neuzeit*, p. 149–156.

Viewed in this way, the theory is then a stronger version of Einstein's intention, insofar as it no longer needs to postulate the equivalence principle, deriving it instead from topology and local Lorentz invariance. Einstein had to assume both in any case. The dualism, in contrast to monism, is preserved in this theory, exactly as in the Einsteinian, due only to two matters of fact:

1. The existence of a topological spacetime continuum and the local validity of the special theory of relativity is in this continuum not an effect of matter but a precondition for a precise definition of matter.
2. Matter does not completely determine the metric. This refers to the above mentioned uncertainty of the Weyl tensor in the context of the vacuum solution. It also pertains to the numerical value of the gravitational constant, the coupling factor in Einstein's equation, determined using real measurement instruments (consisting of atoms). This value does not follow from the theory but must be put in.

Here one further consideration is in order as to the first point. In a hierarchical reconstruction of geometry according to the Erlangen Program one starts with arbitrary point transformations, which one then successively restricts by the requirement of certain relations between points to remain invariant: first topological, then linear (projective or affine), and finally metric relationships. A reverse order would be natural if one were to start with relations between as few points as possible. A metric relationship (distance) exists between two points, a linear one (to lie on a straight line, plane, ...) among at least three points, and a topological one (to be a cluster point etc.) among infinitely many points. Now a metric in fact also defines a topology. In a semantically consistent physics it seems plausible that spatial and temporal distances can be measured. Therefore one perhaps ought to consider the spacetime continuum not primarily as a topological space of "events" onto which a metric is then impressed, but primarily determine it through metrical relations which, at a fictitious absolute precision of measurements, then also define a topology. The metric then does not have to be the pseudo-Euclidean of the Minkowski world, for which points with light-like separation have zero metrical distance, but a metric defined as the positive-definite sum of spatial and temporal distance. Though this metric itself is not Lorentz invariant, we can still then define, by means of the topology defined by it, local Lorentz invariance. We return also to this in Sect. 5.6.

### 2.10.4 Cosmology

The initial discussions of relativity which we described in Sect. 8 dealt with models of the universe, i.e., cosmology. Einstein's theory consequently led back to cosmological questions.

The decisive step beyond the models of the universe passed down from antiquity, whether spatially finite or infinite, was the introduction of the *history of nature*, i.e., *time*. Ancient astronomy treated the world as an everlasting entity. The Christian doctrine of the Creation treated the world as an entity capable of lasting indefinitely, that had once been created by God and one day will be replaced by a new one; this finiteness could be read off the moral history but not the astronomical history of the world. The triumph of mechanics in astronomy forced a rethinking. Newton saw that the planetary system, under the mutual attraction of all its bodies, could not be stable indefinitely. This implied that it could not have existed at all times and could not last forever; Newton had understood the resulting irreversibility for complicated mechanical processes. The enormous mathematical effort in the nineteenth century expended on the stability problem had, on the one hand, made plausible that the system could be stable for so long (about $5 \cdot 10^9$ years, as we know now) but did not rule out irreversibility in principle; stable and unstable solutions are densely packed in parameter space. Newton's idea, that he had thereby proved the creation and repeated restoration of the system through God's "direct" intervention, could not survive the progress of science. It led to theories of the mechanical formation of the system of which perhaps Kant's (*Allgemeine Naturgeschichte und Theorie des Himmels*, 1755) came closest to the present conceptions (closer than their resumption by Laplace).

The expansion of our empirical knowledge of the universe to distances of several billion light-years and back in time by several billion years allows us nowadays to approach the problem of a model of the universe and its history empirically-critically. Also Einstein's cosmological models show a tendency toward instability, and the Hubble effect makes us suspect a beginning of our universe in time. Cosmology has become one of the great scientific fashions of our time. The theory developed in the present book cannot sidestep cosmological questions (Chap. 5.2 and 6). Precisely because of this we first want to make several skeptical remarks concerning cosmology.

Considering the four mathematical forms of the laws of nature, enumerated in Sect. 3 of this chapter, then without question the older, prerelativistic models are to be ascribed to the first type, the morphology. They assume a structure of space (finite or infinite) and a distribution of matter without any causal justification. Ever since the general theory of relativity one has a system of differential equations, and the models of the universe are now considered solutions of those. There one is usually not aware of the paradox that one assumes a differential equation which permits infinitely many solutions and only a single one of these is identified as reality. What is then the meaning of all the other solutions? And what is then the meaning of the differential equation? Differential equations as laws of nature make sense in a world where many different initial conditions occur, and therefore many different solutions can be verified empirically. What empirical or logical argument can we invoke to apply such an equation to the universe as a whole? Perhaps we are merely repeating the mistakes of the Greeks and of Copernicus, who considered that part of the world which was known to them to be the entire universe, and in that way justified a morphological characterization of this part? Perhaps the part of the universe known to us is indeed a solution of Einstein's equations, but merely because it is only a small part of the whole. Perhaps our present concepts are utterly inadequate to describe the entirety of the universe; perhaps the very concept of the "entirety of the universe" is inconsistent.

This will remain here only as a question, to be remembered in later chapters, in order that the magnificent progress of cosmology in our time not be repudiated. Perhaps only in this way will it be taken seriously.

## 2.11 Quantum theory, historical

### 2.11.1 1900–1925. Planck, Einstein, Bohr

Planck's starting point was the thermodynamics of radiation. This starting point was abstract and fundamental. Kirchhoff had shown in 1859 that the requirement of thermodynamic equilibrium implies the existence of a universal spectral distribution law for blackbody radiation. At the end of the nineteenth century, with the victory of the Maxwell–Hertz theory of the radiation field, the question of Kirchhoff's spectrum came within the reach of theory. Planck had concerned himself for more than a decade with the foundations of thermodynamics. He attacked the problem. The measurements of Rubens provided him in 1900 with an opportunity for precise empirical verification. His rapid success was due to the correct representation of Rubens' result. Planck's accomplishments were twofold: first the fundamental but for his contemporaries less conspicuous result that due to the quantum hypothesis the total energy of the spectrum, at finite temperature, became finite; second, the precise description of the intensity distribution of the spectrum.

The first, fundamental result followed from the abrogation of the equipartition principle of statistical thermodynamics. This required a radical break with classical mechanics: precisely this assumption was necessary that for each frequency $\nu$ of oscillation only discrete energy values are possible, their spacing increasing with increasing frequency. Planck was under no illusion that this was inexplicable in classical mechanics. The second was more of a stroke of luck; it depended on the simple quantum theoretical spectrum of the harmonic oscillator, which Planck guessed.

Planck saw the factual necessity for a break with classical physics. That this break was unavoidable was perhaps first seen by Einstein in 1905 (Pais (1982), pp. 372). He saw that classical electrodynamics necessarily leads to the Rayleigh–Jeans radiation law, which implied an infinite energy content of the radiation field at any finite temperature. He then dared to introduce the concept of the light quantum, as physical carrier of Planck's quanta of energy. He thus explained Lenard's observations of the photoelectric effect, but created, in the dualism of the description of light as a wave and a particle a paradox that remained unsolvable in classical physics.

Einstein was aware of the radical nature of his quantum theoretical results and alarmed from the very outset. He wrote to his friend C. Habicht in 1905, about his current paper on relativity only that "its kinematic part will interest you". The article on quantum theory, however, he called in the same letter "very revolutionary" (Pais (1982), p. 30). More than forty years later he wrote: "All my attempts, however, to adapt the theoretical foundation of physics to this new knowledge failed completely. It was as if the ground had been pulled out from under one, with no firm foundation to be seen anywhere, upon which one could have built." (Einstein (1949), p. 44).

From 1905 through 1912, the intellectual leadership on the road to quantum theory lay with Einstein, and from 1913 until 1925 with Bohr.

Bohr realized that the empirically well-founded Rutherford model of the atom was unstable by virtue of Maxwell's electrodynamics, hence impossible. He applied Planck's quantum conditions to this model and thereby achieved two things: one fundamental, not immediately recognized, and one quick, spectacular empirical success.

The fundamental success was the solution and thus the first understanding of a basic problem of chemistry that had never previously been clearly understood. Not only must atoms be extended and practically impenetrable if they are to explain the existence of liquids and solids; the discreteness of chemical substances (see Sect. 2.4 above) requires all atoms of a particular substance to be equivalent to one another, and distinct from atoms of any other substance by a finite amount in the qualities. Rutherford's planetary system did not even explain that all hydrogen atoms were of the same size; any elliptical orbit of the electron around the nucleus was admissible. Only quantum theory implied the equality, stability, and impenetrability of atoms. Only quantum theory, and only through its classically paradoxical features, reconciled chemistry and mechanics.

The immediate empirical success, the quantitative theory of the hydrogen spectrum, was on the other hand more a stroke of good luck. Only for the one-body problem of the Coulomb field (and the oscillator) do the quantum conditions of Planck and Bohr lead directly to the same spectrum as the fully developed quantum mechanics. However, in addition there was the confirmation of discrete energy levels through electron collisions, which also would have made a spectacular impact if the predicted spectrum had quantitatively not been correct.[27] Finally, Bohr could qualitatively explain the entire periodic system of the elements. Certainly Bohr's solution of the problem of the discreteness of chemical substances was only preliminary, inasmuch as he had to assume the existence of discrete elementary particles (proton and electron, to be joined by the neutron later).

---

[27] Gustav Hertz has told me that, as a young experimentalist in 1913, he and Franck had read Bohr's work and that they had both agreed: "This is crazy. But we also have the experimental means to refute Bohr by demonstrating a continuous absorption of energy in collisions with electrons." They did the experiment, and the energy absorption turned out to be discontinuous. The consequences were the three Nobel prizes for Bohr, Franck, and Hertz.

How clearly Bohr saw from the very outset the principal problem can be attested by a quote from his 1913 report to the Danish Academy of Sciences:

> Before closing, I only wish to say that I hope I have expressed myself sufficiently clearly that you appreciate the extent to which these considerations conflict with the admirably coherent group of conceptions that have been properly termed the classical theory of electrodynamics. On the other hand, by emphasizing this conflict, I have tried to convey to you the impression that it can also be possible in the course of time to discover a certain coherence in the new ideas.

### 2.11.2 Quantum mechanics

Heisenberg in 1925, following the line of Bohr, together with Schrödinger in 1926, following the line of Einstein and de Broglie (1924), found the final form of quantum theory. The two versions turned out to be identical. The mathematical formulation commonly used nowadays is due to Neumann (1932). In this general abstract form, based on the mathematical theory of Hilbert space, the theory has remained unchanged ever since. It is nowadays the fundamental theory of physics.

The success of the theory is probably related to its abstractness. It is a model of what Heisenberg called a closed theory: it does not contain any special laws or constants of nature[28] whose change could lead to an "improvement" of the theory. Heisenberg himself had to learn the general validity of his theory. He and Bohr initially saw it only as a good version for the mechanics of the outer shell of atoms but expected a new theory for the atomic nucleus. In actual fact, thus far not a single trustworthy violation of quantum theory has been found. When we proceed to "reconstruct" quantum theory, we will seek to base it on assumptions of the same abstract generality as factually shown by the theory.

### 2.11.3 Elementary particles

Over the course of history, in a retreat, the notion of the atom has constantly been sharpened. Chemists again took up the ancient idea of the atom, and formulated it more precisely, claiming that there are as many sorts of atoms as there are chemical elements. Bohr explained the possible existence of these atoms by means of quantum theory, while at the same time renouncing the elemental "indivisible" nature of "atoms" and describing them as composites of "elementary particles." In doing so, he employed the systematic classification of the elements according to the periodic table. Since the discovery of the neutron (1932) and the positron (1933), the number of known particles

---

[28] The "value" of Planck's quantum of action $h$ is a statement about our system of units, not about the theory.

has constantly increased. They, too, can be classified nowadays according to "internal symmetries," and nobody still believes that they are all to be considered the ultimate building blocks. One tries to distinguish some of them as constituents of all the others. The theory is still under development. Experimentally, the construction of very large accelerators was essential for it. Heisenberg, in 1958 (cf. his 1967), introduced the idea that all particles ought to be attributed to one elementary field which itself, strictly speaking, can no longer be considered a field of particles. His attempt, however, cannot be considered successful.

The group-theoretical perspective has been decisive for the theory of elementary particles. The fourth mode of describing laws of nature (Sect. 2.3) is accepted. The spacetime group of special relativity is fundamental, and additionally there are internal symmetries. The real problem regarding the foundations of physics now appears to be the explanation of why groups at all, and why these groups in particular, determine the laws of nature.

## 2.12 Quantum theory, plan of reconstruction

In the following chapters we attempt to reconstruct quantum theory. We explained the concept of the reconstruction of a theory in the first chapter of this book: we understand it to be the retroactive construction of the theory based upon the most plausible postulates. Postulates that only formulate conditions which make experience feasible would be ideal. We will not attain this ideal, for two reasons. First, it is not easy to formulate precisely such prerequisites for experience. Second, we must propose at least *one* postulate, which we cannot justify at all from the principle of possible experience. The notion of a retroactive system, the re-construction, takes this into account. Science evolved historically, through the interplay of experience and theoretical concepts.

We therefore begin the outline of a reconstruction with a review of the place of quantum theory in the system of theories and its present structure. We only just referred to it as the fundamental theory of contemporary physics. To verify this we again turn to Figure 2.1 of the introductory Sect. 2.1.

The arrows of the diagram converge in two theories: on the right in quantum theory, on the left in the general theory of relativity. We consider the right side first. Four arrows meet in quantum theory: from classical mechanics, from field theories, from chemistry, and from the thermodynamics of the continuum. It combines the particle picture of classical mechanics with the wave picture of classical field theories, mediated by the concept of probability, which played its first fundamental role in statistical thermodynamics. It reconciles the laws of motion of classical mechanics with the observed stabilities of chemistry. It solves the paradox of the possibility of a thermodynamics of the mechanical continuum. In the reconstruction, however, we wish to avoid revisiting these same paths to quantum theory. As already emphasized in the first chapter,

## 2.12 Quantum theory, plan of reconstruction

the latest closed theory offers the best hope of being reconstructed from completely general principles. We will propose postulates which do not utilize at all any of the special concepts of mechanics, field theories, or chemistry. We will use the impossibility of a classical thermodynamics of the continuum as an argument for a suspected impossibility of a fundamental classical physics in general. The correspondence principle, which was so important historically, we will not use at all directly, but only in the reverse sense: quantum theory, formulated without recourse to classical physics, must in a natural way ultimately explain the empirical success of classical physics as a limiting case. Similarly we will not use at all the mathematical process of the "quantization" of a classical theory, but only its converse, the definition of a classical limiting case.

We wish to state this procedure as a methodological principle:

*Methodological principle of the reversal of classical arguments:* For the reconstruction of a theory it can be useful to reverse the historical sequence of arguments and begin with the most abstract notions of the latest and thus most general closed theory.

We turn now to the left side of the diagram. Geometry and classical mechanics lead initially to the science of kinematics, which together with the field theories leads to special relativity, which, combined with Riemannian geometry and the theory of gravitation, then leads to the general theory of relativity. The left side is thus the theory of space and time, while the right side is the theory of matter, in the general sense of "things in space and time." This fundamental dualism of contemporary physics Einstein tried in vain to convert into a monism, on the basis of the theory of space and time, in which matter appeared as a special state of the spacetime continuum. The reverse monism would be the reduction of the spacetime continuum to quantum theory. It is precisely this that we wish to attempt. If this were to succeed, then two more arrows would need to be drawn from the special and general theory of relativity to quantum theory. These would historically denote the relativistic quantum field theory of gravitation. In line with the reversal of the arguments, these arrows could then also be read backwards, in the narrow sense, to the effect that the spacetime continuum is also a possible factual predicate of matter.

The plan of the reconstruction is shown in Fig. 2.2. The main features of the interpretation are on the left. The right side describes in more detail the reconstruction of quantum theory and from it the theory of relativity and elementary particles. The titles, with chapter and sometimes also section numbers, are arranged in 12 rows.

The interpretation of physics has, in this book, its origin in the phenomenon of time, which stands alone at the top of the diagram. The reconstruction is also steeped in time, as it explains what we mean by conditions that make experience possible. But we will see in 3.6 that we need *one* postulate which cannot be explained on the basis of these conditions ( "indeterminism"). Similarly, the reconstruction has *one* starting point, entitled *individuality*, which is not derived from the logic of temporal propositions.

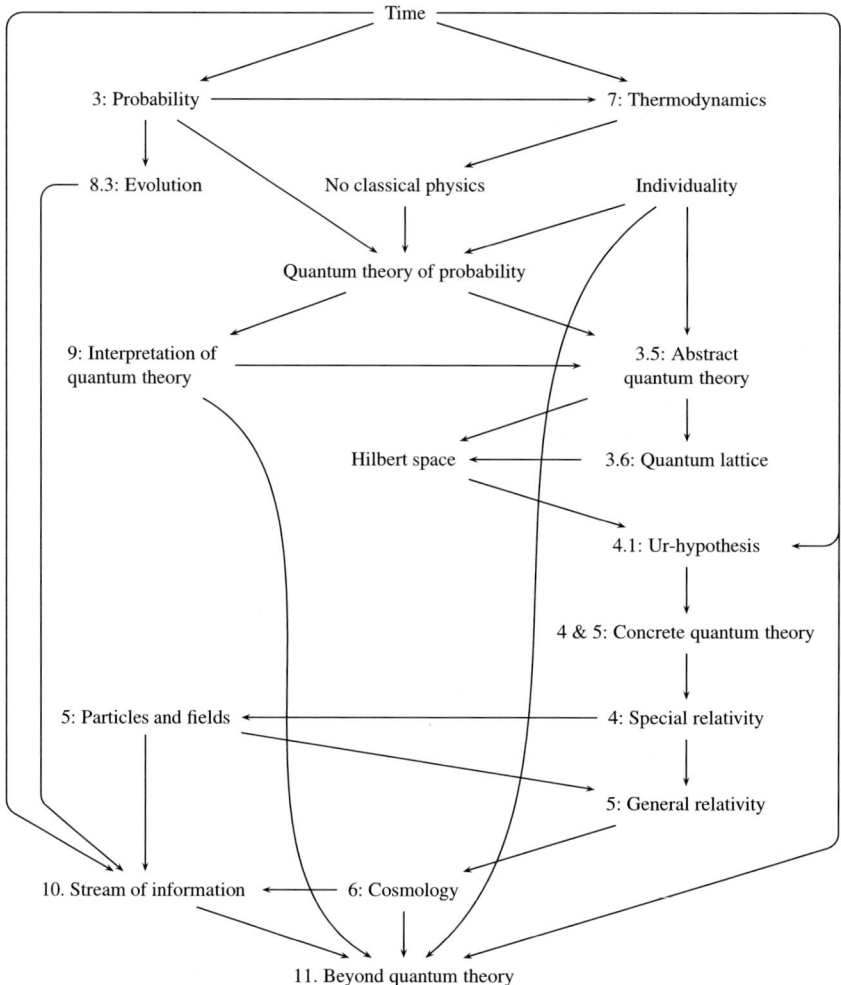

**Fig. 2.2.** Reconstruction of quantum theory

Now we go over these points individually.

In the first row the single topic is the basic phenomenon of *time*. From this topic there originate four arrows; two to the sciences of probability (Chap. 3) and thermodynamics (Chap. 7), and two to the final topics "flow of information" and "beyond quantum theory," hence to the interpretation of our results. Finally, one arrow branches off to the ur hypothesis, which as a matter of fact resumes once more directly the topic of time.

The second row initially contains *probability*. This basic concept continues on three completely different paths. First, to thermodynamics, which offers to quantum theory the negative vantage point of the *impossibility of a*

## 2.12 Quantum theory, plan of reconstruction

*fundamental classical physics*. With this topic we begin Chap. 7. Second, directly to the *quantum theoretical amplification of the theory of probability*, to the theory of *evolution*, which then leads directly to the chapter on the interpretation of the flow of information.

The *third* row contains in addition the only concept not reached by an arrow that directly or indirectly comes from time: the *individuality* of processes. We choose this name introduced by Bohr for the central phenomenon of quantum theory. On the one hand, the concept of a process assumes temporal development. But the concept of individuality denies processes a characteristic that is essential for our notion of time: the difference between facticity and possibility. This might be the deepest reason for the lack of "visualizability" or paradox of quantum theory. From this concept originate therefore two fundamentally different paths. On the one hand it is indispensable for the *constructive* derivation of quantum theory (arrow to 3.4 and 3.5–6), to the extent that we can implement it within our notion of time. On the other hand, there is an arrow leading directly from it to the *critical* chapter "beyond quantum theory," bypassing all chapters that depend on the concept of time.

The *fourth* row mentions preliminary discussions of the quantum theory of probability, which come from the same, somewhat older article as the chapter of this book.

Beginning with the *fifth* row, the diagram definitely branches out into a *right, constructive* and a *left, interpretive* half. On the right begins the *reconstruction*, initially of *abstract quantum theory*, which does not yet employ the concept of space. On the left we refer to the detailed chapter on the *interpretation* of quantum theory, with which the interpretive part of the book opens.

On the reconstruction path we encounter an essentially new idea: the hypothesis of *ur alternatives*. This leads to Chap. 4 and 5 on *concrete quantum theory*, as mentioned in the *eighth* row, i.e., the deduction of the special theory of relativity (*ninth* row) from quantum theory, starting with the concept of space. In this point alone does our book go *physically* beyond generally accepted physics. This proposal should also determine uniquely the theories of particles and fields, but thus far this has remained merely a program.

The *general theory of relativity*, mentioned in the *tenth* row, one should then be able to understand as the quantum theory of the gravitational field. This too is for the time being only a program.

The *eleventh* row contains, on the one hand, cosmology, important for our construction of the theory of particles, and on the other hand, under the heading *the flow of information*, a discussion of how the reconstructed unified physics is to be understood as a physics of *time*.

That all this is not yet the last word is indicated in the *twelfth* row "beyond quantum theory." There begins a new, philosophical task.

# 3
# Probability and abstract quantum theory

## 3.1 Probability and experience

*In memoriam Imre Lakatos*

*This and the following section are from my essay "Probability and Quantum Mechanics" (1973). Imre Lakatos had read this work in the last years of his life, when he was a member of the Scientific Council of the Max Planck-Institute in Starnberg. He viewed it as an example of a "rational reconstruction" and had it published in the* British Journal for the Philosophy of Science, *24, 321–337. I dedicate these sections to his memory.*

The theory of probability had its origin in an empirical question: Chevalier de Meré's gambling problem. Equally, the present-day physicist finds no difficulty in empirically testing probabilities which have been theoretically predicted, by measuring the relative frequencies of the occurrence of certain events. On the other hand, the epistemological discussion on the meaning of the application of the so-called mathematical concept of probability is by no means settled. The battle is still raging between "objectivist," "subjectivist," and even other interpretations of probability. Probability is one of the outstanding examples of the "epistemological paradox" that we can successfully use our basic concepts without actually understanding them. In many apparent paradoxes associated with fundamental philosophical problems, the first step toward their solution consists in accepting the seemingly paradoxical situation as a phenomenon, and in this sense as a fact. Thus we must understand that it is the very nature of basic concepts to be practically useful without, or at least before, being analytically clarified. This clarification must use other concepts in an unanalyzed manner. It may mean a step forward in such an analysis to see whether a hierarchy exists in the practical use of basic concepts, and which concepts then practically depend on the availability of which other concepts, and also to see where concepts interlink in a non-hierarchical manner. I will try to show that one of the traditional difficulties in the empirical interpretation of probability stems from the idea that experience can be treated as a given concept and probability as a concept to be applied to experience. This is what I call a mistaken epistemological hierarchy. I will try to point out that, on the contrary, experience and probability interlink in a manner that will preclude understanding experience without already using some concept of probability. I will offer a particular way to introduce probability in several steps.

We will interpret the concept of probability in a strictly empirical sense. We consider probability to be a measurable quantity whose value can be empirically tested, much like, for example, the value of an energy or temperature. What we need for defining a probability is an experimental situation in which different "events" $E_1, E_2, \ldots$ are the possible results of one experiment. We further need the possibility of meaningfully saying that an equivalent experimental situation (in short "the same situation") prevails in different cases (in different "realizations," "at different times," "for different individual objects" etc.) and that, given this situation, an equivalent experiment (in short "the same experiment") is carried out in each case. Let there be $N$ performances of the same experiment, and assume that event $E_k$ occurred $n_k$ times. In this series of cases we call the fraction

$$f_k = \frac{n_k}{N} \tag{3.1}$$

the relative frequency with which event $E_k$ occurred in the series. Now consider a future series of performances of the same experiment. Let us assume that our (theoretical and experimental) knowledge enables us to calculate a probability $p_k$ of the event $E_k$ in this experiment. Then we will take the meaning of this number $p_k$ to be that it is a prediction of the relative frequency $f_k$ for the future series of performances.[1] Finally, $p_k$ will be empirically tested by comparing it with the values of $f_k$ found in this and subsequent series of the experiment under consideration.

This is the simplistic view of the ordinary experimentalist. I think it is essentially correct and it will only need to be defended against the objections of the epistemologists. Of course we hope to understand it better by defending it.

Let us use a simple example in formulating the main objection. Our experiment will consist in the single cast of a die. There are six possible events. Let us choose the event that a "5" appears as the event of interest. Its probability $p_5$ will be 1/6 if the die is "good." Now let us cast the die $N$ times. Even if $N$ is divisible by 6, the fraction $f_5$ will only rarely be exactly 1/6, and, what is more important, the theory of probability does not expect $f_5$ to be 1/6. The theory predicts a distribution of the measured values of $f_5$ in different series of casts around the theoretical probability $p_5$. The probability is only the *expectation value* of the relative frequency. But the concept "expectation value" is generally defined by making use of the concept "probability." Hence

---

[1] This formulation was proposed by Drieschner (1967).

## 3.1 Probability and experience

it seems impossible to define probability by referring it to measurable relative frequencies, since that definition itself, if rigorously formulated, would necessarily contain the concept of probability. It would, so it seems, be a circular definition.

We will not evade the problem by defining the probability as the limiting value of the relative frequency for long series, since there is no strict meaning to a limiting value in an *empirical* series which is essentially finite. These difficulties have induced some authors to abandon the "objectivist" interpretation altogether in favor of a "subjectivist" one which, e.g., reads the equation $p_5 = 1/6$ as meaning: "I am ready to lay odds of 5 to 1 that a 5 will come up next time." The theory of probability is then a theory of the consistency of a betting system. But this is not the problem of the physicist. He wishes to discover empirically who will become a rich man by his betting system. I am not going to enter into the discussion of these proposals,[2] but instead immediately offer my own.

The origin of the difficulty does not lie in the particular concept of probability but more generally in the idea of an empirical test of any theoretical prediction. Consider the measurement of a position coordinate $x$ of a planet at a certain time; let its value predicted by the theory be $\xi$. A single measurement will give a value $\xi_1$, different from $\xi$. The single measurement may not suffice to convince us whether this result is to be considered a confirmation or refutation of the prediction. Thus we will repeat the measurement $N$ times and apply the theory of errors. Let $\bar{\xi}$ be the average of the measured values. Then, comparing the distance $|\xi - \bar{\xi}|$ with the average scatter of the measured values, we can formally calculate a "probability" with which the predicted value will differ from the "real" value $\xi$ by a quantity $d = |\xi - \bar{\xi}|$. This "probability" is itself a prediction of the relative frequency with which the measured distance $|\xi - \bar{\xi}|$ will assume the value $d$, if we repeat the series of measurements many times. This structure of the empirical test of a theoretical prediction is slightly complicated, but well known. We can compress it into an abbreviated statement: "An empirical confirmation or refutation of any theoretical prediction is never possible with certainty but only with a greater or lesser degree of probability." This is a fundamental feature of all experience. Here I am satisfied to describe it and to accept it; its philosophical relevance is to be discussed in another context.[3] Whoever works in an empirical science has already tacitly accepted it by his practice. In this sense the concept of scientific experience in practical use presupposes the applicability of some concept of probability, even if this concept is not explicitly articulated. Hence the very attempt to give a complete definition of probability by recursing on a given concept of experience is likely to result in a circular definition. Of course it would be equally impossible to define the concept of an empirical test by

---

[2] Cf. *Zeit und Wissen*, Chap. 4.
[3] Cf. *Zeit und Wissen* 3 and 4.

using a preconceived concept of probability. These two concepts, experience and probability, are not in a relationship of hierarchical subordination.

In practice every application of the theory of errors implies that we consider relative frequencies of events to be predictable quantities. In this sense probability is a measured quantity. This implies that our "abbreviated statement" also applies to probability itself: The empirical test of a theoretical probability is only possible with some degree of probability. The appearance of the probabilistic concept of an expectation value in the "definition" of probability is therefore not a paradox but a necessary consequence of the nature of the concept of probability; or it is a "paradox" inherent in the concept of experience itself. Still, probability is not on the same methodological level as all other empirical concepts. The precise measurement of any other quantity refers us to the measurement of relative frequencies, that is, to probabilities; the precise measurement of probability refers us to probabilities again. Due to this higher level of abstraction the predictions of the theory are better defined. The scatter of the measured values of any quantity about its average value depends on the nature of the measurement device; the scatter of relative frequencies about their expectation values is itself defined by the theory.

## 3.2 The classical concept of probability

We have yet to achieve a definition of probability that can avoid the objection of being circular. I will now sketch a systematic theory of probability as an empirical concept, i.e., a concept of a quantity which can be empirically measured. This is not a rigorously developed classical theory of probability, but a sketch for an analysis of its concept of probability that emphasizes aspects of the theory where epistemological difficulties usually arise. I hope that this analysis will suffice for the construction of a consistent classical theory of probability, where for the mathematical details we might use any good textbook. The word "classical" means here only "not yet quantum theoretical."

This is done in three stages. We first formulate a "preliminary concept" of probability. It does not aim for precision; it is meant to describe in comprehensible terms how probability concepts are actually used in practice. Secondly we formulate a system of axioms of the *mathematical theory* of probability. In this section we can adopt Kolmogorov's system. Thirdly we give empirical meaning, *physical semantics*, so to speak, to the concepts of the mathematical theory by identifying some of its concepts with some concepts associated with the preliminary concept of probability. This three-step procedure can also be described as a process of giving mathematical precision to the preliminary concept. The most important part of the third stage is a study of the consistency of the whole process. The interpreted theory of the third stage offers a mathematical model of those structures which were imprecisely described in the preliminary concept. I propose to call a theory *semantically consistent* if it permits one to use the preliminary concepts without which it would not have been given meaning in such a manner that this use is correctly described by the mathematical model offered in the theory itself.[4]

The *preliminary concept* is described by three postulates:

A. A probability is a predicate of a formally possible future event or, more precisely, a modality of the proposition which asserts that this event will happen.
B. If an event, or the corresponding proposition, has a probability very near to 1 or 0, it can be treated as practically necessary or practically impossible. A proposition (event) with a probability not very near to 0 is called possible.
C. If we assign a probability $p$ ($0 \leq p \leq 1$) to a proposition or to the corresponding event, we thereby express the following expectation: out of a large number $N$ of cases in which this probability is correctly assigned to this proposition there will be approximately $n = pN$ cases in which the proposition will turn out to be true.

The language in which we formulated these postulates needs further explanation. We first see that restrictive concepts like "practically," "approximately," "expressing an expectation" are used. Their task is to indicate that our preliminary concept is not precise but should be made more precise. We will see that in this process these restrictive concepts will not be eliminated but be made more precise themselves. The word "correctly" in $C$ indicates that we consider ascribing a probability to an event not an act of free subjective choice, but a scientific assertion subject to test.

The language of the postulates refers to the logic of temporal propositions. For propositions about the future this logic proposes not to use the traditional truth values "true" and "false," but the "future modalities": "possible," "necessary," "impossible." The postulate proposes to use probabilities as a more precise form of future modalities. With respect to the ordinary use of the word "probability" this can be considered a terminological convention; further on, we wish to restrict the use of this word to statements about the future. But behind this convention lies the view that this is the primary meaning of probability and that other uses of the word can be reduced to it. For example, we apply it to the past in saying "it is probable that it was raining yesterday" or "the day before yesterday it was probable that it would rain the following day." But in the second example probability is referred to what was then the future; characteristically we say here "it *was* probable." In the first example we first of all admit lack of knowledge concerning the past; to make the statement operative we must apply it to the future in the sense "It is probable that, if I investigate, I will find out that it was raining yesterday."

---

[4] Cf. 2.3 and *Zeit und Wissen* 5.2.7.

For the *mathematical theory* we can adopt Kolmogorov's text literally, changing only some notation:

"Let $M$ be a set of elements $\xi, \eta, \zeta, \ldots$ which we call *elementary events*, and $F$ a set of subsets of $M$; the elements of $F$ will be called *events*.

I. $F$ is a lattice of sets.
II. $F$ contains the set $M$.
III. To every set $A$ of $F$ we assign a non-negative number $p(A)$. This number is called the probability of the event $A$.
IV. $p(M) = 1$.
V. If $A_1$ and $A_2$ are disjunct, then $p(A_1 + A_2) = p(A_1) + p(A_2)$."

We leave out axiom VI, which formulates a condition of continuity, since we will not discuss its problems here. We need, however, the definition of the expectation value:

"Let there be a partition of the original set $M$,

$$M = A_1 + A_1 \ldots A_r,$$

and let $x$ be a real function of the elementary event $\xi$ which is equal to a constant $a_q$ on every set $A_q$. Then we call $x$ a *stochastic quantity* and consider the sum

$$E(x) = \sum_q a_q p(A_q)$$

the mathematical expectation of the quantity $x$."

We now turn to the *physical semantics*. In order to simplify the expression and concentrate on the essentials, we assume the set $M$ of elementary events to be finite. We call the number of elementary events $K$; in the case of the die $K = 6$. We further consider a finite *ensemble* of $N$ equivalent cases, e.g., of casts in the case of die. To every elementary event $E_k$ (we write $E_k$ instead of Kolmogorov's $\xi$; $1 \leq k \leq K$) we assign a number $n(k)$ which indicates how many times this event $E_k$ (say the "5" on the die) has actually happened in the particular series of $N$ experiments which forms our given ensemble. Correspondingly we assign a $n(A)$ to every event $A$. It is easy to see that the quantities

$$f(A) = \frac{n(A)}{N}$$

satisfy Kolmogorov's axioms I to V if we insert them for $p(A)$. This model of the axioms is, however, not the one intended by the theory of probabilities, but we reach our goal by adding a fourth postulate to the preliminary concept:

D. The probability of an event (of a proposition) is the expectation value of the relative frequency of its happening (its coming true).

The expectation value used in D is not defined on the original lattice of events $F$. It can be defined on a lattice $G$ of "meta-events." We call a meta-event an ensemble of $N$ events belonging to $F$ which happen under equivalent

## 3.2 The classical concept of probability

conditions. We here use the language that the "same" events can happen several times ("it has been raining and it will be raining again"). $G$ is not a subset of $M$ or $F$, but it is a set of elements of $F$ with repetitions. Now we can assign a probability function $p(A)$ to $F$ (it may express our expectation of the events $A$ according to the preliminary concepts). Then the rules of the mathematical theory of probability permit us to *calculate* a probability function for the elements of $G$; it is only necessary to assume that the $N$ events which together form a meta-event can be treated as independent. Assuming the validity of Kolmogorov's axioms for $F$ we can then *prove* their validity for $G$ and the validity of the formula

$$p(A) = E\left(\frac{n_A}{N}\right). \tag{3.2}$$

We can now forget our preliminary ideas of the meaning of the $p(A)$ in $F$. Instead we can apply the three postulates $A, B, C$ to the lattice $G$ of meta-events. After having thus given an interpretation (in the preliminary sense) to the $p$ in $G$, we use (3.2) to deduce an interpretation of the $p$ in $F$. It is exactly what postulate $D$ says: $p(A)$ is the expectation value of the relative frequency of $A$. If we now rememberer again how we would have interpreted the $p$ in $F$ without this construction, we would only have used $A, B, C$. This preliminary concept is now justified as a weaker formulation of $D$. The concepts "practically," "approximately," "expectation" can now be more precisely interpreted by estimating probable errors. The mathematical "law of large numbers" proves that the expectation values of these errors tend to zero as $N$ increases.

What have we gained epistemologically? We have not gotten rid of the imprecise preliminary concepts, we have merely transferred the lack of precision from events to meta-events, i.e., to large ensembles of events. The physical semantics for probabilities rest on the preliminary semantics for meta-probabilities. This is a more precise expression of our earlier statement, that a probability can only be empirically tested with some degree of probability. The solution of the paradox lies in its acceptance as a phenomenon; no theory of empirical probabilities can be meaningfully expected to yield more than this justification, which at least makes its consistency more evident.

If we wish, we can iterate our process and call this ladder of meta-probabilities a "recursive definition" of probability. While a typical recursive definition offers a fixed starting point ($n = 1$) and a rule of recursion from $n + 1$ to $n$, the recursion here can go as high as we like. At some step of the ladder we must halt and rely on the preliminary concepts. Due to the "law of large numbers" it will suffice for this highest step to postulate $A$ and $B$. This will yield $A, B$, and $C$ for the next lower step, and $D$ for the ones below that.

## 3.3 Empirical determination of probabilities

We distinguish the *probability of an event* from the *probability of a rule*, but assume, however, (in contrast to Carnap's mean[5]) that the two quantities are of exactly the same nature but at different levels of the application. The probability of an event $x$ is the prediction (the expectation value) of the relative frequency $f(x)$ with which an event of this type $x$ will occur, upon frequent repetitions of the experiment in which $x$ can happen. The content of a rule (an "empirical law of nature") is the specification of probabilities of events. Rules always specify *conditional* probabilities: "If $y$ then $x$ will occur with the probability $p(x)$." But in the same way, by the conditional probability one also means the probability of an event. After all, the relative probability can only be measured if "the same" experiment is always performed, i.e., if equivalent conditions are realized. One can say that empirically testable probabilities are by nature conditional probabilities. The probability of a rule is now meant as the probability that the rule is "true." An empirical rule is true if it proves itself in experience. Its probability is then the prediction of the relative frequency with which just that rule $R$ proves itself, upon frequent repetitions of the same empirical situation for testing that rule. This initially formal definition we have only to elucidate in detail, and we naturally arrive at an interpretation of Bayes' rule. What we are looking for can simply be called an iterated probability $P(p(x))$. In *Zeit und Wissen* II.4.4b we will see that one does better to speak of a "higher-order probability" $P(f(x))$. For the present discussion such finesse does not matter.

One can (and in general will) approach the empirical determination of a probability from a starting point that expresses the *prior knowledge*. Methodologically we remember that an objective, empirically testable probability is at the same time related to the prior knowledge of a subject. As an example we choose two die, each successively thrown once. Observers $A$ and $B$ are to state the probabilities for obtaining a 12, $A$ before the double throw, $B$, however, after the first die had been cast. $A$ gives $p(12) = 1/36$, $B$ on the average in one sixth of the cases $p(12) = 1/6$, on the average in five sixths of the cases $p(12) = 0$. Empirically both are correct as they refer to different statistical ensembles due to different prior knowledge.

The starting point for the desired rule expresses what one already knows before the test series. For simplicity let us first assume that one can describe the setup conceptually, but has never experimented with precisely this realization of the concepts. For example, one might be about to throw "heads" or "tails" with a coin or a "1...6" with a die, or to draw a ball from an urn containing $w$ white and $b$ black balls. Here is the legitimate application of Laplace's concept of *equal possibilities*, i.e., a *symmetry argument*. One knows which "cases" are possible, i.e., one knows the catalog of all possible events. One does not know what would distinguish one of the elementary events (the

---
[5] Cf. *Zeit und Wissen* II.4.3

## 3.3 Empirical determination of probabilities

atoms of the lattice of events) from any other. In this sense they are all equally possible. *Therefore* one assesses them to be all equally probable, i.e., one predicts an equal relative frequency for their occurrence. The empirically motivated assumption of a symmetry is at this phase of the experiment essentially an expression of ignorance. This is the legitimate meaning of Laplace's approach, as subsequent investigation of the experiment will show.

At any rate, certain relative frequencies will be found in this experiment. Roughly speaking, we can distinguish three cases:

a) relative frequencies arise that are consistent with the starting point
b) relative frequencies arise which correspond to a different starting point
c) relative frequencies arise which do not correspond to any uniform statistical distribution

The phrase "are consistent with" means "in agreement with the expected distribution," within the limits of error the observer has set himself, using the calculus of probability. For the observer there is essentially no certainty, only a probability he can assign to each of the available scenarios; how he does this we will discuss shortly in more detail in connection with Bayes' problem. That the frequencies correspond at all to a unique starting point is by no means self-evident. This is indicated by the listing of case $c$. In this case one suspects that the catalog of events needs to be expanded to bring conditions into view which do not vary statistically but systematically. In view of these possibilities, cases $a$ and $b$ are hardly self-evident, and one could ask by what right they can be expected to occur at all. At the present stage of our epistemological examination we can only recognize Hume's problem in this difficulty and reply that according to our present understanding the occurrence of regular statistical distributions is a condition for the possibility of experience. At a later stage we will recognize in Laplace's symmetries basic symmetries of the world, namely in the equal probabilities of the sides of a uniform coin or a uniform die representations of the group of spatial rotations, realized in objects with negligible interaction with their environment, and a representation of the group of permutations of objects in the equal probabilities for picking any of the balls in the urn. There we must justify, through a discussion of the interaction, why the inclusion of new objects cannot remove the symmetry of the world under this group, such that every deviation of individual objects from the symmetry only stems from their individual interaction with other objects.

The classical model of Bayes' problem involves several urns (say 11) with different mixing ratios of black and white balls (say in the zeroth urn 0 white and 10 black balls, in the $k^{\text{th}}$ urn $k$ white and $10 - k$ black ones). In drawing from each urn we assume with Laplace equal probabilities; each of the urns is thus characterized by the probability $p_1$ of drawing a white ball and $p_2$ of drawing a black ball. According to our assumptions we have for the $k^{\text{th}}$ urn

$$p_1(k) = \frac{k}{10}, \tag{3.3}$$

and always
$$p_1 + p_2 = 1. \tag{3.4}$$
Now one picks one of the urns, without knowing which, and proceeds to draw and immediately return a single ball, $n$ times in succession. If the outcome was $n_1$ white and $n_2$ black balls ($n_1 + n_2 = n$), how probable is it that this had been the $k$-th urn? In other words, one then determines a probability $P_k$. One can interpret $P_k$ as a prediction of a relative frequency in a twofold way. On the one hand, $P_k$ is, according to Laplace's assumption applied to the selection of one urn, the predicted relative frequency with which a particular urn turns out to be the $k$-th one *if* just $n_1$ white and $n_2$ black balls have been drawn from the urn. On the other hand, $P_k$ enables one to predict new probabilities $p'_1$ and $p'_2$ for subsequent drawing from the urn, according to the formulas
$$p'_1 = \sum_k P_k p_1(k), \tag{3.5}$$
$$p'_1 + p'_2 = 1. \tag{3.6}$$
Before the start of the experiment, according to Laplace's assumption, one would set each $P_k$ to 1/11 for the selection of one urn, and compute from it the "a priori probabilities" $p_1^{(0)}$ and $p_2^{(0)}$, which in our case would both be 1/2. The test series of drawing $n$ balls is then the empirical determination of newer, i.e., "a posteriori probabilities." Bayes' procedure thus assigns to each of the 11 possible rules (3.3) a probability of the rule $P_k$ and determines the probabilities $p'_1$ proposed for the practical usage from the probabilities *according* to the rule $(p_1(n))$ and the probability of *the* rule $P_k$ according to (3.5).

Bayes' procedure thus corrects an initially assumed equipartition by means of insight into possible cases leading to different rules, for which again an equipartition is assumed. Naturally this can also be modified. One can introduce unequal a priori probabilities for picking one urn. This again can be reduced to an equipartition by assuming different numbers for each type of urn. The practical value of the procedure depends on the fact that for large values of $n$ the influence of the assumed a priori probabilities gradually disappears. With an ontological assumption that all phenomena are built from equally possible elementary events one can thus even further justify the empirical determination of probabilities. Without such an assumption one can still describe the empirical determination "as if" such an assumption were justified; we need the assumption to *count* cases, and thereby are able to *define* in this way absolute, as well as relative, frequencies.

## 3.4 Second quantization

In Sect. 3.2 we interpreted probability as expectation value of the relative frequency of events in an ensemble. There we defined the expectation value

## 3.4 Second quantization

in an ensemble of ensembles which we also referred to as "meta-events." We repeat now, for a simple model, this conception within quantum theory.[6]

Let us consider the quantum mechanical measurement of a single observable. To simplify the expression we assume again that this observable may only have a finite number of different values. Suppose this number is exactly $R = 2$. We have then a simple (or, put differently, a binary) alternative, e.g., the measurement of the spin of an alkali atom in a Stern–Gerlach experiment. We denote the two possible alternatives by $r = 1$ and $r = 2$. As regards this alternative, thus ignoring its other degrees of freedom, the object has a two-dimensional Hilbert space. We call its state vector $u_r$ ($r = 1, 2$). If $u_r$ is normalized, the probabilities of finding the results 1 and 2 are then

$$p_1 = u_1^* u_1, \qquad p_2 = u_2^* u_2, \qquad (3.7)$$

$$p_1 + p_2 = 1. \qquad (3.8)$$

Now we consider a statistical ensemble of $N$ such objects on which the same alternative can be decided. The result 1 can be found in $n_1$ cases, the result 2 in $n_2$ cases

$$n_1 + n_2 = N. \qquad (3.9)$$

We wish to treat the ensemble as a real ensemble, i.e., as a quantum mechanical object composed of $N$ simple objects. Formally this is possible, even when the measurements are performed at different times, but we leave this case aside, as it would require a complicated description (that includes the problem of symmetry). We treat the measurements as being simultaneous. The general state of the ensemble, in which a given alternative need not be decided, can be described by a wave function in the $2^N$-dimensional configuration space. To simplify the calculation we make a specific assumption about the symmetry of this wave function. Let it be symmetric, i.e., the simple objects will have Bose statistics; Fermi statistics would have restricted us to the uninteresting case $N = 2$. One can then describe the state by a wave function $\varphi(n_1, n_2)$. The set of all (normalized) $\varphi(n_1, n_2)$ for arbitrary $n_1, n_2$ describes then all possible ensembles with finite values of $N$; each definite ensemble has a fixed value of $N$.

If the ensemble consists of $N$ objects in the same state $u_r$, we obtain

$$\varphi(n_1, n_2) = c_{n_1 n_2} u_1^{n_1} u_2^{n_2}. \qquad (3.10)$$

We normalize $\varphi$, considering (3.9):

$$\sum_{n_1} \varphi^*(n_1, n_2) = \sum_{n_1} |c_{n_1 n_2}|^2 p_1^{n_1} p_2^{n_2} = 1. \qquad (3.11)$$

As

---
[6] This section is from the essay Weizsäcker (1973c).

$$(p_1+p_2)^N = \sum_{n_1} \frac{N!}{n_1!n_2!} p_1^{n_1} p_2^{n_2} = 1 \qquad (3.12)$$

we obtain

$$|c_{n_1 n_2}|^2 = \frac{N!}{n_1!n_2!}. \qquad (3.13)$$

The numbers $n_1$ and $n_2$ can be interpreted as eigenvalues of the operators $n_1$ and $n_2$, whose action on $\varphi$ consists in multiplication by $n_1$ or $n_2$, respectively. The expectation value of $n_1$ in $\varphi$ is

$$\begin{aligned}\overline{n}_1 &= \sum_{n_1=0}^N \varphi^*(n_1,n_2)\, n_1\, \varphi(n_1,n_2) \\ &= N p_1^N + N(N-1) p_1^{N-1} p_2 + \cdots + N p_1 p_2^{N-1} + 0 \\ &= p_1 \frac{\partial}{\partial p_1}(p_1+p_2)^N = p_1 N (p_1+p_2)^{N-1} = p_1 N.\end{aligned} \qquad (3.14)$$

Thus we obtain

$$p_1 = \frac{\overline{n}_1}{N}, \qquad (3.15)$$

in accordance with postulate D of Sect. 3.2. This calculation can be easily generalized to larger values of $R$.

This little calculation was none other than the simplest case of *second quantization*. The index $r$ is a quantum mechanical observable capable of two values. The operators $n_r$ can be constructed out of operators $u_r, u_r^\dagger$, which satisfy the commutation relations

$$\begin{aligned} u_r u_s^\dagger - u_s^\dagger u_r &= \delta_{rs} \\ u_r u_s - u_s u_r &= u_r^\dagger u_s^\dagger - u_s^\dagger u_r^\dagger = 0, \end{aligned} \qquad (3.16)$$

such that

$$n_r = u_r^\dagger u_r. \qquad (3.17)$$

Second quantization has typically been regarded as a skillful formal manipulation. One could prove it to be equivalent to the method of configuration space, but it never became quite clear what the iteration of the quantization procedure was actually supposed to mean.

From the standpoint of the correspondence principle the name is indeed paradoxical. Heisenberg had strictly forbidden me as a student to use it; he spoke of the quantization of a classical field theory. From the correspondence principle it is incomprehensible how the same theory, which was obtained through quantization of a classical theory, all at once can be considered a new classical theory. The Schrödinger wave denotes a probability and is not measurable like a classical quantity. Also, the Schrödinger equation of the one-particle problem is not formally identical to the classical ("de Broglie") field equation: the latter may and must contain nonlinear (or multilinear) interaction terms, whereas the former, as a quantum theoretical equation, is strictly linear.

This renunciation of the term "second quantization", however, which is consistent from the standpoint of the correspondence principle, leaves the fact unexplained that a quantum theoretical equation like the Schrödinger equation is after all formally identical to a special (specifically, force-free) classical field equation. Second quantization indeed defines an ensemble of identical objects, each of which, if it occurred separately, would be described by the wave function of first quantization. On the other hand, the formalism of second quantization is a correct quantization procedure. This suggests the conjecture that quantization in general is a process of forming ensembles according to the special rules of probability calculus characteristic of quantum theory. This is now precisely my thesis: quantum theory is nothing but a general theory of probabilities, i.e., of the expectation values of relative frequencies in statistical ensembles.

## 3.5 Methodological: Reconstruction of abstract quantum theory

The title of this section initially suggests three questions:

1. What is meant by reconstruction?
2. What is meant by abstract quantum theory?
3. What ways are there for a reconstruction of abstract quantum theory?

### 3.5.1 The concept of reconstruction

We have already explained this concept in Chap. 1 and recapitulated it at the beginning of Sect. 2.12. We mean by reconstruction its retrospective derivation from the most plausible postulates. We articulate once again the difference between two kinds of such postulates. They may either express conditions which make experience possible, thus conditions of human knowledge; then we call them *epistemic*. Or they formulate very simple principles which we hypothetically, inspired by concrete experience, want to assume as universally valid for the particular area of reality; we call these postulates *realistic*.

We emphasized in the first chapter that our method of a *Kreisgang*[7] does not permit a completely sharp distinction between these two kinds of postulates. We merge in the *Kreisgang* two traditions of thought which, in the history of philosophy, were in hostile opposition most of the time. All our knowledge of nature is subject to the conditions of human knowledge; that is the epistemological question. Humans are children of nature and their knowledge itself is a process in nature; that is the evolutionary question. Even our

---

[7] Cf. p. XXII, fn. 14.

evolutionary knowledge is, as human knowledge, subject to the conditions of such knowledge, as studied in epistemology. Also, the back of the mirror we only see in the mirror. But the mirror in which we see the back of the mirror is also just the mirror with this rear surface; epistemology, like the cognition it investigates, is also an event in nature. In this way every epistemological postulate is at the same time a statement about a process of nature, and every realistic postulate is formulated subject to the conditions of our knowledge.

The historical phenomenon that there are closed theories, however, permits us a relative distinction of epistemic and realistic postulates, as regards a particular theory. "Only theory decides what can be measured" (Einstein to Heisenberg; cf. Heisenberg 1969). We will begin the reconstruction of quantum theory with *one* postulate which, for quantum theory, is epistemic: the existence of separable, empirically decidable *alternatives*. An alternative characterized in this way expresses the quantum theoretical concept of an observable, reduced to its logical foundation. The fact that quantum theory is so successful, and that one is able to succeed with the concept of the alternative in the totality of physical experience known to us, is an empirical fact which a priori does not appear to be certain. In this sense the postulate of alternatives is realistic, but it is also epistemic in another sense. First, as just noted, it is epistemic in the context of quantum theory: it formulates a condition without which the concepts of quantum theory are inapplicable. But second also as a matter of principle: we scarcely can imagine how scientific knowledge might be possible at all without separable, empirically decidable alternatives. The high degree of generality of quantum theory thereby confers upon its basic postulate a position reminiscent of Kant's perception a priori: *that* experience is possible at all we cannot know a priori; we can only know what circumstances must obtain *in order for* experience to be possible.

However, the *second* central postulate for the actual quantum theory, which we call the postulate of *expansion* or *indeterminism*, must also be considered *realistic* in the context of quantum theory. We cannot imagine a theory of probability predictions about decidable alternatives in which this postulate is not applicable. This question, however, we can only discuss after the reconstruction is accomplished. It will follow us into Chaps. 9 and 11.

### 3.5.2 Abstract quantum theory

Terminologically we distinguish abstract and concrete quantum theory. One can characterize abstract quantum theory by means of four *theses*. We use the concept of "thesis" to distinguish it from the reconstructive concept of "postulate." These could be at the foundation of a formally axiomatic deduction of the theory. But they cannot claim to be "evident" as we require it of the postulates. Rather, their explanation is the goal of our reconstruction.

A. *Hilbert space.* The states of every object are described by rays in a Hilbert space.

B. *Probability metric.* The absolute square of the inner product of two normalized Hilbert vectors $x$ and $y$ is the conditional probability $p(x, y)$ of finding the state belonging to $y$, if the state belonging to $x$ is present.
C. *Composition rule.* Two coexisting objects $A$ and $B$ can be considered to be a composite object $C = AB$. The Hilbert space of $C$ is the tensor product of he Hilbert spaces of $A$ and $B$.
D. *Dynamics.* Time is described by the real coordinate $t$. The states of an object are functions of $t$, described by a unitary mapping $U(t)$ of the Hilbert space onto itself.

We call this theory abstract because it is universally valid for any object. One example of an abstract theory we have seen in classical point mechanics (Sect. 2.2). There, Eq. (2.1) characterizes the universally valid law of motion for arbitrary numbers $n$ of mass points, arbitrary masses $m_i$, and arbitrary force laws $f_{ik}(x_1 \ldots x_n)$. Von Neumann's quantum theory is even more abstract, as it does not presume the concept of a point mass and the existence of a three-dimensional configuration space. These concepts enter into quantum theory itself only via the special choices of the dynamics and the selection of certain observables associated with the dynamics. They belong to the *concrete* theory of specific objects.

Concrete quantum theory is the subject of Chaps. 4 and 5.

## 3.6 Reconstruction via probabilities and the lattice of propositions

This reconstruction path was chosen by Drieschner (1967) and described later (Drieschner (1979)) in improved form. It follows most closely Jauch (1968) and the usual axiomatics; it goes beyond these in the way of the justification and the thereby implied choice of postulates. The reconstruction is sketched here to facilitate the connection to existing quantum axiomatics. This offers the opportunity to explain the abstract basic concepts within a familiar context. We follow the layout and sometimes the wording of the presentation in *Die Einheit der Natur* II.5.4, pp. 249–263.

### 3.6.1 Alternatives and probabilities

Physics formulates probability predictions about the outcome of future decisions of empirically decidable alternatives. The concept of probability is described in Chap. 3. Here, however, we will replace axiom I of Kolmogorov by another one; the catalog of events is not the lattice of the subsets of a set.

We describe all possible observations as decision of $n$-fold alternatives. There $n$ means either a natural number $\geq 2$ or the denumerable infinite. An $n$-fold alternative represents a set of $n$ formally possible events which satisfy the following conditions:

1. The alternative is *decidable*; i.e., a situation can be created in which one of the possible events becomes an actual event and subsequently a fact. We then say that this event has happened.
2. If an event $e_k$ ($k = 1\ldots n$) has happened then none of the other events $e_j (j \neq k)$ has occurred. The results of an alternative are *mutually exclusive*.
3. If the alternative has been decided and all events except one, thus all $e_j (j \neq k)$, have not occurred, then the event $e_k$ has happened. The alternative is defined as being *complete*.

*Note about the nomenclature.* Probabilities can be considered to be predicates of possible *events* or of *propositions*. About the philosophical interpretation of the difference between the two expressions see *Zeit und Wissen* 4. Here in this sketch we use both expressions indiscriminately; sometimes the one, sometimes the other is more convenient. This leads to the following expressions.

An alternative is a set of either *events* or of *propositions*. Both we call its *elements*. An event consists of the determination of a formally possible (conditional) *property* of an object at one time. Instead of this we also say that the object is at this time in a certain *state*. The word "state" is in this context not restricted to "pure cases." The *proposition* which asserts the existence of a property or state is formulated in the *present* tense, in the sense of 2.3. This means that one can often decide "the same" alternative. An alternative can also be referred to as a *question*; the propositions are then its possible *answers*.

### 3.6.2 Objects

The elements of an alternative consist of the determination of formally possible properties of an object at one time.

We introduce the "ontological" concept of an object in addition to the "logical" concept of an alternative. The alternatives for an object are, speaking quantum theoretically, its observables. We follow here the mode of thinking customary in all of physics, in particular in quantum theory, which interprets all its catalogs of propositions as propositions about, respectively, an "object" or a "system." These two words are practically synonymous in contemporary physics. "Object" is perhaps the more general concept as it encompasses composite as well as the possibly existing elementary objects, whereas the word "system" is more indicative of compositeness (sy-stema, standing together). In this book we will therefore in general choose the term "object."

In the reconstruction we need the concept of the object to define the lattice of propositions which in each case is determined as the lattice of propositions about a fixed object (or the properties of a fixed object).

The concept of the object, however, contains a fundamental problem which we will discuss subsequently to subsection E.

### 3.6.3 Final propositions about an object

For every object there ought to be final propositions and alternatives whose elements, logically speaking, are final propositions. As a final (conditional) proposition about an object we define a proposition which is not implied by any other proposition about the same object.[8] In the quantum theoretical language this means that there are pure cases. Lattice-theoretically these final propositions are "atoms," i.e., the lowest elements of the lattice; Drieschner (1979) therefore calls them atomic propositions. Drieschner argues for the postulate of the existence of atomic propositions from the requirement that it ought to be possible in principle to completely describe every object in terms of its properties.

### 3.6.4 Finitism

Drieschner (1967) introduced the postulate of finitism, which one might perhaps formulate thus: "The number of elements of an arbitrary alternative for a given object does not exceed a fixed positive number $K$ which is characteristic of that object." In contrast, we have also admitted denumerably infinite alternatives in 3.2a. Furthernore, Drieschner (1979) no longer requires finitism. The technical benefit of the finitism postulate is that it avoids mathematical complications of an infinite-dimensional Hilbert space in the axiomatic reconstruction of quantum theory. Philosophically, behind this is the observation that no alternative with more than a finite number of elements can actually be decided by an experiment.

For convenience we will use here only finite alternatives. We can afford this as we enter this reconstruction only as an avenue leading to Sect. 4.2, where we explicitly introduce infinite alternatives, i.e., with a justification. Physically, the infinite dimensions of Hilbert space become indispensable if we wish to unitarily represent in it the non-compact transformation groups of special relativity. In other words, we need it for relativistic quantum theory. In that regard, the present chapter is restricted to nonrelativistic quantum theory. In Chap. 4 we will define the simplest objects, particles, as representations of relativistic transformation groups, following Wigner; thereby for every object $K = \infty$. The "objects" of finitism, however, retain an assignable meaning as representations of the compact part of the group in finite-dimensional subspaces. We will then call them "sub-objects."

### 3.6.5 Composition of alternatives and objects

Several alternatives can be combined to a *composite alternative*. This is done by "Cartesian multiplication." Given $N$ alternatives ($N$ finite or perhaps denumerably infinite): $\{e_k^\alpha\}$ ($k = 1 \ldots n^\alpha; \alpha = 1 \ldots N$), a combined event means

---

[8] With the trivial exception of the "always false proposition" 0 which, by definition, implies any proposition—*ex falso quodlibet*.

that an event $e_j^\alpha$ from each alternative occurs (not necessarily simultaneously). This is an element of the combined alternative which has $n = \prod_\alpha n^\alpha$ elements.

Now $N$ objects also define a total object of which they are its parts. The Cartesian product of any alternatives of the parts is an alternative of the total object. In particular, the product of always final alternatives of the parts is a final alternative of the total object.

The concept of an object, as we now see, contains some sort of self-contradiction which one cannot eliminate without eliminating all of physics known to us, which is built upon the concept of the object. Objects are known to us only through their interaction with other objects, ultimately with our own body. Completely isolated objects, free of any interaction, are no objects at all to us. The Hilbert space of an object describes just the possible states of only this one object. The introduction of dynamics, as we will perform it afterward, i.e., of a Hamiltonian operator, describes the influence of a fixed environment on the object and, insofar as one considers the object to be composite, the interaction of its parts with one another. To describe its influence on the environment one must combine it with other objects, thereby forming a aggregate object. In the Hilbert space of the aggregate object, however, the pure product states, in which the individual objects are in a definite state, are a set of measure zero. But it is just these definite states in terms of which quantum theory describes the individual objects. It appears that quantum theory could be formulated only approximately, which, if the theory is correct, would practically never be exactly valid. In short, the feasibility of theoretical physics rests upon its character as an approximation.

The philosophical problems herein I have discussed in detail in previous essays.[9] Let us accept here the concept of an object in its common usage.

### 3.6.6 The probability function

Between any two states $a$ and $b$ of the same object there is defined a probability function $p(a,b)$ which gives the probability of finding $b$ if $a$ is necessary. The formulation and content of this postulate depended on the assumption that everything which can be said about an object in an empirically decidable way must be equivalent to the prediction of certain probabilities. The empirical verification of a proposition lies in the future at the time to which this proposition refers. About the future, however, only probabilities can be stated, which of course may approach the values 1 and 0, certainty and impossibility. The formulation of the condition in $p(a,b)$ by means of "if $a$ is necessary" includes the case "if $a$ is present," as $a$ is then, due to the reproducibility, necessary in the future, as well as the case that one knows the necessity of $a$ for other reasons.

The really strong assumption in this postulate remains inconspicuous in the above formulation, namely that this probability function assigns to each

---
[9] *Die Einheit der Natur* II.3.5, IV.6.4; *Der Garten des Menschlichen* II.1.9.

pair of states $a$ and $b$ the value $p(a,b)$ *independent of the state of the environment*. This means at the same time that the states of an object admit an "internal description," consisting only of its relative probabilities without reference to "external" objects. How one can identify the respective states through observation, however, is then only determined in terms of the interaction of the object with its environment.

This strong assumption of independence is the form in which the *identity* of an object with itself expresses itself in this reconstruction, which ought to hold independently of its changing environment. Here is a specification of the concept of an object which we need for the reconstruction but which here we do not justify any further.

### 3.6.7 Objectivity

If a certain object actually exists, then a final proposition about it is always necessary. This, too, is a strong statement. For its justification we refer to Drieschner (1979, p. 115–117). There it is described as being equivalent to the statement: "Every object has at any time as property a probability distribution of all its properties." The premise "If a certain object actually exists" is necessary, because in states of composite objects which are not product states of the partial objects, no final statement about such a partial object is necessary. We then say that this partial object does not actually exist in such a state (compare Sect. 9.3d).

We call this postulated fact the *objectivity* of the properties of actually existing objects. For an actually existing object there is always a final proposition, independently of whether we know it, i.e., it must inevitably be found if one looks for it. In other words, when one says that an object actually exists, one means that in principle one can know with certainty something about the object. Knowledge is not "merely a subjective state of the mind." To know means, putting it tautologically, knowing that the known is as we know it. Here as well we refrain from following up on the philosophical implications of our assertion.

### 3.6.8 Indeterminism

To any two mutually exclusive final propositions $a_1$ and $a_2$ about an object, there is a final proposition $b$ about the same object which does not exclude either of the two. Two propositions $x$ and $y$ exclude one another if $p(x,y) = p(y,x) = 0$.

This is the central postulate of quantum theory. Following Drieschner it is called here the postulate of indeterminism. Within the context of the reconstruction it turns out to be equivalent to, e.g., the principle of superposition formulated by Jauch (1968, p. 106). It is the "realistic" fundamental postulate; for it is at least not immediately obvious that experience without this postulate is not possible.

We can denote this postulate equivalently by the more abstract term *postulate of expansion*. The connection between the two names is as follows. Every alternative of final propositions is expanded through this postulate by final propositions about the same object which are not elements of the lattice of propositions which form the original alternative. The expansion is here formulated as a requirement on the probability function, i.e., on predictions: there are always predictions which have neither the value of certainty nor impossibility. This is juxtaposed with the postulate of objectivity according to which there are always necessary predictions. Both always exist. The requirement is at the same time formulated universally: it holds for *any* pair of mutually exclusive final propositions. It implies that there can be no probability assignment of the catalog of propositions about any object whatever for which every proposition is either true ($p = 1$) or false ($p = 0$). It thus implies the openness of the future as a matter of principle.

### 3.6.9 Sketch of a reconstruction of quantum theory

For the implementation of the reconstruction we refer to Drieschner (1979). Here we merely mention the most important steps.

The catalog of propositions is constructed about an object. Negation, disjunction, and implication are defined in terms of obvious requirements on the probability functions, such that the catalog proves to be a lattice, and, in fact, for the case of finitism, a modular lattice. It can be shown that, with the imposed requirements, it is even a projective geometry. This can be represented as the lattice of the linear subspaces of a vector space. There remains the question of the field of numbers in which the vector space is erected. As a real metric is defined in it by means of the probability function, the field of numbers must contain the real numbers. Following Stückelberg (1960) Drieschner concludes from the uncertainty relations that it specifically must be the field of complex numbers. The dynamics is to be described in it, i.e., the time-dependence of the state, in terms of transformations under which the probability function remains invariant. These must be unitary transformations. In this fashion, abstract quantum theory is reconstructed.

For the time being, we forgo any attempt to examine how close the individual postulates have come to the ideal of an epistemic justification.

### 3.6.10 Historical remark

The first formulation of the ideas utilized here, in my version, is given in the work *Komplementarität und Logik* (1955). To Drieschner's indeterminism axiom, there corresponds, for example, the "theorem of complementarity" (Sect. 6): "To every elementary proposition there are complementary elementary propositions." But only the work of Drieschner transformed this "complementary-logical" way of thinking, together with the quantum axiomatics of Jauch (1968), into a reconstruction of quantum theory. The goal

## 3.6 Reconstruction via probabilities and the lattice of propositions

of the present historical note is to point out the reconstruction of quantum theory previously begun by F. Bopp. Bopp's work of 1954 I quoted in 1955 (Sect. 5). It provided me with essential suggestions for the elaboration of my arguments at that time; see also his more recent work (Bopp 1983). Bopp begins, as we do in Sect. 4.1, with a simple alternative ("Sein oder Nichtsein als Grundlage der Quantenmechanik"). He postulates, as in Drieschner's uncertainty postulate, the existence of additional states defined in terms of relative probabilities and the continuity of this state space to make a continuous kinematics of the states possible. He, however, takes the spacetime continuum for granted and considers the alternative to depend on position ("ur fermion").

# 4
# Quantum theory and spacetime

## 4.1 Concrete quantum theory

In the previous chapter we indicated a path to the reconstruction of abstract quantum theory, i.e., the quantum theory of arbitrary alternatives and objects, and arbitrary forces. Now our goal is the reconstruction of concrete quantum theory, i.e., the quantum theory of actually existing and actually possible objects. For this we need three concepts which did not play any substantive role, being at most used for illustrative examples:

a. Space
b. Particles
c. Interactions

For their preliminary description we go back to the system of theories presented in Chap. 2.

### 4.1.1 Space

We summarize in four theses the role of the concept of space in the system of theories:

$\alpha$. *Spatiality.* All objects, at least all objects of classical physics, are "in space" (2.2d), whether as extended objects, localizable mass points, or fields defined in space. To distinguish it from abstract mathematical space, this space is also termed *position space*.

$\beta$. *Symmetry.* The position space is a real, three-dimensional Euclidean space of points. Its symmetry group is the six-parameter Euclidean group $E(3)$ of rotations and translations.

$\gamma$. *Special theory of relativity.* Position space is combined with time to form a four-dimensional spacetime continuum, also known as Minkowski space. The symmetry group of this "space" is the ten-parameter Poincaré group (inhomogeneous Lorentz group ) consisting of three spatial rotations, the three proper Lorentz transformations which hyperbolically rotate space and time into one another, three spatial translations, and time translation.

$\delta$. *General theory of relativity.* The spacetime continuum has a Riemannian geometry. At every location there is a local Minkowski space as tangent space.

*Comment:* We will not require any of these four theses as epistemic or realistic postulates but will try—as with the four theses of abstract quantum theory in Sect. 3.1b—to justify them from postulates. There we only assume abstract quantum theory and *one* additional, purely quantum theoretical postulate (Sect. 3.4b). In the present preliminary discussion, we wish to visualize the role of the concepts of space, particles, and interactions within the system of theories. First we address the four theses $\alpha$ through $\delta$.

$\alpha$. *Spatiality.* There is absolutely no *a priori* necessity for all objects of physics to be in one common space. We are led to this insight by the fact that we used the assumption of "spatiality" (i.e., being in space) neither in the theses (3.1b) about abstract quantum theory nor in its reconstruction. On the other hand, spatiality is traditionally so self-evident that it requires a certain effort of abstraction to recognize in it a special assumption. It may therefore be useful to briefly survey how the three great epistemological schools of realism, empiricism, and apriorism, as well as the newer behaviorism, have dealt with this assumption.

Realism usually simply accepts the spatiality of objects as a self-evident attribute of the concept of reality. Empiricism realizes that this is not self-evident; but in accepting spatiality it interprets it as an empirical fact. One thinks, however, that which is described as "empirical" might also be different, considered abstractly. Kant was aware of this in his aprioristic approach. He therefore postulated space as the prerequisite for possible experience, not in the sense of a conceptual necessity but as a form of our perception.[1] These are successive steps on the way to clarify the prerequisites for physics. In a *Kreisgang*[2] we now follow the evolutionist epistemology of Lorenz. It acknowledges spatiality as an innate form of perception but seeks to explain it in terms of adaptation; i.e., in its principles of explanation it returns to the most naïve thesis, that of realism.

In our own approach we have no difficulty in acknowledging spatiality as an empirical fact within the range of applicability of classical physics, and thereby

---

[1] Note that he distinguished space as "formal perception," in which we "construct" our geometrical concepts, through a "synthesis" from the mere "form of perception" which only leads to "mere multiplicity" (*Critique of Pure Reason*, B 160, footnote).

[2] Cf. p. XXII, fn. 14.

also evolutionary as the reason for our form of perception.³ But precisely these empirical facts about the spatiality of empirically known objects, as well as about the form of our perception, are what we do not want to presume in quantum theory. If possible, we instead wish to explain them in terms of quantum theory.

This suggests the question of the systematic role of coordinate space in the quantum theory that developed historically. That Heisenberg started with the commutation relation between position and momentum was a sharpening of Bohr's correspondence principle. In terms of the reversal of the historical order of the arguments (Sect. 3.1) one can say that if in quantum mechanics the energy operator $H$ depends on any two operators $p$ and $q$ that satisfy the Heisenberg commutation relations, then this defines a classical limit in which $p$ and $q$ are canonically conjugate variables. The reason why coordinate space is distinguished lies in the fact that all classical laws of interaction depend on position (and possibly on the velocity in coordinate space). As we observe only by means of interaction, every measurement starts out as a position measurement. This dependence of interaction on position one simply must accept as an empirical fact within the correspondence principle. In an abstract interpretation it is again plausible to reverse the order of the arguments. If there is any parameter at all on which all interactions depend, one can expect that the actually observable objects and their states are most directly described in a representation which is based on this parameter as an independent variable.

One can then ask why there should be a state parameter at all on which all interactions depend. We can learn about this from the following theses.

$\beta$. *Symmetry* In the spirit of Felix Klein's Erlanger Program we consider a geometry to be defined by its symmetry group (see Sect. 2.3). In the reconstruction of quantum theory the symmetry of the state space appears to us as an expression of the separability of alternatives. For an $n$-fold alternative, this initially yields the group $U(n)$, thus the complex geometry of Hilbert space. But if all actually occurring interactions depend on a three-dimensional real geometry, this must mean the *all actually occurring dynamical laws have a common symmetry group* which is much smaller than $U(n)$ for large $n$. We consider this to be the *central phenomenon of concrete quantum theory*. We will try to explain it in terms of the postulate of the *ur alternative* (Sect. 3.2b). Coordinate space will then be explained as a homogeneous space of the universal symmetry group of the dynamics.

$\gamma$. *Special theory of relativity.* The special theory of relativity describes the kinematics as a four-dimensional pseudo-Euclidean geometry with the Poincaré group as defining symmetry group. Historically kinematics was understood as a prerequisite of dynamics. Kinematics describes all formally pos-

---

³ In the spirit of the previous footnote we further remark that the psychologically elicited form of our perception is neither Euclidean nor non-Euclidean but fuzzy. Euclidean geometry is an "idealization" of our perception (Sect. 2.2).

sible motions within the structure of space and time; dynamics, by specifying the forces, selects from them the actually possible ones. Both of the mathematically precise advances in theoretical physics of our century, Einstein's special theory of relativity of 1905 and Heisenberg's quantum mechanics of 1925, introduce new *kinematic* laws.

In the spirit of the reversal of arguments we must ask why a universal kinematics is possible at all. We will explain the symmetry group of the kinematics as the common symmetry group of all actually possible dynamics. In that respect it is more comprehensive than the symmetry group of position space as it also encompasses transformations of velocities (and time translations). The restriction to velocities, i.e., uniform motions, is related to the law of inertia (Sect. 4.1$b$). We will subsequently recognize this inclusion of time in the transformations to be quantum theoretically consistent.

To date the relationship between quantum theory and relativity is justifiably considered to be not yet fully clarified. Certainly relativistic quantum field theory with local interactions is empirically very successful. But as long as the singularities that occur in it are only omitted in a relativistically invariant manner, it is not clear whether the theory actually exists in a strict mathematical sense. Even if one hopes that these difficulties are eventually clarified, relativistic quantum theory still remains an instance of two foreign theories that have been "glued together."

Our reconstruction of quantum theory, on the other hand, starts with the expectation that each of these two theories can only be fully understood in connection with the other. That holds in both directions. On the one hand, we will derive position space and relativistic invariance as *consequences* of a quantum theory with variable alternatives by means of the postulate of the ur alternatives. The quantum theory thus constructed is therefore a relativistic quantum theory from the very outset. This is precisely what shows, on the other hand, that only the consideration of relativistic invariance *completed* quantum theory. Only the representation of the non-compact relativistic symmetry group forces us to go beyond finitism to an infinite-dimensional Hilbert space. The fundamental role of the theory of relativity does not show up in the traditional construction of quantum theory according to the correspondence principle, as there one starts with a classical physics which from the outset is formulated in a continuous, infinitely extended coordinate space. The wave functions in this space automatically form an infinite-dimensional Hilbert space. But only our slow, initially finite reconstruction of quantum theory offers the prospect that from the outset no singularities arise in the interaction.

$\delta$. *General theory of relativity.* We have discussed in detail the problems of principle of the general theory of relativity in Sect. 2.10, and will not return to them until Chap. 5.

### 4.1.2 Particles

Here we only remark that the existence of particles which can be described as point masses follows according to Wigner (1939) from relativistic quantum theory: their state spaces are the representation spaces of irreducible representations of the Poincaré group. We address this in Chap. 4.

### 4.1.3 Interaction

*All dynamics is interaction.* We introduce this thesis as a postulate in the next section. Here it is initially meant to be a descriptive statement about present physics, and will be explained as such.

The thesis is by no means self-evident. In mechanics (see Sect. 2.2) one distinguishes between force-free motion, motion under the influence of an external force, and motion under the influence of interaction. That there are force-free motions we referred to in Sect. 2.2 and 2.3 as a causal paradox; in Sect. 4.3c we return to this question. It does not affect our present arguments if we understand by dynamics only the effect of forces. The above thesis says then that all "external" forces are in reality interactions. Interaction between two objects (or alternatives) means that if their motion (or evolution) is affected at all, the effects are always mutual. If the action is negligible in one of the two directions, then we speak of an external force. One can perhaps add that most physicists believe in this thesis. Newton's Third Law formulates if quantitatively. We used it implicitly above when we related the concept of space not to forces but to interactions.

## 4.2 Reconstruction of quantum theory via variable alternatives

### 4.2.1 Variable alternatives

#### α. Three postulates

***1. Establishment of possibilities.*** *The actual possibilities are governed by the actual facts.*

Comment. We first explain the terminology. The word *actual* will denote something that exists in the immediate present. Using the language of temporal logic, we distinguish between presentic, perfectic, and futuric statements. An actual fact is that which is described in the immediate present by a true presentic proposition; we distinguish it from a perfectic fact, which is similarly described by a true perfectic proposition. We express its relationship to the *immediate* present by means of the term *actual*, rather than *presentic*. An actual possibility is then something that just became possible in the immediate present, which then posits a true futuric proposition concerning the immediate future. Not everything we can describe as being formally possible, nor everything that is possible in the more distant future, is possible now. The postulate means that the actual facts govern what is now correspondingly possible. For examples we refer to the discussion of irreversibility (Sect. 7.1) and evolution (Sect. 8.1).

We specify time in this and the following chapter using a real coordinate, where the immediate present is represented by a corresponding point. This is a mathematical idealization. The phenomenon of the present is neither a instant of time nor a time interval, nor is it measured along a scale. We abide by this idealization in modern quantum theory and relativity theory; only in Chap. 11 will we inquire beyond that. However, we emphasize here that the present, as indicated by the word *actual*, is the present of a *single* observer, or in any case a group of people communicating with one another in a mutually consistent present.

We understand this postulate to be *epistemic* in the sense in which we understand time in its modalities as prerequisites for experience. How else should actual possibilities be recognizable if not from actual facts? Even if somebody assumed facts from the distant past were "directly" influencing actual possibilities, he would thereby declare the facts acting nowadays to a certain extent to be actual facts. The postulate can be considered a present-day version of Kant's reply to Hume that the principle of causality is a prerequisite for possible experience, save that in our version the actual facts determine mere possibilities, not the future facts.

**2. Open finitism.** *All alternatives that are decidable in real terms are finite, but no upper bound can be specified for the number of elements.*

*Comment.* We have already explained this in Sect. 3.6D. Initially, the postulate simply describes the role of natural numbers when dealing with experience. Every set whose elements one can actually enumerate is finite. But for every finite cardinal number, one can always find a larger one. The enumeration of all natural numbers (counting) takes time to occur, and has no natural end.

For measurements in the continuum it is initially unclear how exactly open finitism is to be applied; frequently the experimentalist himself does not know how accurate a measurement is, and thus how large the resolved alternative is.

Let us consider, in anticipation of a well-known theory which, however, we here first want to justify, the observables of a force-free particle. The eigenfunctions of operators with a continuous spectrum like position and momentum are not in Hilbert space. The Hilbert space may, however, be erected over a discrete basis from the eigenfunctions of angular momentum and Laguerre polynomials of the magnitude of momentum. To every fixed total angular momentum there belongs a finite-dimensional subspace which is defined by a finite alternative. From now on we term such a space the state space of a subobject. Then the state space of a free particle is the direct sum of infinitely many finite-dimensional state spaces of subobjects. We will interpret the open finitism so that the state space of every free object admits such a formal decomposition. For justification we do not draw on the spatial interpretation, as we just did for angular momentum, but instead derive it from the initially abstract postulate of open finitism.

**3. Actual alternative.** *The actual possibilities are, in the approximation of the separability of alternatives, in each case given by the state space of* one *alternative.*

*Comment.* The actual possibilities are thus understood in the sense of separable alternatives, the postulate of expansion being assumed to hold for each individually. All actual possibilities, each in the present of an observer, can then always be described by the state space of *one* alternative, which is the Cartesian product of all independent alternatives just then decidable by him. We call this the actual alternative. It is, as the basis in state space, naturally only defined up to a coordinate transformation in state space.

We will see that our approach explicitly leads us to consider the separability of finite alternatives to be only an approximation.

## β. Three conclusions

**1. Determinism of possibilities.** *The actual possibilities determine their own change in time.*

*Comment.* We initially describe possibilities quantitatively by specifying the probabilities of formally possible events. Actual possibilities are formally possible events with a nonvanishing actual probability. Our conclusion asserts that these events on their part will determine the subsequently existing actual possibilities, and so on.

Those who know quantum mechanics will understand that there is an ambiguity hidden in this apparently simple deduction. An actual fact has irreversibly happened. According to Bohr it must be described classically; we will discuss this in detail in Sect. 9.2. For an actually possible event to become an actual fact another irreversible process must happen, the "measurement process." In classical physics one disregards this problem. A fact $a$ at time $t_0$ determines in the classical theory the conditional probabilities of all facts $b$ at time $t_1$, and in terms of the latter the conditional probability of a fact $c$ at an even later time $t_2$ according to

$$p_{ac} = \sum_b p_{ab} p_{bc} \tag{4.1}$$

(see Feynman (1948)). If, however, one does not make a measurement, then according to quantum theory (3.1) must replaced by the combination of amplitudes

$$\psi_{ac} = \sum_b \psi_{ab}\,\psi_{bc}. \tag{4.2}$$

We can then draw our conclusion if not the probabilities $p_{ab}$ for actually occurring facts but the amplitudes $\psi_{ab}$ for formally possible events determine the subsequent development of the actual possibilities. We can understand this as a continuation of the epistemic justification of the postulate on establishing possibilities: How else should future actual possibilities be recognizable now if not as mediators of the possibilities lying between the present and then. At the moment I am unable to formulate this argument any more sharply.

**2. Variable alternatives.** *The temporal change of the actual possibilities can signify the transition to another actual alternative (according to postulate 3).*

*Comment.* We start with the experience of time in which actual possibilities can emerge and pass on. This will happen not only in the state space of a fixed alternative; greater and lesser alternatives will emerge as well. In the terminology we introduced in the commentary to postulate 2, the temporal variation will transform one subobject into other subobjects. We devise a quantitative theory for this in the next subsection with the help of binary ur alternatives. The third conclusion may serve as a qualitative explanation:

**3. Growth of possibilities.** *As a statistical average, the multiplicity of actual possibilities increases.*

*Comment.* We discuss in Chap. 8 the sense in which information increases, in the statistical mean, upon every emergence of forms; biological evolution provides the most striking example. Picht (1958) expressed this growth, without scientific theory and without using the concept of information, with the statement "The past does not pass. The multiplicity of possibilities increases." (For elucidation, see *Der Garten des Menschlichen*, II.7: *Mitwahrnehmung der Zeit*). The sentence "The past does not pass" describes the facticity of time. The sentence "The multiplicity of possibilities increases" can then be interpreted as follows. If the past does not go away, i.e., if everything that was a fact once remains a fact, then the number of facts will continually increase, as new facts are constantly emerging (one calls these *events*!). Hence the multiplicity of possibilities determined by these facts should also grow.

Now certainly only the number of past facts increases, but not necessarily also the number of actual facts and thus of actual possibilities. Every actual event eliminates certain possibilities and creates different possibilities instead. But the open suggests that the number of actual possibilities also increases, at least in the statistical mean. Let us begin with an actual alternative of order $n$. According to the variability of alternatives this will evolve into alternatives of lower or higher rank. But $n$ cannot go below the value 1 if something observable is still to be left; it can, however, grow beyond all bounds. Here the theory of ur alternatives will also provide us with a quantitative model.

### 4.2.2 Ur alternatives

**1. Theorem of the logical decomposition of alternatives.** *An n-fold alternative can be mapped into the Cartesian product of k binary alternatives with $2^k \geq n$.*

*Comment.* The theorem is logically trivial. Every finite alternative can be decided by successive yes/no decisions.

**2. Theorem of the mathematical decomposition of state spaces.** *An n-dimensional vector space can be mapped onto the tensor product of k two-dimensional vector spaces, with $2^k \geq n$, such that its linear and metric structure are preserved.*

*Comment.* Take, e.g., $k = n - 1$, and retain only symmetric tensors of rank $k$. These constitute an $n$-dimensional irreducible representation of SU(2).

Physically one can interpret the theorem as a radicalization of classical atomism that only became possible in terms of abstract quantum theory. Heretofore, in chemistry and physics, *relative atomism* had proven itself. All objects consist of smaller and smaller objects that can be arranged into a few classes (chemical atoms, elementary particles); the objects of any single class are mutually equivalent. Atomism is relative: one does not know whether the smallest objects that happen to be known are further divisible or not. Quantum theory has made atomism more precise. The composition of objects does not in general have to be visualized as a spatial coexistence; it consists in the formation of the tensor products of their Hilbert spaces. The necessary sacrifice of visualizability we read as an indicator that in quantum theory the spatial extent of objects is only a derived property.

From our theorem we can now develop the hypothesis of *radical atomism*. In the imprecise language of classical atomism one might cast it in a form that claims that any object can be decomposed into the very smallest possible objects. However, one immediately sees that the language of classical spatial extent is inadequate for the proposed idea. What are the "smallest possible objects" supposed to be? One plunges into the difficulties of a classical continuum theory, as discussed in Sect. 2.2b and 2.4. Quantum theory avoids this problem. We have developed quantum theory from the concept of the alternative. Alternatives are discrete. The smallest alternative that will represent a decision is the binary (twofold) alternative[4], $n = 2$. It is the "smallest" alternative in the sense of the amount of information; its decision provides, assuming no prior information, just one bit. Clearly, any association here with spatial smallness must be held at bay: to localize a particle in a very small space requires very many yes/no decisions.

---

[4] In previous texts I have called it, following the colloquial expression, the "simple alternative" (*one* yes/no decision).

**Definition.** *We call* ur alternatives *the binary alternatives from which the state spaces of quantum theory can be constructed. The subobject associated with an ur alternative we call one* ur.

*Comment.* Quantum theory also modifies the perceptions of the temporal behavior of elementary objects. The atom of classical atomism was assumed to be a substance in the strict sense: never created, indivisible, everlasting. The elementary particles of present-day physics still have some sort of associated temporal identity. But elementary particles can be transformed into one another. The ur is defined in terms of a simple alternative which, according to the above postulate 2 of the variable alternatives, must be capable of emerging and passing on. That its states can be recognized again requires, however, that its state space be invariant under the dynamics as long as the defining alternative actually lasts. This will be ensured by the following postulates.

*3.* **Postulate of interaction.** *All dynamics is interaction.*

*Comment.* We have already discussed the postulate in Sect. 4.1c as the description of a widespread perception of physics. In the context of a reconstruction of quantum theory from empirically decidable alternatives we will call it *epistemic*. An empirically detectable external force can be described by alternatives. It should therefore itself be an object in the sense of quantum theory. One can combine it with the object upon which it acts, into a combined object which in itself one can then decompose into urs. For the action of the force not to be one-sided, but an interaction (*actio = reactio*), it should suffice that there be no difference between the urs that make up the two parts of the combined object. We formulate this as a separate postulate.

*4.* **Postulate of the indistinguishability of urs.** *Urs are momentarily indistinguishable.*

*Comment.* The word "indistinguishable" is used here in the same way as in quantum statistics of identical particles. Below (4.7d) we make this mathematically more precise. For the justification of the postulate let us say that a distinction of the urs would again be an alternative, which itself should be describable by urs.

We should, however, remark at once that the existence of urs is deduced from the mathematics of the time-independent vector space, and affects only those alternatives that follow from this "momentary" structure of the state space. According to the determinism of the alternatives, this momentary structure ought to govern subsequent development. But this also requires the specification of a dynamical law. From the postulate of the interaction we will deduce one conclusion about the *symmetry group* of this law. But from the outset it is not ruled out that the interaction permits the definition of "temporal" alternatives that have not yet appeared in the momentary structure, and which allow a distinction between types of urs. Only with this reservation is the ur hypothesis an epistemologically justified consequence of abstract quantum theory.

*About the terminology:* The two postulates 3 and 4, in their application to Theorems 1 and 2, we summarize under the designation *postulate of the ur alternatives* or *ur hypothesis*.

### 4.2.3 The tensor space of urs

The question now is how the composition of subobjects from urs is to be described. According to the composition rule, the state space of an object consisting of various parts is the tensor product of their state spaces. This rule we apply here. All possible states of $n$ urs thus lie in the space $T_n$ of all tensors of rank $n$ over the vector space $V^{(2)}$ of the ur. The totality of all possible states of an arbitrary but finite number of urs then lies in the direct sum of all $T_n$:

$$T = \sum_{n=0}^{\infty} T_n. \tag{4.3}$$

A question that arises is whether the full tensor space is utilized. Urs are assumed to be indistinguishable. This naturally suggests that urs have Fermi or Bose statistics. Fermi statistics is out of the question as then only two urs could exist in the world, one in each of the states defining the alternative. Bose statistics is possible. This implies restriction to symmetric tensors in $T$; we denote the space of symmetric tensors by $\overline{T}$. For simplicity we adopt the arguments of the present chapter in $\overline{T}$. In the next chapter we show, however, that the statistics adequate for urs is the para-Bose statistics, which utilizes all possible symmetry classes of tensors.

We now define states and operators in $T$.

The ur is the subobject to a binary alternative. The two outcomes of the alternative we label with an index $r$, which can assume the values 1 and 2. In Sect. 4.1 we have occasion to introduce urs and anti-urs, which we formally combine into a single fourfold alternative. Thus we also admit that $r$ may assume not just two but also four values. The number of possible values of $r$ we denote by the letter $R$. The vector space with $R$ dimensions we call $V^{(R)}$, the tensor space erected upon it $T^{(R)}$. The vectors in $V^{(R)}$ we call $u$, their components relative to the basis $u_r$. The choice of the letters $u_r$ is intended to suggest the ur.

A symmetric basis tensor of rank $n$, normalized to 1, is identified by the number $n_r$ of urs in the state $r$. We initially choose $R = 2$, and thus speak of "double-urs." Then the basis tensor of rank $n$ is characterized by the two numbers $n_1$ and $n_2$, with the condition

$$n = n_1 + n_2. \tag{4.4}$$

We write such a basis tensor as $|n_1, n_2\rangle$. All tensors of rank $n$ have a basis of $n+1$ basis tensors. We supplement the tensor space with a "vacuum"

$$\Omega = |0, 0\rangle. \tag{4.5}$$

The basis states of a single ur are then written

$$|1, 0\rangle = (n_1 = 1, n_2 = 0), \tag{4.6}$$
$$|0, 1\rangle = (n_1 = 0, n_2 = 1). \tag{4.7}$$

In the tensor space $\overline{T}$ we define a metric according to which $|n_1, n_2\rangle$ and $|n'_1, n'_2\rangle$ are always orthogonal, except for $n_1 = n'_1$ and $n_2 = n'_2$. Tensors of different rank are connected, according to the well-known rules of Bose statistics, by means of ladder operators $a_r, a_r^\dagger$ which satisfy the commutation relations

$$[a_r, a_s^\dagger] = a_r a_s^\dagger - a_s^\dagger a_r = \delta_{rs}, \tag{4.8}$$
$$[a_r, a_s] = [a_r^\dagger, a_s^\dagger] = 0. \tag{4.9}$$

The effect of these operators on tensors is described by the equations

$$a_r |n_r\rangle = \sqrt{n_r}\, |n_r - 1\rangle, \tag{4.10}$$
$$a_r^\dagger |n_r\rangle = \sqrt{n_r + 1}\, |n_r + 1\rangle. \tag{4.11}$$

We return to these operators in Sect. 3.4, where we discuss second quantization. We discuss the meaning of this expression in the context of tensor spaces in Sect. 5.2e.

We now ask what Lie groups we can represent by means of these operators.

It is easiest to construct the Lie algebra of the desired group *linearly* from the $a_r$. The self-adjoint operators

$$p_r = \tfrac{1}{2}\left(a_r + a_r^\dagger\right), \quad q_r = \tfrac{1}{2\mathrm{i}}\left(a_r - a_r^\dagger\right) \tag{4.12}$$

have the Heisenberg commutation relations

$$[p_r, q_s] = \delta_{rs}, \tag{4.13}$$

thus define an $R$-dimensional Heisenberg group. This group we cannot immediately interpret. We will subsequently see that even numbers $n$ belong to integral spin, odd $n$ to half-integral spin. The group linear in the $a_r, a_r^\dagger$ then indicates a supersymmetry which we can only interpret in a fully developed theory of particles.

The second step is the construction of a Lie algebra from *bilinear* expressions in the $a_r, a_r^\dagger$. We restrict attention to those in the following. Higher multilinear expressions will in general have commutators of even higher degree, and thus will not lead to a finite-dimensional Lie algebra. This question we have not investigated further.

## 4.3 Space and time

> *Du siehst, mein Sohn, zum Raum wird hier die Zeit.*
>
> Wagner, Parsifal, 1. Akt[5]

### 4.3.1 Realistic hypothesis

If in the context of abstract quantum theory, especially from the ur hypothesis, there should arise a universal symmetry group for the laws of temporal variations that is isomorphic to an empirically derived universal symmetry group of spacetime processes, then the processes governed by both groups will heuristically be considered identical.

*Comment.* Isomorphic mathematical structures, occurring in different physical circumstances, are identical as abstract structures but can concretely stand for completely different things. Thus any binary alternative defines in quantum theory a two-dimensional state space with the symmetry group $U(2)$; there the alternatives may stand for such different decisions as the two directions of spin of an electron, the two states of polarization of a light quantum, the passage through one of two holes in Young's interference experiment, two values of a component of isospin, or the decision whether one fermion state is occupied or unoccupied (Weizsäcker 1958 ) . None of these decisions will one interpret as an ur alternative. Ur alternatives are what we call alternatives

---

[5] I have asked Mr. Martin Gregor-Dellin whether there were here an influence of Schopenhauer on Wagner, thus indirectly also an influence of Kant. With his permission I quote from his reply in a letter of 5/26/1984: " 'Du siehst, mein Sohn': there are no immediate forerunners or sources to be found... Richard Wagner has much discussed with Cosima the problem 'time' and 'space', there are a few places in Cosima's diaries which I, however, cannot quote (due to a still missing subject register)—they also say no more than that according to Wagner's conjecture there must exist a connection. And then he works on the performance of his Parsifal where now the following happens: Parsifal, while still immature, is led by Gurnemanz to the Grail and thereby to a way of life—how is the 'dramatist' Wagner going to present such a long development which breaks the unity of place and time, and how to bridge a necessary change of scenery? Now at the first performance a wound-up canvas was unrolled—Parsifal is thus scarcely moving, yet enormous distances are overcome, clearly to symbolize Parsifal's development, the temporal component. Zum Raum wird hier die Zeit—space turns here into time—such, and I also think in a most practical way for the theater, I can understand this sentence. But it is always this way: Wagner, when living entirely in the *idea of a work*, he also gets it right philosophically."

To this a quote from a letter to Mathilde Wesendonck from August 1860: "Thus all of life's terrible tragedy is but to be found in time and space being asunder: as, however, time and space are only our modes of perception but have otherwise no reality, then one must be able to explain to a perfectly clear-sighted person that the most tragic pain is due solely to the mistake of the individual: I think it is so!"

into which any decidable alternative can be decomposed. If there are ur alternatives, then this implies a *universal* symmetry group. If we now know from experience an isomorphic universal symmetry group, then the *hypothesis* suggests itself that the empirical symmetry group is thereby *explained* as a deduction from abstract quantum theory with the ur hypothesis.

Let us explain the nature of such a "realistic hypothesis" with the example of Columbus' voyage. If somebody sailing westward across the ocean reaches land that corresponds to the known description of India, then the realistic hypothesis says that he has actually reached India. One can be mistaken with such identifications. Columbus thought he had landed in India, but instead had discovered a new continent. Nevertheless, his hypothesis was correct. Magellan's ships proved this by first finding a country beyond America that corresponded exactly to all descriptions of India, and from there sailing back to Europe following the then already known route around the Cape of Good Hope.

We might experience the same. If we abstractly find a universal symmetry group, we will try to identify it with the empirical symmetry groups of relativity theory and particle physics. But it can be that we first reach a thus far empirically still unknown intermediate station between abstract theory and present-day experience. And whether the whole enterprise is justified we will only see if we manage to close the *Kreisgang*[6] when the ships come home again, i.e., when we can reconstruct the known experience from our theory.

"Zum Raum wird hier die Zeit" ("here space turns into time"). Assume our undertaking to be successful. We started with the analysis of time in its modalities. From that we reconstructed abstract quantum theory, where binary alternatives appear to be fundamental. These alternatives define a symmetry group which will turn out to be isomorphic to the symmetry group of the relativistic spacetime continuum. "Space is the plural": it is the entirety of relations that govern the quantum theoretical interactions of multiple objects. Space only exists in the approximation in which we can mentally separate several objects as being distinct.

### 4.3.2 The Einstein space: A model of space

The continuous symmetry group of the ur is $U(2)$; its continuation by means of complex conjugation we discuss below in Sect. 4.3d. $U(2)$ contains the two commuting subgroups $U(1)$ and $SU(2)$. $U(1)$ is the group of the temporal changes of the state, this we will address under c. $SU(2)$ is locally isomorphic to $SO(3)$. It is thus natural to interpret it as the rotation group in a three-dimensional real space (Weizsäcker (1955); Weizsäcker et al. (1958)).

The simple argument for this reads: all dynamics is interaction. All interactions are in the last analysis interactions between urs. Therefore it will be invariant if the state of all urs is simultaneously transformed with the same

---

[6] Cf. p. XXII, fn. 14.

elements of the symmetry group of the ur. Consequently position space ought to be a homogeneous space of $SU(2)$.

The most natural homogeneous space of $SU(2)$ is $SU(2)$ itself. It is an $\mathbb{S}^3$, thus isomorphic to the spatial part of the Einstein space. Hence for us the Einstein space turns out to be the simplest model of the coordinate space implied by quantum theory. Naturally, the Einstein model of the universe by no means follows from this. We still have no theory of length and time measurement. In the sense of the general theory of relativity, we only have an admissible coordinate system in the spatial part of the spacetime continuum.

The general element of $SU(2)$ is given by

$$U = \begin{pmatrix} w + iz & ix + y \\ ix - y & w - iz \end{pmatrix} \quad (4.14)$$

with

$$w^2 + x^2 + y^2 + z^2 = 1. \quad (4.15)$$

The two column vectors of $U$ are functions on $\mathbb{S}^3$ as defined by (4.15), which we can interpret as two mutually orthogonal spinor representations $u_r^{(s)}(w, x, y, z)$ $s = 1, 2$, of the two basis states of the ur. Both states are extended over the entire universe. The ur is, as we have said above, not localizable; it "does not yet know the difference between particle physics and cosmology."

An arbitrary element $U'$ of SU(2) with special values $w', x', y', z'$ acts on these two spinor functions in $\mathbb{S}^3$ as a right-handed Clifford screw. It is well known that in a spherical space the two operations of translation and rotation differ only locally. For example, on $\mathbb{S}^2$, like the surface of the Earth, the same operation which is defined as rotation around the axis passing through North and South pole is a translation along the equator. The SO(4) which leaves (4.15) invariant is the direct product of two SO(3) that everywhere on $\mathbb{S}^3$ act in the sense of right- or left-handed screws, i.e., at the same time as rotations and translations. Both can be represented in terms of the action of some $u'$ on the general matrix $U$, taking the form $u'U$ or $Uu'$ (left and right multiplication, respectively). The eigenfunctions of the generators of the left-handed screws are the row vectors.

The unit element of SU(2) is denoted in (4.1) by the coordinates $w = 1, x = y = z = 0$. In the Einstein space, this is the place we will call "here" (origin of the coordinate system). There the column spinors have the form

$$u^{(1)} = \begin{pmatrix} 1 \\ 0 \end{pmatrix}, \quad u^{(2)} = \begin{pmatrix} 0 \\ 1 \end{pmatrix}. \quad (4.16)$$

### 4.3.3 Inertia

In Sect. 2.2 and again in Sect. 2.3 we remarked that the law of inertia represents something of a causal paradox in classical mechanics: a motion without any force acting. It is this paradox that separates modern from Aristotelian physics; that is why in all recent times one was interested in suppressing its paradoxical character. In Sect. 2.9 we argued in some detail to what extent the special theory of relativity depends on the law of inertia.

Our reconstruction of quantum theory also reverses the historical order of the arguments in this question: We begin with the concept of time in its three modalities. Present, future, and past would be meaningless concepts if change were not always present. This concept of change came systematically earlier than the distinction between forced and force-free motion. We simply assume that the state constantly varies in time, and derive from this the complex character of the state space.

For the single ur in the Einstein space we obtain from this the time dependence

$$u_r = u_r^0 \, e^{-i\omega t}. \tag{4.17}$$

Explicitly this is the time dependence

$$\begin{aligned} u^{(1)}(\boldsymbol{x}, t) &= \begin{pmatrix} w' + iz' \\ ix' - y' \end{pmatrix}, \\ u^{(2)}(\boldsymbol{x}, t) &= \begin{pmatrix} ix'' + y'' \\ w'' - iz'' \end{pmatrix} \end{aligned} \tag{4.18}$$

with

$$\begin{aligned} w' &= cw - sz, \ z' = cz + sw, \ x' = cx + sy, y' = cy - sx \\ w'' &= cw - sz, \ z'' = cz + sw, x'' = cx - sy, \ y'' = cy + sx \end{aligned} \tag{4.19}$$

and

$$c = \cos \omega t, \quad s = \sin \omega t. \tag{4.20}$$

The points where the $u^{(s)}$ have the forms (4.16) move along the $w$-$z$-axis in opposite directions, and the $x$ and $y$ perform the corresponding rotations.

In $\mathbb{S}^3$ this is a natural inertial motion, not singling out any particular point. Our approach to quantum theory thus leads directly to the law of inertia as the simplest subobject.

Symbolically one might say that inertia is the simplest manifestation of time.

In doing so, we have not ruled out the possibility that in a worked-out theory of interaction, the free motion will prove to be the influence of the universe on the approximately isolated object.

### 4.3.4 Special theory of relativity in binary tensor space[7]

In the approach above we considered only one single ur. In particular, we have thus abstained from adopting the variability of the alternatives. We now

---

[7] Editors' note: A more refined introduction of ur operators than given in this and the following subsection obeying common sign conventions and including a vacuum state definition can be found in Sect. 6.2.

## 4.3 Space and time

wish to describe this in the tensor space of the urs. As the simplest model we consider $\overline{T}^{(2)}$, the space of symmetric tensors over the binary vector space $V^{(2)}$. We consider the largest Lie group for which unitary representations in $\overline{T}^{(2)}$ can be obtained by means of bilinear expressions in the $a_r, a_r^\dagger$ ($r = 1, 2$). These representations were studied in particular by Heidenreich (1981).

The group has ten independent generators. With the notation

$$a_{rs} = a_r a_s, \quad a_{rs}^\dagger = a_r^\dagger a_s^\dagger, \quad \tau_{rs} = a_r^\dagger a_s, \quad n_r = \tau_{rr}, \quad n = \sum_r n_r \quad (4.21)$$

they are

$$M_{12} = i/2\,(n_1 - n_2)$$
$$M_{13} = 1/2\,(-\tau_{12} + \tau_{21})$$
$$M_{23} = i/2\,(\tau_{12} + \tau_{21})$$
$$M_{45} = i/2\,(n + 1)$$

$$N_{14} = 1/4\,(\alpha_{11} - \alpha_{22} - \alpha_{11}^\dagger + \alpha_{22}^\dagger)$$
$$N_{24} = i/4\,(\alpha_{11} + \alpha_{22} + \alpha_{11}^\dagger + \alpha_{22}^\dagger)$$
$$N_{34} = -1/2\,(\alpha_{12} - \alpha_{12}^\dagger) \qquad (4.22)$$
$$N_{16} = i/4\,(\alpha_{11} - \alpha_{22} + \alpha_{11}^\dagger - \alpha_{22}^\dagger)$$
$$N_{26} = -1/4\,(\alpha_{11} + \alpha_{22} - \alpha_{11}^\dagger - \alpha_{22}^\dagger)$$
$$N_{36} = i/2\,(\alpha_{12} + \alpha_{12}^\dagger).$$

We denote by $M_{ik}$ the generators of a compact subgroup, and by $N_{ik}$ the generators of a non-compact subgroup. The following relations hold:

$$[M_{ik}, M_{k\ell}] = M_{i\ell}, \quad [N_{ik}, N_{k\ell}] = M_{i\ell}, \quad [M_{ik}, N_{k\ell}] = N_{i\ell}$$
$$M_{ik} = -M_{ki}, \quad N_{ik} = N_{ki}. \qquad (4.23)$$

These operators generate the group SO(3,2), the so-called anti-de Sitter group, which leaves the expression

$$F = x_1^2 + x_2^2 + x_3^2 - x_4^2 - x_5^2 \qquad (4.24)$$

invariant. It characterizes a four-dimensional hyperplane in a five-dimensional real space. Let $F < 0$. Then this hyperplane can be interpreted as a four-dimensional "world," with a time coordinate $t$ related to the $x_4$ and $x_5$ via the equations

$$x_4 = x_0 \cos \omega t, \quad x_5 = x_0 \sin \omega t. \qquad (4.25)$$

This is called the anti-de Sitter space. At the point $x_1 = x_2 = x_3 = 0$ we have $x_0 = -F = |F|$. The time $t$ recurs cyclically with the period $2\pi \omega^{-1}$. But one can also choose the infinite covering group of SO(3,2), for which the time runs without periodicity from $t = -\infty$ to $t = +\infty$ and thereby creates infinitely many "sheets," each of which is an anti-de Sitter space. For that space we have

$$x_0^2 - x_1^2 - x_2^2 - x_3^2 = |F|, \qquad (4.26)$$

i.e., the position space is a hyperbolic space.

The space $F = $ const. is the factor space SO(3,2)/SO(3,1). The SO(3,1) which occurs as the stability group of the points of this space is the homogeneous Lorentz group. Thus we have automatically obtained a special case of the special theory of relativity, namely Lorentz invariance in an anti-de Sitter space. We have mentioned this representation of Lorentz invariance here only as the simplest model case. We will not study in more detail, as in several ways it does not yet adequately represent our physical intentions.

First, the time transformation generated by $M_{45}$ is compact. Thus it keeps the number $n$ of urs invariant. In other words, it does not quite yet describe the anticipated variability of the alternatives.

Second, it does not contain the above model of position space as $\mathbb{S}^3$. The SU(2) is generated by the three operators $M_{ik}$ ($i,k = 1,2,3$) and is here simply the local rotation group (or its covering group). Position space is hyperbolic; translations on it are non-compact. If this were a necessary consequence of the ur hypothesis we would have to accept it. It is, however, as we shall see, a consequence of our special approach; in Sect. 5.1d we will hypothetically adopt the compactness of position space as an important element of the theory.

Third, the assumption that $\omega$ in (4.25) takes only *one* sign is indeed arbitrary. $M_{45}/\mathrm{i}$ is positive definite, as one desires for an energy operator. But a fixed sign for $\omega$ does not follow from our approach thus far. It will be assumed that $\omega$ is a frequency that can take either possible sign, as in the quantization of fields. To describe this will be our next goal.

### 4.3.5 Conformal special theory of relativity[8]

We denote by $Q$ the full compact symmetry group of a binary alternative. It originates from the three subgroups SU(2), U(1), and the complex conjugation $K$. SU(2) commutes with the other two subgroups, but they do not commute among themselves. We describe the state of the ur by a complex column vector

$$u = \begin{pmatrix} u_1 \\ u_2 \end{pmatrix}. \qquad (4.27)$$

The action of the three subgroups can be visualized as a mapping of $u$ to a real three-vector or a null four-vector:

$$k^\mu = \bar{u}_r \, \sigma^\mu_{rs} \, u_s, \qquad (\mu = 0,1,2,3). \qquad (4.28)$$

---

[8] The arguments in this subsection are essentially needed for the development of ur theory in the fundamental paper by Castell (1975).

Here the bar denotes complex conjugation and the $\sigma^\mu$ are the Pauli matrices with $\sigma^0 = 1$. The $k^\mu$ satisfy the relation

$$k_\mu k^\mu = (k^0)^2 - (\vec{k})^2 = 0 \qquad (4.29)$$

with $\vec{k} = (k^1, k^2, k^3)$. SU(2) rotates the vector $\vec{k}$, two elements of opposite sign producing the same rotation of $\vec{k}$; SU(2) is a two-valued representation of SO(3). $k^0$ is invariant. U(1) multiplies $u$ by $e^{i\omega t}$, where $t$ is a real number; $k$ thus remains invariant.

Complex conjugation we represent by an operator K whose square is $-1$:

$$K \begin{pmatrix} u_1 \\ u_2 \end{pmatrix} = \begin{pmatrix} -\bar{u}_2 \\ \bar{u}_1 \end{pmatrix}. \qquad (4.30)$$

The pure complex conjugation follows from K through an additional rotation of SU(2). K acts on $k$ as a spatial reflection,

$$K\vec{k} = -\vec{k}. \qquad (4.31)$$

For the single ur we can put $k^0 = 1$, i.e., work with normalized vectors. $\vec{k}$ then defines a direction in space. If we consider the SO(3) (or, if we wish, the SU(2) itself) as coordinate space, then $\vec{k}$ denotes a Clifford screw, which puts the unit element of the group in the direction of $\vec{k}$. In the SO(3,2) representation, $\vec{k}$ is a half-integral angular momentum. For a free ur, it will keep $\vec{k}$ constant in both interpretations. Thus we will consider U(1) to be the dynamics of the free ur, and may interpret $t$ to be time.

Note that here, in the spirit of the realistic hypothesis, we take the decisive step in the derivation of the theory of special relativity from quantum theory. Time was given to us in quantum theory before the beginning of the axiomatic reconstruction. Then the postulate of the dynamics, *in* the reconstructed quantum theory, represented time as a real variable. In that sense we now interpret $t$ in the expression $e^{i\omega t}$ as "time." We see then that precisely this $t$ in the theory now to be reconstructed acts as the parameter of a one-dimensional subgroup, and is related to the parameters of three additional subgroups by a Lorentz transformation. Thus the theory turns out to be *isomorphic* to the historically known special theory of relativity. We endow our theory with physical semantics by *postulating* it to be *identical in content* with the historically given theory.

Now $K$ changes the expression $e^{i\omega t}$ into $e^{-i\omega t}$. In the language of particle physics one would say that $K$ exchanges urs with anti-urs. We thus ask whether we should assume that in addition to urs there are also anti-urs. We will not introduce any interaction that transforms an ur into an anti-ur or vice versa, i.e., that contains the discrete operator $K$ as an element or generator of a continuous group. But a priori we cannot rule out that both types of motion occur. Should the ur hypothesis, in one of the meanings discussed above, reveal itself to be trivial, it is to be suspected that it is trivial only *with*

100   4 Quantum theory and spacetime

anti-urs. We will see in the following that we can build particles and antiparticles only from urs and anti-urs, and, as antiparticles are described quantum theoretically, the ur-theoretic version of general quantum theory must also contain anti-urs.

It is convenient to express the antilinear transformation (4.30) linearly by means of a doubling of the state space. Thus we introduce four-urs with a four-dimensional vector space ($r = 1, 2, 3, 4$). Let $A^{(2)}$ be any matrix of SU(2); then the four-dimensional representations are defined by the block matrices

$$A^{(4)} = \begin{pmatrix} A^{(2)} & 0 \\ 0 & A^{(2)} \end{pmatrix},$$

$$K = \begin{pmatrix} 0 & 1 \\ -1 & 0 \end{pmatrix}.$$

(4.32)

To the new vector space $V^{(4)}$ there corresponds a tensor space $T^{(4)}$ or $\overline{T}^{(4)}$. If there is no danger of confusion, we write again only $T$ or $\overline{T}$. Now we ask anew which groups can be represented on $T^{(4)}$. We now choose four pairs $a_r, a_r^\dagger$ ($r = 1, 2, 3, 4$). For arbitrary dimensions $2f$ ($r = 1, 2, \ldots, 2f$) one can generate bilinearly the symplectic group Sp(2f,$\mathbb{R}$) from the raising and lowering operators. Its dimension is $2f(4f+1)$. The SO(3,2) for $f = 1$ is the 10-dimensional Sp(2,$\mathbb{R}$). For $f = 2$ we could also construct the Sp(4,$\mathbb{R}$) with 36 dimensions. Castell considered only its 15-dimensional subgroup SU(2,2) or SO(4,2), whose generators can be written

$$M_{12} = i/2 \, (n_1 - n_2 + n_3 - n_4)$$
$$M_{13} = 1/2 \, (-\tau_{12} + \tau_{21} - \tau_{34} + \tau_{43})$$
$$M_{23} = i/2 \, (\tau_{12} + \tau_{21} + \tau_{34} + \tau_{43})$$
$$M_{15} = i/2 \, (\tau_{12} + \tau_{21} - \tau_{34} - \tau_{43})$$
$$M_{23} = 1/2 \, (\tau_{12} - \tau_{21} - \tau_{34} + \tau_{43})$$
$$M_{35} = i/2 \, (n_1 - n_2 - n_3 + n_4)$$
$$M_{46} = i/2 \, (n + 2)$$

$$N_{14} = i/2 \, (\alpha_{13} + \alpha_{13}^\dagger - \alpha_{24} - \alpha_{24}^\dagger)$$
$$N_{24} = 1/2 \, (-\alpha_{13} + \alpha_{13}^\dagger - \alpha_{24} + \alpha_{24}^\dagger)$$
$$N_{34} = i/2 \, (-\alpha_{14} - \alpha_{14}^\dagger - \alpha_{23} - \alpha_{23}^\dagger)$$
$$N_{16} = 1/2 \, (-\alpha_{13} + \alpha_{13}^\dagger + \alpha_{24} - \alpha_{24}^\dagger)$$
$$N_{26} = i/2 \, (-\alpha_{13} - \alpha_{13}^\dagger - \alpha_{24} - \alpha_{24}^\dagger)$$
$$N_{36} = 1/2 \, (\alpha_{14} - \alpha_{14}^\dagger + \alpha_{23} - \alpha_{23}^\dagger)$$
$$N_{45} = 1/2 \, (\alpha_{14} - \alpha_{14}^\dagger - \alpha_{23} + \alpha_{23}^\dagger)$$
$$N_{56} = i/2 \, (\alpha_{14} + \alpha_{14}^\dagger - \alpha_{23} - \alpha_{23}^\dagger)$$

(4.33)

The choice of this group can be justified as follows. It is the largest subgroup of Sp(4,ℝ) that leaves invariant the operator

$$s = \tfrac{1}{2}(n_1 + n_2 - n_3 - n_4). \tag{4.34}$$

($is$ itself belongs as generator to SP(4). Thus, strictly speaking, SO(4,2)$xe^{is}$ is the largest such subgroup). $2s$ is now the difference of the number of urs and anti-urs. Assuming that urs and anti-urs cannot be transformed into one another, the restriction on our group means that an ur corresponds to a binary and not a quaternary alternative.

In the notation of the generators, the indices $r, s = 1 \ldots 4$ correspond to the interpretation as a Lie algebra of SU(2,2), the indices $i, k = 1 \ldots 6$ to the interpretation as SO(4,2). As SU(2,2), the group expresses its character as symmetry group of the ur theory. As SO(4,2), it can be described in the language of special relativity, thus also of particle physics. We can say that by virtue of its derivation, the *full theory of special relativity is well-founded ur-theoretically in quantum theory.*

We say the *full* theory of special relativity, as the conformal group does not prejudice the structure of the four-dimensional "space" beforehand, while the theory of the mere binary ur does. It leaves the expression

$$G = x_1^2 + x_2^2 + x_3^2 - x_4^2 + x_5^2 - x_6^2 \tag{4.35}$$

invariant. This defines Minkowski space as the homogeneous space, if $x_5$ and $x_6$ are held fixed; the (3,2)-de Sitter space (anti-de Sitter space) if $x_5$ is held fixed; and the (4,1)-de Sitter space (anti-de Sitter space) if $x_6$ is held fixed. One obtains the Poincaré group as subgroup with the coordinates

$$y_\mu = \frac{x_\mu}{x_5 - x_6} \quad (\mu = 1, 2, 3, 4). \tag{4.36}$$

Then $M_{ik}$ $(i, k = 1, 2, 3)$ are the angular momenta, $N_{i4}$ $(i = 1, 2, 3)$ generate the special Lorentz transformations, and the momenta $P_\mu$ $(\mu = 1, 2, 3, 4)$ are defined by

$$\begin{aligned} P_i &= M_{i5} + N_{i6} \quad (i = 1, 2, 3) \\ P_4 &= N_{45} + M_{46}. \end{aligned} \tag{4.37}$$

The $K_\mu$,

$$\begin{aligned} K_i &= M_{i5} - N_{i6} \\ K_4 &= N_{45} - M_{46} \end{aligned} \tag{4.38}$$

generate the special conformal transformations. $s$ is the helicity, a SO(4,2)-invariant operator; the mass

$$m^2 = P^\mu P_\mu \tag{4.39}$$

is the Casimir operator of the Poincaré group. For the definition (4.37) one finds in general from (4.33)

$$m^2 = 0, \tag{4.40}$$

i.e., the representations of SO(4,2) chosen here by means of symmetric tensors belong to rest mass zero. We will furthermore find representations in Sects. 4.9 and 5.1 belonging to finite rest mass in $T^{(4)}$, but not in $\overline{T}^{(4)}$.

One ought to emphasize that in our derivation of special relativity the *homogeneous Lorentz group* SO(3,1) necessarily occurs, whereas the inhomogeneous group, i.e., the representations of translations, remains to be chosen arbitrarily; only the containing conformal group SO(4,2) is fixed. Now the homogeneous Lorentz group contains only local transformations, whereas the selection of the translation group implies a corresponding global model of the world. This remark will be important in the transition to general relativity.

The representation of the ur in the $\mathbb{S}^3$ of subsection *b* can now be repeated in the (4,1)-de Sitter space. The $\tau_{rs}$ $(r, s = 1, 2)$ generate right-handed rotations in its spherical coordinate space, and the $\tau_{rs}$ $(r, s = 3, 4)$, left-handed rotations. The $M_{ik}$ $(i = 1, 2, 3)$ are rotations about $w = x_5 = 1$; the $M_{i5}$ $(i = 1, 2, 3)$ are translations there.

### 4.3.6 Relativity of urs

In Chap. 9 we discuss at length the importance of the observer in quantum theory, but even at this point we can use the customary language according to which the probabilities of quantum theory refer to future measurements by an observer. There, in nonrelativistic quantum theory, it suffices to speak of *one* observer. The time $t$ which we use in the axiomatic construction of the theory is then, speaking relativistically, the time registered by a clock at rest in the same reference frame as the observer. Nothing then stands in the way of also admitting several observers who use the same time, i.e., relativistically speaking, are at rest relative to one another.

The theory of special relativity is, expressed in the language of quantum theory, the theory of communication among *several* observers moving relative to one another. Einstein himself, however, found it important to point out that he had not described the behavior of moving observers but that of moving rulers and clocks. In our context, however, we can speak of observers, as we have introduced relativity as a theory of *quantum theoretical* symmetries from the outset.

We now consider the effect of the transformations introduced in Sects. 4.1d and 4.1e on a particular ur. Compact subgroups, like the rotations $M_{ik}$ $(i, k = 1, 2, 3)$ in (4.33), for example, rotate the ur into itself. That is, the alternatives $r = 1, 2$ and $r = 3, 4$ transform into different alternatives within the same state space. The non-compact subgroups, however, do not leave the number of urs invariant. They create or destroy urs, hence transform *one* ur into a superposition of arbitrarily many urs. This holds in general for the special Lorentz transformations $N_{i4}$, and in (4.38) for the special conformal transformations $K_i$. In Minkowski space it also holds for the space and time translations $P_i$ ($1 = 1, 2, 3, 4$), in the anti-de Sitter space for the space translations, and in the de Sitter space for the time translation $N_{45}$.

Let us consider a Lorentz transformation as an example. $N_{14}$ transforms $u$ into

$$u' = e^{N_{14}\beta u} = \sum_{k=0}^{\infty} \alpha_k \overline{T}'_{2k+1} \tag{4.41}$$

where $\beta = v/c$ is the relative velocity of the reference frames and $\overline{T}'_{2k+1}$ a symmetric tensor of rank $2k+1$. In addition, $u'$ is a binary alternative, since $u$ and $k^\mu$ transform like vectors in $\mathbb{C}^2$ and $\mathbb{C}^4$, and in $\mathbb{R}^3$, respectively. We can then say that $u'_r$ ($r = 1, 2$) is the ur alternative $u_r$ as it is perceived from the moving reference frame. $u'_r$ itself is thus not an ur alternative but a superposition of arbitrarily many ur alternatives, to be decided simultaneously. Now, however, we regard all inertial frames as equivalent. Thus the concept of the ur with which we started turns out to be merely a nonrelativistic concept, pertaining to the rest frame of an observer. Every observer has a different definition of the ur, and the theory of relativity is precisely the theory of how to map these definitions into one another. Here one sees very clearly that an ur is not an object, but only the state space of an alternative which, however, is measured on an object. It is the state space belonging to an observable, and observables depend on the reference frame. In this way the "radical atomism" of the ur hypothesis completely dissolves the concept of the "smallest object," the atom, into "elementary information."

The same holds for space translations, at least in the Minkowski and anti-de Sitter space. Thus, observers at rest relative to one another but separated in space also have in these non-compact coordinate spaces different individual definitions of the ur. In the compact spherical de Sitter space $\mathbb{S}^3$, however, the transformation rotates the directions of the Clifford screws corresponding to the two alternatives (1,2) and (3,4), but does not change the number of ur alternatives.

In the Minkowski and de Sitter space, the time translation is also ultimately non-compact. That is, for the observer the number of urs also varies over time. If at time $t_0$ there exists a state which contains a finite number $N$ of urs, or at least a finite expectation value $\overline{N}$ of that number, then the number will grow in the statistical mean with increasing time. This is because the time generator, i.e., the energy operator, contains a term that increases $N$ by 2, and another term that decreases $N$ by 2. Repeated application of the operator will spread the values of $N$ out, by analogy with Galton's board. But at $N = 0$ the terms vanish, while they may grow without bound for large $N$. The statistical reasoning is in principle the same as in statistical mechanics far from equilibrium, with the "equilibrium" here being at $N = \infty$.

We thus have reason to believe that the number of urs will increase over time. We can fully interpret the operational meaning of this assertion only in the context of the theory of measurement (Sects. 9.2b,c). Here we can only say that according to the present arguments, only the number of *decidable* alternatives increases, rather than the number of those decided. The ur at time $t_0$ is a binary alternative (in its state space); the tensor space that will have arisen from it by time $t > t_0$ is still a tensor space of exactly one binary alternative. This is a special case of the fact that the wave function in quantum theory evolves deterministically as long as no measurement is made. The growth in $N$ only means that the number of measurements that are possible in principle increases. So long as nothing is measured, the state space evolving from a 2-ur always remains two-dimensional; it only becomes the subspace of an ever-growing tensor space. One can still describe the state not subjected to a measurement at any time in terms of (or "as") the state of *one* ur at time $t_0$. Finally, in anti-de Sitter space the time translation is compact; here also the number of "momentary" urs does not increase.

# 5
# Models of particles and interaction

## 5.1 Open questions

### 5.1.1 Recapitulation

We begin with a review of the system of theories. What in addition have we learned by reconstructing quantum theory and the special theory of relativity? How does the unity of physics appear to us now? What questions are there still open in physics? What questions await us beyond physics?

The questions beyond what nowadays is called physics we postpone to the third part of the book. The open questions in physics we articulate with the concepts we have gained in the meantime. As a first guideline we choose the pair of concepts of the *whole* and the *part*. Roughly, and with abridged names, we can subdivide the objects of physics, as it appears to us at the present state, into five *"orders of magnitude"*:

A. The universe
B. Heavenly objects
C. Things
D. Particles
E. Alternatives

*C. Things.* We start with the most familiar, thus in the middle. As things we denote the sphere of human life, as it first appears to physics. Stones and trees, water and bread, tables and chairs are *things* to the physicist.

Soon there arises anew the question of the narrowness of the physical horizon. Are plants and animals, are humans, i.e., men, women, and children, families, peoples, and cultures things? Are thought and emotion, are name and form things? Initially, however, we ask in an apparent opposite direction. We ask about changes within physics itself, and the limits, within physics, of the concept of a thing.

We consider once more the diagram in Sect. 2.1, p. 14. There we started with classical physics. Compared to the talk about stones, tables, and bread, classical physics is already an enormous *abstraction*. Classical mechanics speaks of bodies, forces, space, and time (Sect. 2.2). On the one hand, these highly abstract concepts can be demonstrated in everyday experience: a falling stone is a body which a force moves through space in time. On the other hand, this abstraction reaches far beyond the sphere of human life. It reaches into the whole, of which this Earth of humans is a small part, and it reaches to the small, invisible particles that we believe make up the things of our sphere of life.

B. *Heavenly objects.* Sun, moon, and stars in the firmament belong to the world of human experience. Astronomy has subjected the stars to mathematics since antiquity, to mechanics since modern times. At first it appeared that in this way the heavenly bodies were describable in terms of terrestrial mechanics, and thus stripped of their mystery. Actually, however, mechanics had begun at the same time to differ in structure from its terrestrial models. The general law of gravitation could only be found in the realm of the stars; it somehow subjugated the terrestrial motion of falling objects to a cosmic law. The appropriate version of geometry and mechanics for the orders of magnitude of the stars up to the distant galaxies is according to our present understanding the *general theory of relativity*, at the bottom left corner of the diagram. Here we meet the *reversal of the historical order of the arguments* (Sect. 2.12): geometry and mechanics are abstractions from our everyday life, and primarily provide us with the vocabulary with which we interpret the general theory of relativity, the semantic preconceptions (Sect. 2.7). But subsequently the general theory of relativity appears to be the simpler and, because of this, more abstract theory, whose content can only approximately, for the small dimensions of our sphere of life, be subdivided into physical geometry and mechanics.

A. *The universe.* Between the heavenly objects and the universe lies, at least today, the outermost limit of our knowledge. That there is an all-embracing whole was in the history of human thought an unavoidable as well as an unreachable idea. In the context of modern physics and astronomy, the general theory of relativity offered for the first time a model that enabled one to think hypothetically of the universe as the "all-encompassing" thing. We have formulated the necessary skeptical follow-up questions in Sect. 2.10d. It would be surprising if there were not a reversal of the arguments here as well. What is a whole? For the necessary modification of this concept we can draw a still partial instruction from quantum theory.

D. *Particles.* We come to the main concept of the present chapter. The historical path is sketched in the right half of Fig. 2.1 on p. 14 and, in more detail, in Fig. 2.2 on p. 56 (Sect. 2.1 and 2.12). Ancient philosophy contemplated the smallest parts of bodies, modern chemistry and statistical thermodynamics postulated them with great empirical success. The attempt to extend the range of applicability of mechanics to these parts, however, led to a scientific revolution: quantum theory. The reversal of the arguments arises here as well:

only quantum theory explains the always-assumed stability of bodies. We ask at once:

1. What approach explains the universal validity of quantum theory?

We initially remain with the historical development. What today terminologically is called a particle one can only say in relativistic quantum theory: an object whose state space permits an irreducible representation of the Poincaré group. Yet from this definition there arise two further questions:

2. Two interacting particles are, according to this definition, no longer particles. How then is the interaction to be described, and how, in appropriate context, the concept of a particle?

3. The definition does not rule out a further divisibility of a particle into smaller interacting "particles." Is there a limit to divisibility?

E. *Alternatives.* Between alternatives and particles lies today the border between the consensus of physicists and the approach of this book. We attempt to answer the above three questions; for this we must exchange the order of questions 2 and 3 above.

1. The reconstruction of the third and fourth chapter attempts to explain the universal validity of quantum theory from the concept of an alternative. The reconstruction is based on an analysis of the conditions which make experience possible and on *one* "realistic" "*expansion postulate.*" This corresponds to the traditional superposition principle and implies the quantum theoretical composition rule. It explains in what sense the whole is more than the sum of its parts.

3. If we reduce objects, thus also particles, to alternatives there arises a logical limit to the divisibility in the binary alternative. This implies the existence of a three-dimensional coordinate space and Lorentz invariance. We learn that divisibility is primarily not a spatial but a logical concept. alternatives are large or small according to their content of information. The "ur" is, spatially, as big as the "universe" (Sect. 4.3b) This should open for us a new approach to the general theory of relativity.

2. This is the topic of the present chapter. To my regret this chapter is only programmatic. The reconstruction of abstract quantum theory and special relativity we can certainly put into question once more; but from the given postulates it ought to be consistent. The open questions relating to interaction are sketched in the following subsection.

## 5.1.2 Program

1. Strictly speaking there are no separable alternatives, thus also no separable objects. Yet our conceptual process begins with the assumption of separable alternatives and objects which manifest themselves as particles in relativistic quantum theory. For a physics based on ur alternatives there arise then three questions of semantic consistency:

A. How is one to describe the fictitious single, free particle?

B. What approximation admits the theoretical decomposition of an object into disjoint parts?

C. How is the correction to this approximation manifested as interaction?

2. The ur hypothesis requires that all physical states be representable in the *tensor space of the urs*. We define general operators in this space and show in particular that the postulate of the indistinguishability of the urs is satisfied through operators according to the requirements of the para-Bose statistics.

One can describe single free particles as irreducible representations of the Poincaré group, and massless particles in particular as representations of the para-Bose order $p = 1$. Suitably symmetrized products of such $p = 1$ representations again lie in tensor space and describe composite objects. They are representations with $p > 1$ and can be considered products of $p$ representations being with $p = 1$. For para-Bose representations, the order $p$ is in general additive for products. Critical for the *interaction* is now that the decomposition of a representation with $p > 1$ into factors with $p = 1$ is not unique. In the language of scattering theory, this can be interpreted as the existence of nontrivial channels in the S-matrix. Here there are no longer free parameters. If the ur theory is correct, it should *uniquely* determine the types of possible particles and their laws of interaction. Due to its considerable mathematical difficulties however, the present chapter is only a sketch of a *program* of such a theory.

3. A simple model is the description of a particle in a *homogeneous space* of a group acting in $T$, as described in Sect. 4.3. As an example we choose the massless particle in Einstein space (Sect. 4.3b) and in the globally defined Minkowski space (Sect. 4.3c). As the most important tool for the following sections we introduce a "local Minkowski space" tangent to an Einstein space at some location.

4. The section title *model of quantum electrodynamics* is deliberately ambiguous.[1] Quantum electrodynamics is a model of an interaction, but the section only sketches a still incomplete model of quantum electrodynamics.

5. The present book does not attempt to catch up with the rapidly progressing theory of *elementary particles*. It formulates only three proposals for this theory: (*a*) its *systematics* should be representable in the tensor space of the urs; (*b*) *local gauge invariance* appears to be a natural requirement in tensor space; (*c*) the existence of *sharp rest masses* is perhaps not sufficiently well understood as the basic problem in present particle physics. In the context of the ur theory this problem is presumably to be associated with questions of cosmology.

6. The *general theory of relativity* should initially be constructed in $T$ as a theory of the gravitational field analogous to quantum electrodynamics. Along the lines of Gupta (1950, 1954) and Thirring (1961) this automatically yields the approach to a Riemannian geometry. Here, in contrast, there is no need

---

[1] Equally ambiguous is the title of Kant's book *Critique of pure reason*. It is a *genitivus objectivus* and *subjectivus*. Pure reason both criticizes and is criticized.

## 5.2 Representations in tensor space

### 5.2.1 Basic operations in $T_n$

We lift the previous restriction to the space $\overline{T}$ of symmetric tensors and consider the full tensor space. We wish to consider two- and four-urs, i.e., coordinates in the vector space of the single ur $r = 1, \ldots R$ with $R = 2$ or $R = 4$. We denote the vector space with given $R$ by $V^{(R)}$, and the tensor space by $T^{(R)}$. For the basis vectors in $V^{(R)}$ we simply write the numerals $r$ ($r = 1 \ldots R$). In $V^{(R)}$ a hermitian metric is defined by

$$\langle r, s \rangle = \delta_{rs}. \tag{5.1}$$

A basis in the space $T_n^{(R)}$ of tensors of rank $n$ consists of the $R^n$ monomials, each of which we describe as an ordered sequence of numerals:

$$x = r_1 r_2 \ldots r_n \qquad (1 \leq r_\nu \leq R). \tag{5.2}$$

Two different monomials of the same rank $n$ are orthogonal according to (5.1); every monomial of the form $x$ has norm 1. Tensors of different rank are always orthogonal.

Let GL(R) be the full linear group in $R$ complex dimensions. Its vector representation lies in $V^{(R)}$. We first consider its representations in a $T_n^{(R)}$ with fixed $n$. The Lie algebra of GL(R) has the $2R^2$ basis elements $\tau_{rs}, i\tau_{rs}$ ($1 \leq r, s \leq R$) with the commutation relations

$$[\tau_{rs}, \tau_{tu}] = \tau_{ru}\delta_{st} - \tau_{ts}\delta_{ru}. \tag{5.3}$$

The generators of the unitary subgroup U(R) are

$$\begin{aligned} a_{rs} &= 1/2\,(\tau_{rs} - \tau_{sr}) \\ b_{rs} &= i/2\,(\tau_{rs} + \tau_{sr}). \end{aligned} \tag{5.4}$$

A basis of GL(R) consists of $a_{rs}, b_{rs}, ia_{rs}, ib_{rs}$. In $V^{(R)}$ $\tau_{rs}$ acts according to

$$\tau_{rs}\, t = r\, \delta_{st}, \tag{5.5}$$

i.e., $\tau_{rs}$ changes $s$ into $r$ and $t \neq s$ into zero. In the tensor space $T^{(R)n}$, $\tau_{rs}$ acts as a derivative according to Leibniz's rule: $\tau_{rs}$ transforms a monomial into a sum of monomials, in each of which at *one* place a basis vector $s$ is now replaced by $r$:

$$\tau_{rs} x = \sum_{\nu=1}^{n} \delta_{sr_\nu} \tau_{rr_\nu} x, \qquad (5.6)$$

For example,
$$\tau_{12} \, 122 = 112 + 121. \qquad (5.7)$$

$\tau_{rs}$ simply multiplies each monomial by the number $n_r$, which indicates how often $r$ occurs in the monomial. Thus in $T^{(R)n}$,

$$\tau_{rr} = n_r, \qquad (5.8)$$

$$\sum_r n_r = n. \qquad (5.9)$$

The theory of the representations of GL(R) in $T_n$ by means of the so-called Young tableau is well known (see, e.g., Boerner 1955). Here we review those features we will utilize later. A Young *diagram* of rank $n$ is a figure of $\ell$ rows of lengths $m_1, m_2 \ldots m_\ell$ with

$$\sum_{i=1}^{\ell} m_i = n \qquad (5.10)$$

and the condition
$$m_1 \geq m_2 \geq \cdots \geq m_\ell. \qquad (5.11)$$

A standard *tableau* arises by inserting the numbers $1 \ldots n$ into the boxes of the diagram, subject to the rule that the numbers in each row must increase toward the right, and downward in every column. We distinguish the different possible diagrams for a given $n$ by an index $k$. For every diagram $k$ there are a number $f_k$ of different standard tableaus. Every diagram $k$ defines, as it turns out, $f_k$ equivalent representations of the symmetric group $S_n$, all of dimension $f_k$. We have

$$\sum_k f_k^2 = n! \qquad (5.12)$$

Every diagram $k$ defines in $T_n^{(R)}$ exactly $f_k$ irreducible representations of GL(R), one to each standard tableau. The basis tensors for each of these representations are described by the possible standard schemes. A standard *scheme* arises by inserting the numbers $r$ ($r = 1 \ldots R$) into the boxes of the diagram according to the rule that the numbers in each row are non-decreasing toward the right, and increase downward in each column. Hence a standard scheme cannot have more than $R$ rows. To every scheme there are $f_k$ different tensors in $T_n^{(R)}$, corresponding to the $f_k$ representations of GL(R).

### 5.2.2 Basic operations in $T$

We now look for operators that raise or lower the rank of a tensor. We restrict attention to changes from $n$ to $n \pm 1$; we assume that we can build all the other relevant operators from them. The operators are to be defined over the entire tensor space, irrespective of the specification of the group to be represented afterward. The simplest operators of this kind we call the operators of *stuffing* and *ripping*.

Let the *stuff*-operator $S_r$ act on a monomial

$$x = r_1 r_2 \ldots r_n \tag{5.13}$$

according to

$$S_r x = r\, r_1 r_2 \ldots r_n + r_1\, r\, r_2 \ldots r_n + \cdots + r_1 r_2 \ldots r\, r_n + r_1 r_2 \ldots r_n\, r. \tag{5.14}$$

That is, $S_r$ stuffs an $r$ at any location between $r_\nu$ and $r_{\nu+1}$, or before or after the entire monomial, and adds up all the $n+1$ monomials generated in this way. The *rip*-operator $R_r$ acts according to

$$R_r x = \sum_\nu \delta_{rr_\nu}\, r_1 r_2 \ldots r_{\nu-1} r_{\nu+1} \ldots r_n. \tag{5.15}$$

That is, $R_r$ rips an $r$ from a location and adds up all the monomials obtained in this way. If $x$ does not contain any $r$ then $R_r x = 0$.

The following commutation relations hold:

$$[R_r, S_s] = \tau_{sr} + (n+1)\delta_{rs}, \tag{5.16}$$

$$[R_r, R_s] = [S_r, S_s] = 0, \tag{5.17}$$

$$[\tau_{rs}, S_t] = S_r\, \delta_{st}, \tag{5.18}$$

$$[\tau_{rs}, R_t] = -R_s\, \delta_{rt}. \tag{5.19}$$

It can be shown that $R_r, S_R$ and $T_{sr}$ together define representations of the Lie algebra GL(R+1) and a unitary representation of SU(R,1) in $T$.[2] Every irreducible representation of this kind is a "tower" in $T$. As "foundation" it has a subspace $T'_n$ of $T_n$ such that all elements of $T'_n$ are nulled out by all $R_r$, and all other elements of the tower originate from those of $T'_n$ by repeated application of the $S_r$. Provided that the ur hypothesis can be extended to a theory of elementary particles, it would seem that these representations, in any event SU(R,1), will have some physical significance. This ought to be more in the direction of a "supersymmetry" (compare Sect. 4.2c). For as we have seen in Sect. 4.3e, even $n$ means even spin and odd $n$ odd spin, at least in the representations of the relativistic groups in $\overline{T}$; we will see in a moment that the same holds over the entire tensor space $T$. Thus we expect that besides the Lie algebra linear in $R, S$ there is also a Lie algebra bilinear in them, mediating representations of other—namely the relativistic—groups.

Note that there can be other relatively simple basic operators in $T$. For example, one could define instead of the "symmetric" stuffing and ripping also corresponding "skewed" operators, where the terms in (5.14) and (5.15) acquire an algebraic sign defined by some rule. We return to this point in (5.37), continuing for the time being with the symmetric operators.

---

[2] K. Drühl, Tech. Report, MPI Starnberg, 1977.

## 5.2.3 Bose representations

The operators $R_r, S_r$ always transform symmetric tensors into symmetric tensors. We generate the entire space $T$ by repeatedly "stuffing the vacuum." The vacuum $\Omega$ (tensor of rank zero) satisfies the equations

$$T_{rs}\,\Omega = R_{rs}\,\Omega = 0 \quad \text{for all } r,s. \tag{5.20}$$

All symmetric tensors are linear combinations of

$$y(n_1 \ldots n_R) = S_1^{n_1} \ldots S_R^{n_R}\,\Omega. \tag{5.21}$$

Thus all symmetric tensors form *one* tower as regards the $R_r, S_r$.

When looking for a Lie algebra bilinear in the $R_r, S_r$ we run into the difficulty that bilinear expressions of the enveloping algebra of a Lie algebra have in general trilinear commutators, etc. That is, most of the time such expressions do not close to a finite-dimensional Lie algebra. This works, however, if the commutators of the initial Lie algebra are pure numbers. We accomplish this via renormalization:

$$a_r^\dagger = \frac{1}{\sqrt{n+1}} S_r, \qquad a_r = \frac{1}{\sqrt{n}} R_r. \tag{5.22}$$

In addition, we define the unit-norm tensors

$$|n_1 \ldots n_R\rangle = \frac{1}{N} y(n_1 \ldots n_R). \tag{5.23}$$

This yields

$$N = \sqrt{n!\, n_1! \ldots n_R!} \tag{5.24}$$

and

$$\begin{aligned} a_r |n_r\rangle &= \sqrt{n_r}\, |n_r - 1\rangle, \\ a_r^\dagger |n_r\rangle &= \sqrt{n_r + 1}\, |n_r + 1\rangle, \end{aligned} \tag{5.25}$$

and in the unit tensors the $n_r$ remain unchanged for $s = r$. There follow the commutation relations

$$[a_r, a_s^\dagger] = \delta_{rs}, \quad [a_r, a_s] = [a_r^\dagger, a_s^\dagger] = 0. \tag{5.26}$$

That is, the $a_r, a_r^\dagger$ are the raising and lowering operators of the Bose statistics defined in Sect. 4.3b, which for $R = 2$ generate the SO(3,2), for $R = 4$ the SO(4,2).

## 5.2.4 Parabose representations

"Para-statistics" was invented by Green (1953)[3] and was introduced into the ur theory by Castell and his coworkers.[4] Green was looking for generalized

---
[3] Cf. also Greenberg and Messiah (1965)
[4] Jacob (1979), Heidenreich (1981), Künemund (1982, 1985).

commutation relations for field operators which admitted the most general representation of indistinguishable particles. We have postulated the indistinguishability of the urs, hence face the same problem as Green. We first recall the significance of the problem as a matter of principle, and its solution by Green.

For particles one tends to justify the occurrence of Bose or Fermi statistics on the basis of the indistinguishability of particles. This statement is not wrong but imprecise, as it is so often with verbalizations of mathematical structures in physics. Let us refer to Ehrenfest's game of urns (see Sect. 7.2), in which we are given $n$ balls of which $n_1$ are in the first urn and $n_2$ in the second. Classically (or, as formulated today, in Boltzmann statistics) one considers it meaningful to ask *which* balls are in the first urn and which in the second. One can give the balls additional attributes by, e.g., writing a number on each. Abstractly speaking, this means that we have expanded the alternatives. Initially we could only distinguish one ball from the first urn from another ball in the second urn ("local" discrimination); now we can distinguish each ball from every other ("individual" discrimination).

Quantum theoretically one formally sets up the state space of $n$ particles as if one could distinguish them (tensor product of the individual state spaces). In the tensor space $T_n$ one distinguishes the different *symmetry classes* by Young diagrams. To a fixed standard scheme there corresponds in each case an irreducible representation of the permutation group $S_n$ of the $n$ basis vectors contained in it. One now interprets the indistinguishability of the particles, and analogously of the urs, by requiring that all tensors of such a fixed irreducible representation space of the permutation group describe the same physical state. In every irreducible representation of the linear group GL(n), or one of its subgroups which contains the respective standard scheme, there occurs exactly *one* of these tensors of the associated representation space of $S_n$; the corresponding state space is then represented by this vector. Formally this is exactly analogous to the ray representation of a linear group in a vector space. In quantum mechanics all vectors of a ray in Hilbert space belong to the same state. Any of these vectors may represent the state; one usually chooses vectors with unit norm. These then still admit an arbitrary phase factor $e^{i\alpha}$, which irreducibly represents the gauge group U(1) in the ray. One can call the group representations in terms of representatives of the symmetry classes *generalized ray representations*.

The indistinguishability of the particles is justified by the fact that the law describing the interaction of the particles is invariant under a permutation of the particles, i.e., that a symmetry class is conserved by the interaction. Thereby the choice of the correct symmetry class is only shifted to the initial conditions. There is no immediately obvious reason why only fully symmetric or fully antisymmetric tensors will occur, i.e., only Bose or Fermi statistics. Green's operators are now so defined that one can use them to generate from the vacuum exactly *one* representative of every possible standard scheme in tensor space.

## 5 Models of particles and interaction

Empirically, only particles with Bose or Fermi statistics have been found thus far. For urs, however, we certainly cannot restrict attention to Bose or Fermi statistics. For Fermi statistics the world would actually consist of just two or four urs. That we cannot restrict attention to symmetric tensors is clarified by a simple argument, anticipating the concept of particles. We will interpret an irreducible representation of the Poincaré group as the state space of *one* particle. Several particles will only be described by a (suitably symmetrized) tensor product of irreducible representations. Now such a product of symmetric tensors is not again a symmetric tensor. Thus a multi-particle theory necessarily must use more general symmetry classes in tensor space. We begin by sketching the mathematical structure, and discuss its physical interpretation afterwards.

We introduce new operators $a_r, a_r^\dagger$ for which we postulate Green's commutation relations ("CR")

$$[\tfrac{1}{2}\{a_r, a_s^\dagger\}, a_t] = -\delta_{st} a_r \tag{5.27}$$

$$[\{a_r, a_s\}, a_t] = [\{a_r^\dagger, a_s^\dagger\}, a_t^\dagger] = 0. \tag{5.28}$$

By means of these operators, whose effect in $T$ we are still looking for, we can initially define the abstract operators

$$\tau_{sr} = \tfrac{1}{2}\{a_r, a_s^\dagger\} \tag{5.29}$$

which, as follows from (5.27)-(5.28), satisfy the CR (5.3). These then yield the CR

$$[\tau_{rs}, a_t^\dagger] = a_r^\dagger \delta_{st}, \qquad [\tau_{rs}, a_t] = -a_s \delta_{rt}. \tag{5.30}$$

These CR have the same form as (5.18) and (5.19); one can therefore assume that $a_r, a_r^\dagger$ can be represented by raising and lowering operators in $T$ which transform as spinors, like the $R_r, S_r$ and the Bose operators for GL(R).

Green has given a representation of $a_r, a_r^\dagger$ which satisfy the trilinear CR (5.27), (5.28) by means of operators $b_r, b_r^\dagger$ whose CR are bilinear:

$$a_r = \sum_\alpha b_r^\alpha, \qquad a_r^\dagger = \sum_\alpha b_r^{\alpha\dagger}, \tag{5.31}$$

with the CR

$$[b_r^\alpha, b_s^{\alpha\dagger}] = \delta_{rs}, \qquad [b_r^\alpha, b_s^\alpha] = [b_r^{\alpha\dagger}, b_s^{\alpha\dagger}] = 0 \tag{5.32}$$

$$\{b_r^\alpha, b_s^{\beta\dagger}\} = \{b_r^\alpha, b_s^\beta\} = \{b_r^{\alpha\dagger}, b_s^{\beta\dagger}\} = 0 \quad \text{for } \alpha \neq \beta. \tag{5.33}$$

Here $p$, the "order" of the operators $a_r, a_r^\dagger$ is an integer $\geq 1$. For a fixed $\alpha$, the $b_r^\alpha$ thus commute like Bose operators, and for $\alpha \neq \beta$ they anticommute. Furthermore, there will be a "vacuum" $\Omega$ with

$$b_r^\alpha \Omega = 0 \quad \text{for all } \alpha \text{ and } r. \tag{5.34}$$

It follows that
$$a_r a_s^\dagger \Omega = p\delta_{rs}\Omega \quad \text{for all } r \text{ and } s. \tag{5.35}$$

This "Green decomposition" is not the only possible representation of the $a_r, a_r^\dagger$ (Heidenreich 1981; Künemund 1982, p. 12). We will, however, only use this decomposition, as it is the only one for which we can provide a general representation in $T$.

We begin by defining, for fixed $p$, an augmented tensor space $T^{(R,p)}$ as the direct sum of $p$ spaces $T^{(R)}$. Thus $T^{(R,p)}$ is erected over a vector space $V^{(R,p)}$ with the $R \cdot p$ basis vectors $r^\alpha$ ($r = 1\ldots R, \alpha = 1\ldots p$). The $b_r^\alpha, b_r^{\alpha\dagger}$ will in $T^{(R,p)}$ act in a subspace $\overline{T}^{(R,p)}$ of suitably symmetrized tensors. These are not completely symmetric, but symmetrized with alternating signs according to the following rule. All basis tensors of rank $n$ arise from the vacuum according to the prescription
$$\psi = b_{r_1}^{\alpha_1\dagger} \ldots b_{r_n}^{\alpha_n\dagger} \Omega. \tag{5.36}$$

$\psi$ is, with a suitable normalization factor, a sum of $n!$ terms. The first ("leading") term is
$$v = r_1^{\alpha_1} \ldots r_n^{\alpha_n}. \tag{5.37}$$

All other terms originate from $v$ through permutation of the positions $1\ldots n$, subject to the sign rule that for every transposition of two neighboring factors the sign is negated, but only when their upper indices differ.

From $\overline{T}^{(R,p)}$ one returns through a projection into a newly defined $T^{(R)}$. Its basis vectors are the "center of mass coordinates" of the $r^\alpha$:
$$r = \sum_{\alpha=1}^{p} r^\alpha. \tag{5.38}$$

One can now write
$$r^\alpha = r + \sum_{\beta=1}^{p-1} r^{[\beta]}. \tag{5.39}$$

Here the $r^{[\beta]}$ are any linear combinations of the $r^\alpha$, independent of $r$ and of one another, and thus "relative coordinates." The parallel projection along the $r^{[\beta]}$ onto the $r$ occurs by expressing each vector $r^\alpha$ in every tensor using (5.39), and then putting all $r^{[\beta]} = 0$, i.e., in short, replacing $r^\alpha$ with $r$. Thus far this mapping has turned out to have a unique inverse, i.e., the projected tensor in $T^{(R)}$ uniquely determines, for a given $p$, its original tensor in $\overline{T}^{(R,p)}$.

For every $p > 1$, the set of tensors generated in this manner in $T^{(R)}$ also contains tensors of other than fully symmetric symmetry classes. For $p \geq R$ the procedure generates for every standard scheme exactly *one* of the $f_k$ linearly independent tensors contained in it. The other $f_k - 1$ tensors of this scheme are obtained by permutation of the sites, but they cannot be generated by the para-Bose operators alone. The number $n_r$ of basis vectors $r$ is

$$n_r = \tfrac{1}{2}\{a_r, a_r^\dagger\} - \tfrac{1}{2}p. \tag{5.40}$$

Substituting $\tfrac{1}{2}\{a_r, a_s\}$ and $\tfrac{1}{2}\{a_r^\dagger, a_s\}$ for $a_r a_s$ and $a_r a_s^\dagger$, respectively, into (4.21), the $M_{ik}, N_{ik}$ are then again generators of SO(4,2). Every irreducible representation belongs to a fixed $p$. It turns out that $p > 1$ describes particles with nonvanishing rest mass.

This result again affirms the interpretation of para-Bose operators as an extended expression of the indistinguishability of the urs. The extension of the tensor space to $T^{(R,p)}$ initially means that the index $\alpha$ expresses a possible distinction among the urs. The symmetrization to $\overline{T}^{(R,p)}$ implies that one cannot say *which* ur carries the particular index $\alpha$, but only *how many* urs $n^\alpha$ there are to one value of $\alpha$. In other words, the index is the analog to a local distinction. Inasmuch as particles are built from urs, one will be able to say, for noninteracting particles and with each particle in a definite state, of how many urs it consists. For an ur, affiliation with a particle is then analogous to the affiliation of a particle with a volume in space ("urn"). We pursue these questions in the next section, where we refer to "particles" defined by a given value of $\alpha$ as *quasiparticles*.

### 5.2.5 Multiple quantization in ur theory

In Sect. 3.4 we justified the concept of second (and multiple) quantization from the theory of probability. In the article Weizsäcker et al. (1958) it was already attempted to reconstruct the entire quantum theory through multiple quantization of a binary alternative. There one succeeded to reconstruct the quantum field theory of free fields but not to introduce an interaction. This became understandable after we had found the probability-theoretic interpretation of multiple quantization. The "ensembles of ensembles" used in the development of probability theory (Sect. 3.1–3.3) are just not any real collections whose elements interact with one another, but independent repetitions of always the same case. Actually they can only be generated in the limit in which the interaction becomes arbitrarily small.

Precisely because of this, the concept of the "ur" as an ur object (now: ur subobject) was introduced, via the concept of the primary alternative (later called the ur alternative). The totality of the states of the urs could then be described in the tensor space $T$. Now the question arises as to how multiple quantization is expressed in this tensor space.

As "formally strict" multiple quantization we will denote here the schematic iteration of the quantum theory of an alternative. It would proceed as follows.

Step 1: the vectors $u_r$ with $R$ basis states

Step 2: the normalizable functions $\psi_r$ with denumerably infinitely many basis states, in each case denoted by the $R$-number $n_r$ ($r = 1 \ldots R; n_r = 0 \ldots$).

Step 3: the functions $\varphi(u_r$ with denumerable many basis states which are described by all possible number functions $N(n_r)$ $(N(n_r)_{=0...\infty})$ for every function $n_r$ $(r = 1...R)$.

Etc.

For $R = 2$ and $R = 4$, the first step denotes the states of a single ur. The second step is explained in Sect. 3.4, where the functions $\psi_r$ correspond to Bose statistics of the urs, thus make up the space $\overline{T}$ of symmetric tensors. The third step initially signifies a distinguishability of types $\alpha$ of urs, such as they were generated by the Green's operators $b_r^{\alpha\dagger}$ ($\alpha = 1...\infty$). If, however, one uses only suitably symmetrized products, it leads directly to the para-Bose representations above. A fourth step ought the again to distinguish such representations from one another and collect them symmetrically.

## 5.3 Quasiparticles in rigid coordinate spaces

For further applications we need representations of the "building blocks" of all composite particles, i.e., our quasiparticles, in an explicitly defined spacetime continuum. We initially restrict attention to the "rigid" homogeneous spaces of the symmetry group, introduced in Sect. 4.3, and select here only three special cases.

### 5.3.1 Einstein space

We take Sect. 4.3b as our starting point, but introduce urs and anti-urs as in Sect. 4.3e. The full rotation group SO(4), i.e., the group of all global rigid motions in $\mathbb{S}^3$, is generated by the 6 generators $M_{ik}$ ($i, k = 1, 2, 3, 5$) of (4.33). To this is added $M = M_{46}/i$ as the energy operator. All these operators are sums of the form $\mu_{12} + \mu_{34}$, where $\mu_{12}$ acts only on the coordinates $r = 1, 2$, and $\mu_{34}$ acts only on $r = 3, 4$. The $\mu_{12}$ are right-screws, the $\mu_{34}$ are left-screws in $\mathbb{S}^3$. The $M_{ik}$ are formulated such that $M_{12}, M_{13}, M_{23}$ are a rotation group about the point $w = 1$, the $M_{15}, M_{25}, M_{35}$, however, act as translations at $w = 1$. The screws are defined globally, rotations and translations only locally. All operators commute with $n$ and $s$, thus leave the cosmological energy and helicity invariant. We distinguish integer and half-integer $s$, i.e., bosons and fermions, and put in both cases

$$\begin{aligned} \text{boson:} \quad & n = 2\nu, \quad s = \mu \\ \text{fermion:} \quad & n = 2\nu + 1, \quad s = \tfrac{1}{2}(2\mu + 1). \end{aligned} \quad (5.41)$$

For $p = 1$, with fixed $n$ and $s$, the number of linearly independent states is

$$\begin{aligned} \text{boson:} \quad & N_{ns} = (\nu + \mu + 1)(\nu - \mu + 1) = (\nu + 1)^2 - \mu^2 \\ \text{fermion:} \quad & N_{ns} = (\nu + \mu + 2)(\nu - \mu + 1) = (\nu + 1)(\nu + 2) - \mu(\mu + 1). \end{aligned} \quad (5.42)$$

118    5 Models of particles and interaction

The states
$$|n_1, n_2, n_3, n_4\rangle \tag{5.43}$$
form a basis.

For a concrete discussion we choose instead of this basis the "quasi-basis" of pure *momentum states*. As "initial state" we choose the state $\psi_{\max}$ with maximum component of momentum in the $z$-direction,
$$M_{35}/i = \tfrac{1}{2}(n_1 - n_2 - n_3 + n_4). \tag{5.44}$$
This is the state
$$\psi_{\max} = \left|\tfrac{1}{2}(n+s), 0, 0, \tfrac{1}{2}(n-s)\right\rangle. \tag{5.45}$$
All other momentum states are obtained from it via rotations. (We remark that for $p = 1$, i.e., "massless" particles, the direction of momentum also determines the direction of spin.) The number of linearly independent momentum states must be set to $N_{n,s}$. Their number will in any case be the number of pure angular momentum states obtained by rotations from one state
$$\psi'_{\max} = \left|\tfrac{1}{2}(n+s), 0, \tfrac{1}{2}(n-s), 0\right\rangle. \tag{5.46}$$
This, however, must be given by (5.42), due to the composition of the independent angular momenta of all urs and anti-urs. But in contrast to the states (5.43), the pure momentum states are in general not orthogonal to one another. To every pure momentum state belongs the energy
$$M = \tfrac{1}{2}(n+2). \tag{5.47}$$
The specification of the concept of energy we will further discuss in Sect. 4.8c.

### 5.3.2 Global Minkowski space

The usual description of particles does not take place in Einstein space but in Minkowski space. Here we first summarize the representations of massless particles in Minkowski space as given by Castell (1975). For $R = 4$ the space $\overline{T}^{(4)}$ of symmetric tensors contains for every $s$ exactly one representation of SO(4,2), i.e., exactly one massless particle for each helicity. $s = 0$ defines a representation which must be interpreted as an empirically unknown spinless and massless particle. $s = \pm\tfrac{1}{2}$ corresponds to the neutrino and anti-neutrino, $s = \pm 1$ to right- and left-hand polarized photons, and $s = \pm 2$ to gravitons.

Following Castell, we illustrate these arguments with the example $s = +\tfrac{1}{2}$, which describes a single neutrino. There it is formally assumed that the neutrino actually has rest mass zero; otherwise this particle with $m = 0, s = \tfrac{1}{2}$ would be called something else. The identification of these representations with names of particles is in any case fictitious as only in para-Bose representations can we describe more than one individual particle. Our representations in $\overline{T}^{(4)}$ only give the relativistic symmetry of one free massless particle, not how it interacts.

## 5.3 Quasiparticles in rigid coordinate spaces

One can characterize the states in a representation by means of four quantum numbers

$$s^{(1)} = \tfrac{1}{2}(n_1 + n_2) \quad s_3^{(1)} = \tfrac{1}{2}(n_1 - n_2)$$
$$s^{(2)} = \tfrac{1}{2}(n_3 + n_4) \quad s_3^{(2)} = \tfrac{1}{2}(n_3 - n_4). \tag{5.48}$$

The notation corresponds to an interpretation of $s^{(1)}$ as "total helicity" of the "half" of the particle consisting only of urs in the states 1 and 2, $s_3^{(1)}$ as their component in the direction of the third coordinate axis; and analogously as regards the states 3 and 4. For the "neutrino" we have

$$s = s^{(1)} - s^{(2)} = \tfrac{1}{2}. \tag{5.49}$$

The basis states of they neutrino can be characterized by

$$n = n_1 + n_2 + n_3 + n_4 = 2\left(s^{(1)} + s^{(2)}\right) = 2j \tag{5.50}$$

and

$$j_3 = s_3^{(1)} - s_3^{(2)}; \tag{5.51}$$

in (5.50), farthest to the right, we have used Castell's notation of $2j$, according to which $j_3$ is just the $x_3$-component of $j$.

In the lowest state of the neutrino $n = 2s^{(1)} = 1, s^{(2)} = 0$. This state consists of a single ur. It is doubly degenerate: $s_3^{(1)} = \pm\tfrac{1}{2}$. As the generators of SO(4,2) are bilinear in the $a_r, a_r^\dagger$, all states of the neutrino have odd $n$; they contain $(n+1)/2$ urs and $(n-1)/2$ anti-urs, respectively. Castell denotes the operator

$$M = M_{44}/\mathrm{i} \tag{5.52}$$

as energy, more precisely as "cosmological energy" (see also Segal 1976). Its eigenvalue, denoted $m$ by Castell, is related to $n$ and $j$ by

$$m = j + 1. \tag{5.53}$$

This relationship is represented by the solid line in the following figure.

The dashed line represents the value $m$ would have if for every $j = n/2$ exactly $n$ urs with total energy $m = n \cdot m_1$ were present, with $m_1$ being the energy of the free ur, thus also of the lowest neutrino state. The $m$ of all higher neutrino states is, according to (5.53), less than $n \cdot m_1$. Castell interprets this as a binding energy of the urs in the neutrino.

Describing the neutrino in Minkowski space, its energy is, according to (4.37), given by the sum of $M_{46}$ and the non-compact generator $N_{45}$. The discrete states are then not eigenstates of energy; actually, energy in Minkowski space has a continuous spectrum of eigenvalues. In the Minkowski coordinates $y_\mu$, according to (4.36), the states of the discrete basis are wave packets which at a certain time have a smallest spatial extent; prior to that, they contract,

## 5 Models of particles and interaction

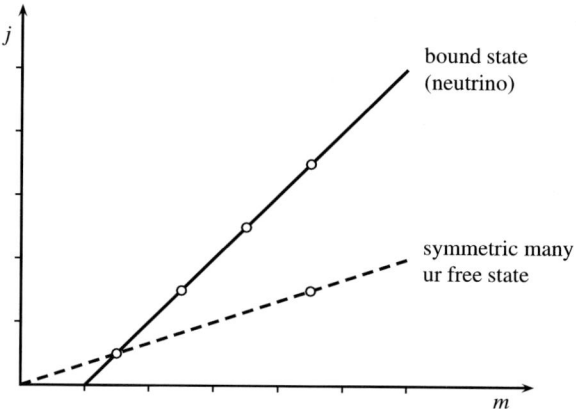

and afterwards they disperse. For the single ur Castell specifies the Minkowski wave function as

$$\psi = \frac{y_k \sigma_k + y_4 - i}{(y_k^2 - (y_4 - i)^2)^2} \begin{pmatrix} \psi_1 \\ \psi_2 \end{pmatrix} \tag{5.54}$$

with fixed complex numbers $\psi_1$ and $\psi_2$.

### 5.3.3 Local Minkowski space[5]

Our third model is a Minkowski space tangent to an Einstein space at some point. There the Einstein space is assumed to be the physically real space, and the Minkowski space merely provides the language in which observers in the vicinity of the tangent point describe the phenomena. This corresponds to Einstein's postulate in the general theory of relativity that the Riemannian space of the real universe should, everywhere locally, be approximately a Minkowski space. In this model we will see that here the Minkowski notation is obviously not only a first approximation, but also that in it only a discrete manifold of states occurs, which is embedded in the continuous manifold of possible states in the global Minkowski space.

From a different standpoint, however, the Einstein space is perhaps a worse approximation than the Minkowski space. The Einstein space is assumed to be invariant over time, in its spatial structure as well as in the amount of matter contained in it; in addition it is not Lorentz invariant. With regard to the constancy in time. astronomical evidence suggests the use of an expanding model of the universe, and in ur theory it is probable that the total number of urs varies in time and increases on the average. In Sect. 5.5d we will see that the radius of the universe, measured in atomic units of length, must arguably be a monotonically increasing function of the number of urs. Our

---

[5] This section owes essential ideas to a discussion with Th. Görnitz.

present model does not take this into account. Thus, at most, it can describe only a cosmologically short period of time.

Conceptually even more incisive is the question of the relationship between the Einstein space and Lorentz invariance. In the general theory of relativity that relationship is simple: the Einstein space is a special solution of Einstein's field equations, and special solutions need not be invariant under the symmetry transformations of the equation; such a transformation in general changes solutions into other solutions. In ur theory we first arrived at the Einstein space as the coordinate space for the representations of a single, free ur. It can then also be the coordinate space for a fixed number of free, i.e., noninteracting urs. As Einstein space is not Lorentz invariant, there is no natural unit of energy defined in it. But such a unit can be used in a tangent Minkowski space. Thereupon, however, local Lorentz invariance becomes fundamental for the physics in the Einstein space. A local Lorentz transformation changes the Einstein space into another (3+1)-dimensional space. Now the question that arises is whether both of these universes satisfy a common condition—are they perhaps solutions of the same basic equation?

In the spirit of the reversal of the sequence of arguments, in ur theory initially there is given neither a differential equation nor a group of motions, but a tensor space and a notion of time. If a group can be represented in tensor space, and, among others, it contains a parameter one can interpret as time, then, secondly, one must expect the invariance of the laws under this group. Thirdly, a formulation of such laws can then be a differential equation for a function in the homogeneous space of the group. We return to these arguments when we deal with the general theory of relativity.

Minkowski space is initially explained as a parallel projection from Einstein space. The coordinates of a point in the Einstein space are determined by the four numbers $w, x, y, z$ subject to the subsidiary condition (4.15)

$$w^2 + x^2 + y^2 + z^2 = 1. \tag{5.55}$$

As point of contact we choose the location $w = 1, x = y = z = 0$. In this Minkowski tangent space the coordinates will simply be $x, y, z$. The time $t$ used in (4.17) will also be the time coordinate in Minkowski space. In Sect. 4.3b we could leave it open whether time is meant as open ($\mathbb{R}^1$) or cyclical ($\mathbb{S}^1$), i.e., whether the Einstein space is the hyper-cylinder $\mathbb{S}^3 \times \mathbb{R}^1$ or the hypersphere $\mathbb{S}^1 \times \mathbb{S}^1$. In the first case the Minkowski space touches the Einstein space along the world line $w = 1, t$ arbitrary, in the second case only at the point $w = 1, t = 0$.

Instead of using this apparently somewhat primitive procedure, one mostly describes the transition from a curved to a flat tangent space in terms of a so-called contraction. In this procedure one describes the curved space by the radius of curvature $r$ and then takes the limit $r \to \infty$. In this way one transforms functions, differential equations, and group representations defined in curved space into the "corresponding" ones of flat space. There one often must perform at the same time a limiting procedure with certain parameters of these functions, etc., such that they are not reduced to a trivial case. Thus there are strictly speaking no massless particles in the Einstein space. Following this contraction, all particles with finite rest mass $M_0$ in the Einstein universe go over into particles of rest mass zero; this happens as in Minkowski space, according to (4.37), and instead of $M_{46}$, $M_{46} + N_{45}$ is the energy operator. Particles with finite rest mass in Minkowski space are obtained in the contraction only if $M_0$ also goes to infinity with $r$.

The procedure of a contraction obviously has a completely different physical meaning than that of a parallel projection. It demonstrates the mathematical analogues and differences between two different physical assumptions; namely between the assumption that the universe is "in reality" an Einstein space, or "in reality" a Minkowski continuum. Our approach, however, starts with the working hypothesis (to be refined later) that the objects built from urs are spatially and temporally correctly described in an Einstein space, and asks how such objects would look in the merely approximate description of a locally tangent Minkowski continuum.

In particular, the finite and discrete rest masses as described in Einstein space must also be described *in* the tangent Minkowski space, although they cannot be explained *from* Minkowski space. If all actually decidable alternatives are finite, then every actually observable state of a particle can only contain a finite number of urs. It is for precisely this reason we have chosen the compact Einstein space for the spatial description. For the implications of this choice for cosmology and the sharpness of rest masses, see Sect. 5.5d.

In local Minkowski space the $M_{ik}$ ($i,k = 1, 2, 3$) will not only generate the local rotation group at $x = y = z = 0$, as in global Minkowski space, but there will also be, in contrast to the global interpretation, the $M_{i5}$ ($i = 1, 2, 3$), the measurable momenta, and $M_{46}$, i.e., $M$ from (5.47) the measurable energy. Actually the three $M_{i5}$ do not commute with one another, thus are strictly speaking not momenta in flat space. But this only means that the flat space approximation can only be carried out as long as the commutators of the $M_{i5}$ remain unmeasurably small. According to the general theory of relativity, this is also the condition for the description of our experience in flat space to be valid.

We thus simply carry over the description of quasiparticles of Sect. 5.3a into the language of the local Minkowski space. One then finds that quasiparticles with $p = 1$ are also no longer massless. The boson with $s = 0$ has a rest state with energy 1, and is without spatial structure, i.e., $\psi(x, y, z) =$ const., to the extent that the Minkowski approximation is valid. $\psi$ remains normalizable, as the flat description fails at large distances. The fermion with $s = \frac{1}{2}$ has momentum $\frac{1}{2}$ in its lowest state, but energy $\frac{3}{2}$, thus a velocity markedly different from the speed of light. We take this question up again in Sect. 5.4c.

## 5.4 Model of quantum electrodynamics

As mentioned in Sect. 5.1b, we discuss here quantum electrodynamics, on the one hand, as a model of more general field theories. Yet on the other hand, I am not able to present a developed theory of quantum electrodynamics, only the outline of a model. For this and the following Sect. 5.1 and 5.2, I had the choice of three options: 1) to further postpone the publication of this book indefinitely while attempting to consistently work out all the details; 2) to publish this book without a discussion of these problems; 3) to include mere model discussions in this book. I have chosen the third option, and present here a program of work still unfinished.

### 5.4.1 Old project, new program

From the quantum theory of the binary alternative, Scheibe, Süßmann and I (Weizsäcker et al. 1958) derived the force-free equations for massless and massive leptons and for transverse photons. At that time, this work had found little resonance, for various reasons. It suffered from three significant flaws:

1. The conceptual foundations of this approach were unclear.
2. For this reason we were unable to bring the formally derived equations into a physically transparent connection with their prerequisites as well as experience; thus there were several inconsistencies in the interpretation of our formulas.[6]
3. We did not succeed in describing interaction.

The clarification of the conceptual foundations has taken about 25 years of reworking; the present book reports on it. In the meantime I believe I have found a consistent approach to interaction. I am again taking up the old project, eliminating the inconsistencies, and formulating from it a new program for fully deriving quantum electrodynamics.

The approach of the article Weizsäcker et al. (1958) (in the following quoted as KL III, namely "Komplementarität und Logik III") consisted of applying multiple quantizations to a binary alternative. I explain the procedure forthwith, based on the simplest example, the derivation of the Weyl equation.

### 5.4.2 Massless lepton

We begin with a simple alternative

$$r = (1, 2). \tag{5.56}$$

---

[6] About 1972 I gave this paper to J. Jauch to read, who had not known about it but in the meantime had become very interested in our later work (with Drieschner and Castell). To my question after his reading, "Would this text have convinced you then?" he replied with a friendly smile, "No!"

In the "first quantization" one assigns complex numbers $u_r$ to both elements of the alternative; one obtains a "pure spinor"

$$u = \begin{pmatrix} u_1 \\ u_2 \end{pmatrix}. \tag{5.57}$$

$u$ in turn is now treated as a classical quantity and subjected to a "second quantization." This quantization we had not yet carried out in full rigor; that is one of the inconsistencies mentioned above. In return the procedure was very simple and quickly led to the result. I first describe the procedure and then the corrections in the present context.

We initially represented the complex 2-vector $u_r$ by means of a real 4-vector $k^\mu$:

$$k^\mu = \overline{u}_r \, \sigma^{\mu \, rs} \, u_s. \tag{5.58}$$

$\sigma^\mu$ ($\mu = 0, 1, 2, 3$) are the Pauli matrices; in particular $\sigma^0$ is the unit-matrix. The norm of $u$ is

$$k^0 = \overline{u}_1 u_1 + \overline{u}_2 u_2. \tag{5.59}$$

The $k^\mu$ determine only three real parameters, as they satisfy the algebraic identity (in the notation of the Minkowski metric)

$$k_\mu k^\mu = 0. \tag{5.60}$$

The upper spin indices are defined by

$$u^1 = -u_2, \qquad u^2 = u_1, \tag{5.61}$$

such that

$$u^r v_r = u_1 v_2 - u_2 v_1 \tag{5.62}$$

is the invariant of all affine transformations of the $u$. The $k^\mu$ determine the $u_r$ up to a phase factor $e^{i\alpha}$; $\alpha$ is the fourth real parameter of $u$.

The $u_r$, considered to be classical quantities that are to be quantized again, are not normalized; i.e., for $k^0$ one initially admits arbitrary values. In the second quantization one must then treat the $u_r$ and $k^\mu$ as operators. It is natural to postulate the canonical commutation relations

$$[u_r, u_s^\dagger] = \delta_{rs}, \qquad [u_r, u_s] = [u_r^\dagger, u_s^\dagger] = 0; \tag{5.63}$$

then the $k^\mu$ ($\mu = 1, 2, 3$) have the commutation relations of angular momentum, and $k^0$ behaves like the magnitude of an angular momentum. We had considered this (KL III, 9a), but chose the simplified procedure to treat the $u_r$ and thus the $k^\mu$, as commuting. For large values of $k^0$ this is an acceptable approximation.

One can then define a wave function of the second quantization level

$$\varphi(k^\mu). \tag{5.64}$$

$\varphi$ may only be different from zero for quadruples $k^\mu$ that satisfy the condition (5.60). This can be accomplished with the requirement

$$k_\mu k^\mu \varphi(k^\nu) = 0. \tag{5.65}$$

Performing now a four-dimensional Fourier transformation from the space of the $k^\mu$ into a space of four coordinates $x^\mu$, (5.65) assumes in $x$-space the form

$$\Box \psi = 0. \tag{5.66}$$

Formally this is the classical wave equation.

In addition, there is an algebraic identity between the $k$ and $u$

$$k_\mu \sigma^{\mu r s} u_s = 0. \tag{5.67}$$

This means that $k_\mu \sigma^\mu$ is the projector in $u$-space onto the direction orthogonal to $u$. We define the spinorial wave function

$$\varphi_r(k^\mu) = u_r \varphi(k^\mu). \tag{5.68}$$

Here the $u_r$ and $k^\mu$ are to be connected via (5.58), i.e., represent the same state. Then we have

$$k_\mu \sigma^{\mu r s} \varphi_s(k^\nu) = 0, \tag{5.69}$$

which goes over into the Weyl equation

$$\partial_\mu \sigma^{\mu r s} \psi_s(x_\nu) = 0 \tag{5.70}$$

by means of a Fourier transformation.

In the spirit of the "realistic hypothesis" (Sect. 4.3a), we had at the time assumed that the $x_\mu$ can be interpreted as coordinates in physical Minkowski space, and thus (5.66) and (5.70) as second-order wave equations. This would be the Weyl equation for a massless lepton, for instance a neutrino. This same assumption will also be made now. But it reveals several inconsistencies of the previous terminology which we must clarify next.

$k^\mu$ appears in this interpretation as the momentum vector, $u_r$ as the spin vector of a particle. One must therefore interpret $k^\mu$ in (5.65) and (5.66), along with (5.69) and (5.70), as the generator of the four translations in Minkowski space. According to (5.58) the $k^m$ ($m = 1, 2, 3$) are then, as components of the spin vector, the three components of angular momentum. The first answer is that for a massless particle, momentum and angular momentum are parallel and proportional, but not equal. The vector $k^\mu$ denotes both quantities, up to a real factor which does not show up in the equations. Thereby, however, the inconsistency is not resolved. In the second step of the quantization, $k^m$ is an operator which, if we still retain the approximation in which its components commute, has arbitrary triplets of real numbers as eigenvalues. On the other hand, $u_r$ is also a Hilbert vector of the first level of quantization in (5.68); it denotes the state of *one* ur. This is indispensable if one wants to interpret the $x$ as positions in Minkowski space. A state of the second level to a fixed vector $k^m$ implies ur-theoretically a large number $k^0$ of urs on the same state. If, however, $k^m$ is an angular momentum, then together they must also have a large magnitude of angular momentum $k^0$. Thus, at that time, we had inconsistently assumed $k^\mu$ to be the momentum for all urs occurring in the particle, but as angular momentum only for a single ur.

One can, however, retain the realistic hypothesis for both wave equations by quantizing the $u_r$ and $k^\mu$ according to (5.63) and using, as introduced by Castell, urs and anti-urs according to Sect. 4.3e. One then arrives at the theory of Sect. 5.3a. $k^\mu$ becomes the generator of a right-screw in Einstein space, thus is *at the same time* momentum and angular momentum. In the same way the two components of the anti-ur, $r = 3, 4$ define the generator of a left-screw. The massless lepton has $s = \frac{1}{2}$, thus $\nu + 1$ urs and $\nu$ anti-urs. The momentum is the sum, the spin angular momentum the difference of the two eigenvalues of the screw. The approximation of commuting $k^\mu$ is the projection in a tangent Minkowski space described in Sect. 5.3c. In the following we will therefore consider the "massless lepton" as the approximate description of a particle with $p = 1, s = \frac{1}{2}$ in Minkowski space.

### 5.4.3 Relativistic invariance

How are we to consider, from the present point of view, the relativistic invariance of the earlier project? Equations (5.65), (5.66) and (5.69), (5.70) are formally Lorentz invariant. Indeed, at that point we made the claim to have placed the special theory of relativity on a quantum theoretic basis: *"The special theory of relativity, insofar as it is a mathematical theory of space and time, is already the quantum theory of a deeper, simple alternative. The Lorentz group is a (unfaithful) representation of the group of complex linear transformations of the quantum state of that alternative."* (KL III, p. 708) Independent of us, Finkelstein (1968) founded the three-dimensionality of space and Lorentz invariance on the quantum theory of the binary alternative.[7] His implementation of this idea was then different from ours; roughly speaking, he interpreted $k^\mu$ not as a momentum four-vector but a position four-vector. Here we do not follow the subsequent development of his theory, but rather our own.

The introduction of Lorentz transformations in KL III, which I have just quoted literally, is different from the one given here in Sects. 4.3d-e, following Castell. It is precisely the difference between the local and global introduction of Minkowski space. The global introduction is accomplished by means of a

---

[7] As I got to know him in 1971, we both did not know about our respective investigations. I told him that in my opinion one could derive the special theory of relativity from the quantum theory of the alternative. He answered:"You are the only man in the world to say such a thing. Of course you are right." His answer was so prompt because it was unfounded: he was the other one who asserted this.

unitary, infinite-dimensional representation in the full tensor space. In KL III we used the fact that the algebraic identities, from which the equations of motion had been derived, do not depend on the metric of the urs but are invariant under SL(2,$\mathbb{C}$). In the first level of quantization we therefore obtained in this way only the homogeneous Lorentz group. It wa preserved under the second quantization with commuting $k^\mu$. Therefore we consistently interpreted the $k^\mu$ as momenta. The four translations arose only in the Fourier transformation, where four additive imaginary parameters remain arbitrary in the exponent. Therefore it was consistent to interpret the $x_\nu$, for which now the full Poincaré group was valid, as space and time coordinates.

The commutation relations (5.63), however, break the SL(2,$\mathbb{C}$)-invariance. They conserve the number of urs and Einstein space. Our former Lorentz invariant theory thus holds only in the tangent Minkowski space. A question arises as to the justification for describing as *massless particles* the $p = 1$ representations of the Poincaré group in tensor space (see the end of Sect. 5.3c). The state space of these objects has a discrete basis, which, e.g., is represented by the momentum eigenfunctions in Einstein space according to Sect. 4.8a. Assume their time dependence to be given by the expression (5.47) of the energy. One can project these eigenfunctions onto a reference frame in the tangent Minkowski space, which is at rest relative to the Einstein space. In Minkowski space, however, these functions, however, do not form a basis for a unitary representation of the Lorentz group (only Castell's functions defined in 5.3b do this in global Minkowski space). Thus the object has no well-defined rest mass in the relativistic sense and is not a particle in the sense of Wigner's definition. For large numbers of urs, however, it goes over into a particle to a good approximation, and this particle can be described as massless in just this approximation.

### 5.4.4 The Maxwell field

In KL III, Sect. 6, we developed the theory of free massless particles with $s = 1$ in analogy to the theory of free massless particles to $s = \frac{1}{2}$. According to the realistic hypothesis this was the theory of the free transverse right-circular polarized Maxwell field; the ($s = 1$)-particles were interpreted as right-circular photons.

The ground state of the ($s = 1$)-photon consists of two urs in the states $r = 1, 2$. Denote by $D_2$ the irreducible representation of SL(2,$\mathbb{C}$) by one ur, then the Lorentz group of the $k^\mu$ is the product $D_2 \otimes D_2^*$, which is also irreducible. On the other hand, the product of $D_2$ with itself is reducible: $D_2 \otimes D_2 = D_1 \oplus D_3$, where $D_1$ is the identical representation in terms of the antisymmetric tensor $u^r u_r$. $D_3$ consists of the three symmetric tensors

$$f^{k0} = \tfrac{1}{2} u^r (\sigma^k)^{rs} u_s, \tag{5.71}$$

which describes the lowest states of the photon built from symmetrical tensors of urs. Explicitly we have

$$f^{10} = \tfrac{1}{2}\left(u^1 u_2 + u^2 u_1\right) = \tfrac{1}{2}\left(u_1^2 - u_2^2\right),$$
$$f^{20} = -\tfrac{1}{2}\left(u^1 u_2 - u^2 u_1\right) = \tfrac{1}{2}\left(u_1^2 + u_2^2\right), \qquad (5.72)$$
$$f^{30} = -\tfrac{1}{2}\left(u^1 u_1 - u^2 u_2\right) = -u_1 u_2.$$

$f^{k0}$ defines a skew-symmetric self-dual tensor

$$f^{ik} = -f^{ki}, \qquad (5.73)$$
$$f^{k0} = \mathrm{i}\, f^{\ell m} \qquad (k, \ell, m \text{ cyclic}). \qquad (5.74)$$

Again defining the $k^\mu$ according to (5.58), there follow the two algebraic identities

$$k^\lambda f^{\mu\nu} + k^\nu f^{\lambda\mu} + k^\mu f^{\nu\lambda} = 0, \qquad (5.75)$$
$$k^\nu f_{\mu\nu} = 0, \qquad (5.76)$$

which we could denote "first-level Maxwell-equations" by analogy with (5.60) and (5.67). The $f^{\mu\nu}$ are "first-level field strengths," which through second quantization become the field strengths of the space-dependent Maxwell field.

Analogously one can define "first-level potentials." We interpret the $f^{\mu\nu}$ as complex combinations of the real field strengths:

$$e^k = \mathrm{Re}\, f^{k0}, \qquad h^k = \mathrm{Re}\, f^{\ell m} \qquad (k, \ell, m \text{ cyclic}) \qquad (5.77)$$

These fields we determine at the second level by differentiation, and thus at the first level by multiplying potentials by $k^\mu$. In particular we require

$$e^k = k^0 a^k - k^k a^0, \qquad h^k = k^\ell a^m - k^m a^\ell \qquad (5.78)$$

It follows that
$$a^k = e^k / k^0, \qquad a^0 = 0. \qquad (5.79)$$

Instead of this gauge one can choose a gauge transformation

$$a^{\mu'} = a^\mu + \alpha\, k^\mu, \qquad (5.80)$$

as according to (5.78), $a^\mu = \alpha\, k^\mu$ implies vanishing field strengths.

The second step of quantization follows as in the case of the lepton. From the function $\varphi(k^\mu)$ one forms

$$\varphi^{\mu\nu}(k^\lambda) = f^{\mu\nu}\, \varphi(k^\lambda) \qquad (5.81)$$

and the real field strengths

$$F^{\mu\nu} = \mathrm{Re}\, \varphi^{\mu\nu}. \qquad (5.82)$$

Due to the self-duality (5.74), the real field strengths $F^{\mu\nu}$ contain as much information as the complex $\varphi^{\mu\nu}$. For the potential

$$A^\mu = a^\mu\, \varphi(k^\lambda) \qquad (5.83)$$

one has
$$k_\mu k^\mu A^\nu = 0. \tag{5.84}$$

In KL III explicit special solutions were given for propagating right-circular plane waves.

Now we must again introduce anti-urs. With them, the right-circular photon with $n = 2\nu$ always has $\nu + 1$ urs and $\nu - 1$ anti-urs, the left-circular one $\nu - 1$ urs and $\nu + 1$ anti-urs. The geometrical arguments are the same as for the lepton.

If electrodynamics is also to describe static fields, then it is well known that longitudinal and scalar photons must also be defined. I suspect that they are generated by the $s = 0$ irreducible representations. Substituting in (5.71) for one of the two spin indices $r, s$ one of the values 1 or 2, and for the other 3 or 4, one then obtains a scalar and a vector. At the time of completion of the manuscript, the mathematical structure generated in this way had yet to be investigated.

### 5.4.5 Electromagnetic interaction

The interaction between leptons and the Maxwell field ought to be uniquely determined. Also this calculation has not been carried out yet. We enumerate here the problems to be solved and sketch the assumed solutions. There are three classes of problems:

$\alpha$) the form of the equations
$\beta$) the separability of quantum electrodynamics from other interactions
$\gamma$) the values of the fundamental constants

$\alpha$) *The form of the equations.* After an approach to reconstruct a relativistic field theory had actually been found, it is not very surprising that we obtained in KL III the correct form of the force-free equations of motion. The symmetry group and the quantum number $s$ determine the form of the equations of lowest order. If our approach actually determines an interaction uniquely, its form presumably will be the one known from quantum electrodynamics. According to our present understanding, besides relativistic invariance, a gauge group representing the "internal symmetries" is essential for the interaction. See also Sect. 5.1c.

Two questions remain open in this argument, namely $\beta$) and $\gamma$). The question $\beta$) is whether one can find at all an approximation in which quantum electrodynamics can be considered separately from other interactions; only in this approximation will one be able to define the form of the electromagnetic interaction uniquely. $\gamma$) is the question of the values of the dimensionless coupling constants not determined by the gauge group, i.e., the fine structure constant and the leptonic rest mass.

$\beta$) *Separability.* We sketch the problem for the simplest example. Assume two interacting quasiparticles. $p$ is additive, so $p$ of the entire system has the value 2. Also $s$ is additive. We write $s_1$ and $s_2$ for the two components and

130    5 Models of particles and interaction

$$S = s_1 + s_2. \tag{5.85}$$

Let the first particle be a fermion, the second a boson. Let $S = \frac{1}{2}$. This can be realized with the following combinations, where corresponding pairs of values are written beneath one another:

$$
\begin{array}{c|cc|cc|cc|c}
s_1 & \frac{1}{2} & -\frac{1}{2} & \frac{3}{2} & -\frac{3}{2} & \frac{5}{2} & -\frac{5}{2} & \\
s_2 & 0 & 1 & -1 & 2 & -2 & 3 & \cdots
\end{array} \tag{5.86}
$$

The solid vertical lines mark the presumed borders between two different theories. The first pair is a lepton and a scalar or longitudinal photon (if the hypothesis at the end of Sect. 5.4d is correct). The second pair is an anti-lepton and a right-circular photon. The two following pairs probably combine a left-circular photon and a massless gravitino with an anti-gravitino and a graviton. The following pairs lead to even higher spins, etc. All these pairs interact with one another, i.e., are in a sense identical. Their separability will be given as far as the dashed vertical lines indicate weak interaction constants. Thus the answer to question $\beta$) depends on that to question $\gamma$).

We also sketch the domain of quantum electrodynamics for $p = 2$:

$$
\begin{array}{c|cccccc}
S & \frac{3}{2} & \frac{1}{2} & \frac{1}{2} & -\frac{1}{2} & -\frac{1}{2} & -\frac{3}{2} \\
s_1 & \frac{1}{2} & \frac{1}{2} & -\frac{1}{2} & \frac{1}{2} & -\frac{1}{2} & -\frac{1}{2} \\
s_2 & 1 & 0 & 1 & -1 & 0 & -1
\end{array} \tag{5.87}
$$

$\gamma$) *Constants.* According to our hypothesis the coupling constants of the interaction should also be uniquely determined by the composition of the urs. The question is initially what that might mean. The values of dimensional constants are initially statements about our system of units. One will expect only dimensionless constants to be determined by the laws of nature. In certain cases one can perhaps formulate it such that there are systems of units distinguished by laws of nature; the pure numbers are then the values of certain quantities expressed in such natural units. For the ur hypothesis only natural numbers characterize the basis states. Thus one might argue that all constants of nature could be derived from integers.

The simplest value of a constant that does not require any special units is zero. An interaction factor zero means *no interaction*. We explain such a result with an example.

Consider a massless lepton ($s_1 = \frac{1}{2}$) interacting with a right-circular photon ($s_2 = 1$). This is the case $S = \frac{3}{2}$ above. The lowest possible state belongs to $n = 3$. The three urs may have the basis states 1,1,2. There are two possible combinations of the two particles: $1 \cdot \overline{12}$ and $2 \cdot 11$; there $\overline{12}$ means the symmetric tensor $12 + 21$. Similarly there are two admissible para-Bose states of the three urs:

$$112 \quad \text{and} \quad \frac{11}{2}.$$

Thus the ambiguity of the representations of para-Bose states by products of particle states does not occur in this case. This means that in this state of the whole object the two particles do not have any interaction. By explicit calculation one can show that this holds for all product states of these particles; it simply follows from the fact that para-Bose states are defined by Green's decomposition which gives exactly the same manifold as the product. If this were normally the case we could say that our massless lepton is a neutrino, which does not have any interaction with the Maxwell field.

There are, however, three circumstances that nevertheless lead to interaction:

1. Ambiguity of the decomposition of the whole object into various kinds of particles
2. States of a part with $n = 0$
3. higher para-Bose order $p$ of a part

Case 1. We have illustrated this in (5.86) for the case $S = \frac{1}{2}$. In the example $S = \frac{3}{2}$ another decomposition would be, e.g., $s_1 = -\frac{1}{2}, s_2 = 3$. This decomposition presumably leads to the theory of gravitation and would thus introduce a gravitational interaction between neutrinos and the Maxwell field.

Case 2. An example would be $S = \frac{1}{2}, s_1 = \frac{1}{2}, s_2 = 0$, i.e., interaction of a massless lepton with a scalar photon. States with $n = 0$ do not occur as summands in Green's decomposition. Thus, it can be expected that scalar photons are bound to massless leptons.

Case 3. A lepton to which many photons were bound would be an object with large $p$. Thus one must expect that a massive lepton will show an electromagnetic interaction with the Maxwell field, i.e., charge.

In this way we come to the question of the magnitude of the charge and the mass of the electron. The present argument suggests that the charge will be a function of the mass. Thus one must first determine the mass. Let us assume the electron has $p_{\text{el}}$ scalar photons bound to it; then its mass in "natural" units would be $p_{\text{el}} + 1 \approx p_{\text{el}}$. There remains the task of finding an approach to $p_{\text{el}}$. This leads us to the general theory of elementary particles.

## 5.5 Elementary particles

### 5.5.1 Prehistory: Atomism or unified field theory

The present theory of elementary particles sounds in its terminology like a continuation of atomism in chemistry (see Sect. 2.4 and 4.1). The last building blocks of nature are understood as localizable small bodies or point masses called "particles," preserving their identities in time. Among them are assumed to be ultimate "elementary" particles which are not further divisible, "atoms" in the Greek sense of the word. Only what one considers elementary varies in time. First there were the atoms of hydrogen, oxygen, etc., then the photon, electron, neutron etc., nowadays leptons, quarks, photons, gluons, etc. Beyond the particles just mentioned one naturally looks for even smaller particles.

It is, however, by no means obvious that there are particles at all. Their empirical existence demands a theoretical explanation. Why should material reality actually be subsumed in localizable, permanent entities? The field theories (Sect. 2.6) offered an alternative in principle. In Maxwell's electrodynamics, the electron is initially something foreign, leading to insoluble theoretical difficulties: if it is a point mass with a Coulomb field, then it must carry infinite self-energy; if it is extended then non-electric forces are necessary to hold it together.[8]

Einstein in his later years devised the idea of a unified field theory (Sect. 2.10). The particles should be only special solutions of the field equations. He did not, however, succeed in finding such solutions.

Quantum theory has changed the appearance of these problems but has not solved any of them. It presents particles and fields as two aspects of the same reality but with an asymmetry which is conveniently expressed in the scheme of multiple quantization. When nonrelativistic quantum theory starts with the concept of a particle, it is just as naïve and unjustified as classical point mechanics (Sect. 2.2). The relativistic quantum theory of fields on the other hand assumes without justification the concept of a field. It leads then, following Wigner, to a justification of the particle concept as an irreducible representation of the Poincaré group. But this only holds for free particles. Why and to what degree interacting particles should retain particle properties at all initially remains completely unexplained.

The divergences of the classical electron theory have not been overcome. Renormalization is only a procedure to discard divergent terms in the equations in a Lorentz invariant way. It offers no more than a pragmatically successful method of calculation, together with the hope of being able to justify the omission of these terms at a later stage of development of the theory. Even if it were possible, as some authors hope, for the divergent terms to cancel one another out in a more comprehensive theory of particles (e.g., supersymmetry), we would still not understand the deeper reason for the resulting convergence of the theory.

Heisenberg in 1958 (cf. his 1967) attempted what was in principle a deeper approach to a unified quantum field theory. He chose a massless spinor field ($s=\frac{1}{2}$) whose equation of motion contained a nonlinear self-interaction term. The solutions one obtained while neglecting the interaction term, formally free massless particles with $s=\frac{1}{2}$, were assumed to be physically irrelevant. The solutions with interaction were to correspond directly to particles with sharp,

---

[8] From the biography of Sommerfeld (Eckert et al. 1984) I have learned that the mathematician Lindemann objected in 1907 to the appointment of Sommerfeld in Munich on the grounds that he adhered to the mathematically contradictory theory of electrons.

finite rest masses. Here the field concept is fundamental and the cohesion and rest mass of a particle are generated by the self-interaction. Heisenberg expected a consistent theory of self-interaction never to lead to divergent results, and that all finite terms in the dynamical equations of the resulting particles would yield a correct description of finite quantities (energies, rest mass, coupling constants). But naturally the solution of Heisenberg's equation was difficult. The theory was not accepted as, among other things, Heisenberg neither wanted to assume the quark symmetry that in the meantime had been successfully applied, but was unexplained by his theory, nor could he derive it from the latter. It appears to me, however, that the mathematical problems encountered by Heisenberg (like those encountered by Einstein in his attempt at a classical unified field theory) are at least symptomatic of the questions that must necessarily be confronted in fundamental physics. The pragmatically successful continuation of the particle concept only pushes these questions off, unasked, and thus unsolved.

### 5.5.2 The offer of ur theory

We have introduced the ur hypothesis as the only consistent atomism possible in quantum theory. There are no smallest particles, just smallest alternatives. The spacetime continuum, particles, and fields are to emerge from them as approximate descriptions. The proof of the program can only be in its implementation.

We pose here as an orientation two preliminary questions about its chances:

1. Can it overcome the conceptual difficulties of the present theories?
2. Does it offer the prospect of explaining the empirical systematics of particle physics?

Question 1. The evident difficulties are the divergences. Our program of open finitism is designed so that such divergences never arise. Every real state is built from tensors of finite rank. Divergences occur if one takes seriously the particle concept with its infinite spectrum of states, or the concept of local fields with arbitrarily sharp localization, and one constructs real states based on interacting particles and fields, preserving their identity in the interaction. Here, infinitely many fictitious intermediate states are introduced. That is, the concepts of particles and fields, which from the outset were only convenient approximations for states in tensor space, are taken seriously precisely in their fictitious properties. One can hope that in a strict ur theory none of these infinities will ever arise.

The same hope underlay Heisenberg's spinor-field theory. I would assume that this theory also represents an approximation to the ur theory. It is necessary to introduce, at least formally, a field (or an infinite-dimensional representation in tensor space) to represent the special theory of relativity. In its language one usually expresses causality, i.e., in our interpretation, the sequence of temporal modes. Individual finite tensors of urs do not give an account of the temporal sequence. Hence it is plausible to expect that between urs and particles there is an "intermediate floor" (Dürr 1977) which describes the relativistic causal dependencies but not yet the individual particles. Precisely this would be Heisenberg's "ur-field."

Our theory thus far has taught us an analog to it in the $p = 1$ representations which we called "quasiparticles." There remains, however, an essential difference with Heisenberg's approach. In common is the starting point of an elementary reality with the transformation properties of a spinor. Only from this one can hope to build up all the various particles. With this argument Heisenberg explicitly required the spinorial behavior. We have derived it, at least for momentary states (Sect. 4.2b2), from abstract quantum theory and justified it for time-dependent states from a postulate of the preservation of identity. Heisenberg immediately had to endow his spinor with the properties of a field in the spacetime continuum as he had no concept available for an informationally smaller unit. We build from urs a fundamental $p = 1$ field for every $s$, and thus, as it were, an infinite sequence of virtual Heisenberg fields of different helicity. Thereby we also avoid the problems that occur for Heisenberg with the composition of higher objects from one *field* (indefinite Hilbert space, etc.). The combination of several $p = 1$ fields takes place for us not in coordinate space but in tensor space with discrete indices.

Question 2. The problem breaks down into two essentially different types of questions about the *discrete* and *continuous* attributes of particles.

Discrete quantum numbers arise from compact groups. The distinction of particles according to their spins follows via special relativity directly from ur theory. We discuss in subsection 5.5c the chances of deriving the internal symmetries of particle physics.

Continuous characteristics of particles are rest masses of free particles and coupling constants. The word "continuous" means here that the symmetry groups we know about permit a continuous range of values. Experience, however, shows that these quantities have sharp values; rest masses are sharp for stable particles and have a precise center for unstable particles, i.e., resonances; the coupling constants are to our knowledge also completely sharp, e.g., $e^2/\hbar c = 1/137.036$. Again observe the meaning of the word "sharp." That an individual stable particle like a free electron has a sharp rest mass is relativistically trivial: it is the value of the Casimir operator "mass" for the irreducible representation of the Poincaré group of this one particle. Mathematically, however, one could easily imagine that electrons, while indistinguishable in their other properties, might all have different rest masses, characteristic of each individual particle. That there belongs to the type of particle "electron" a single, fixed value of rest mass is a phenomenon in need of explanation. Only because one started without justification with the particle concept and accepted the empirical equality of the masses as seemingly self-evident has this problem scarcely entered the awareness of physicists. We deal with it below in subsection d. It will turn out to be a central problem of the ur theory.

In Sect. 5.4e we found arguments for the conjecture that, at least in electrodynamics, the value of the coupling constant depends on the value of the rest mass. This question, however, exceeds the scope of the present stage of our program.

### 5.5.3 Systematics and gauge groups[9]

Before a complete ur-theoretic reconstruction of electrodynamics takes place, we can only make educated guesses about the subsequent systematics.

One main feature of the present systematics of particles appears to be the distinction between fermions on the one hand and bosonic fields on the other. Here the classical dichotomy between particles and fields finds its explanation in the two forms of quantum statistics for particles: fermions repel one another, appearing instead as isolated particles; bosons easily merge into a field. Ur theory reconstructs Pauli's connection between spin and statistics and explains why the mathematically possible para-statistics occurs for urs but not for particles.[10]

Between fermions there appears to be no direct interaction (apart from the statistically implied repulsion); their interaction is mediated through boson fields. Also this seems to follow from ur theory. In Sect. 5.4e we have seen that two massless particles with $s \neq 0$ have no interaction with one another. There it was the example of the neutrino with a transverse (circular) photon. But the same also holds for two neutrinos or two photons. The latter ought to be the reason for the linear character of Maxwell's equations. There the path to the electromagnetic interaction appeared to be going exclusively via the ground state of scalar photons. A pure fermion field, without any ($s = 0$)-admixture, should therefore in principle have no self-interaction, different from Heisenberg's approach. Of course, interaction may also be generated by dropping the separation of electrodynamics from other fields (see Eq.(5.86) in Sect. 5.4).

In the class of fundamental fermions, the present standard model distinguishes two types: leptons and quarks. Each of these two types is presently known in three different "families" which are distinguished mainly by different values of the rest mass. This classification into families is thus far unexplained. We do not attempt their explanation but restrict attention to the lowest families of both types.

The systematics of particles is determined by *local gauge groups*. The simplest example is U(1) of electrodynamics. The matter wave function $\psi(x)$

---

[9] I am indebted to H. Joos for extensive and instructive talks about these questions.
[10] Editors' comment: The method of Heidenreich products of ur tensors of arbitrary para-Bose orders $p$, based on the representations given in Sect. 5.2, was proposed by Heidenreich (1981) (see also Heidenreich's contributions in QTS). Its implications—concerning the spin-statistics connection for instance—have not been developed further and therefore the original section 10.3 is omitted from this edition.

always permits a global phase factor $e^{i\alpha}$ ($\alpha$ real, $0 \leq \alpha < 2\pi$). But one can also introduce a gauge depending on position with a factor $e^{i\alpha(x)}$. This is a redefinition of the $\psi$-function. The electromagnetic interaction remains invariant if one adds at the same time $e\partial_\mu \alpha(x)$ to the potential of the Maxwell field. For other interactions one chooses higher, non-Abelian gauge groups, e.g., SU(3) for quarks. There the fundamental requirement is that the law of interaction be invariant under the gauge group.

If now in ur theory the interaction is uniquely determined, then the invariance of ur theory itself must be, in each relevant approximation, a consequence of the ur hypothesis, thus a property of the tensor space of urs. In principle this appears to be plausible for our model of interaction. One can define in the space $\overline{T}$ of symmetric tensors a "local" gauge transformation by applying to each basic tensor defined by the four numbers $n_1, n_2, n_3, n_4$ a factor $e^{i\alpha(n_1,n_2,n_3,n_4)}$. A tensor of a composite object ($p > 1$) is understood as a symmetrized product of tensors to $p = 1$. One may re-gauge each of these factors such that the product remains invariant. Upon transforming to coordinate space, this will reveal itself as a local gauge transformation in the usual sense. The transformed gauge phase will depend not only on position, but also on other coordinates, e.g., spin components.[11] Hence one will obtain non-Abelian gauge groups.

Now there arises the question of the difference between leptons and quarks. Both have spin $\frac{1}{2}$. In ur theory one would expect only *one* type of such particles. It is therefore natural to assume that they are the same particle in two different states, separated by metastability. One characteristic difference is that leptons have large interactions at small distances and are asymptotically free at large distances; quarks have large interactions at large distances and are asymptotically free at small distances. For quarks one can probably also speculate they have large interactions at small relative momenta and are asymptotically free at large relative momenta. We now consider the representation of SO(4,2) of Sect. 4.1e, Eqs.(4.33) and (4.37). There the translations in coordinate space, up to the compact $M_{i5}$, are the operators $N_{i6}$ ($i = 1, 2, 3$); the translations in momentum space are the $N_{i4}$. The $N_{i4}$ and $N_{i6}$, however, are for every $i$ the two possible skew-symmetric linear combinations of a certain combination of raising and lowering operators, thus can be transformed into one another by means of a reflection. The idea suggests itself that one can transform a lepton into a linear combination of quarks, through a certain reflection in the tensor space of the urs, and vice versa (linear combinations, to transform integer charges into integer charges).[12]

So much for the presently available conjectures about the ur-theoretic systematics of particles.

---

[11] See analogous arguments in Bopp (1983).

[12] Barut (1984) follows a similar idea by considering quarks as normal modes of oscillations of a system of three leptons at small distances, under the influence of their magnetic dipole interaction.

## 5.5.4 Rest masses

It will be a test of the conceptual consistency of ur theory whether it succeeds in explaining the sharpness of the rest masses. Why doesn't each particle of a particular type, for instance every individual electron, have a slightly different value of the rest mass, a continuum of possible values? As mentioned above (end of subsection $b$), this does not even arise as a problem in present theories. The classical assumption of the existence of fixed types of particles postulates the equality of all individuals of one type. Classical chemistry did this for atoms. Bohr was the first to show in detail by means of a physically analyzable atomic model, i.e., Rutherford's, that (as philosophers had long anticipated) atoms with internal structure cannot be at all individually identical to one another according to classical mechanics. His solution was quantum theory. But in that case, he found that he must simply postulate the individual equivalence of all electrons, and likewise for all protons. In quantum field theory the problem is again invisible. Assuming the validity of a field equation with a fixed mass term implies that all particles which correspond to the solutions of *this* field equation automatically have the same mass.

A unified field theory, however, like that due to Einstein or Heisenberg, does not from the outset contain an arbitrarily postulated mass term. One must demand of it that it implies, in the approximation of free particles of one type, the validity of a certain equation with a fixed mass term. First it ought to explain by itself the existence of sharply separated types of particles. Einstein's classical field theory did not even come close to such a result; presumably this is only possible in quantum theory. But also for Heisenberg's theory, this was the decisive problem that he, in any event, did not conclusively solve. In about 1965, when in the ur theory I came across the problem of the individual identity of particles of one type, I asked Heisenberg whether this problem did not also exist in his theory. He said: "No. The theory has to explain the existence of the types. To this then also belongs, for instance, the numerical value of the proton–electron mass ratio. But when one has then found a wave equation for the electron, all electrons must automatically have the same mass." The answer was correct. But it postulated that one would succeed in deriving a wave equation for the electron. As Heisenberg in fact started with a fundamental wave equation, he had also eliminated the problem, as it must appear to ur theory, through his basic assumptions.

In ur theory the existence of types of particles and the validity of field equations have first to be derived. Both can only hold approximately. We can initially characterize types of particles by means of the two numbers $s$ and $p$. For $p = 1$ we call $s$ the helicity. For larger $p$, $|s|$ becomes the spin, and the sign of $s$ distinguishes particles and antiparticles. In electrodynamics we derived field equations in the approximation of the Minkowski space for free fields with $p = 1$. The problem of the quantum numbers belonging to internal symmetries we discussed above, but did not solve. It is to be expected that $p$ or $p-1$ assumes the role of rest mass. One must therefore expect representations

138    5 Models of particles and interaction

with large $p$. The sharpness of the rest mases ought then to manifest itself as the distinction of certain values of $p$. Only if such values are distinguished can one hope to actually find field equations for distinct types of particle, each with fixed rest mass. This is our problem, still unsolved.

Thus far, I have only been able to find an approach to the solution of this problem in the model of a *cosmic finitism* (Weizsäcker 1971a, 1973b, 1974a, 1975). This model assumes that in the universe there is a finite sum total of information, i.e., a definite total number $N$ of urs. Here we only consider the possible consequences of this assumption. Its criticism or justification is beyond the scope of the present discussion.

We begin with the representation of the single ur in Einstein space. A single ur "does not yet know the difference between universe and elementary particles." Its finite amount of information will naturally be represented in a coordinate space of finite volume. Localization of an event in a small subspace of this cosmic space is only possible if many yes/no decisions are made, i.e., through many urs. Assume that there are $N$ urs in Einstein space. How accurately will one be able to localize events on them?

The question is still ambiguous. Assuming that we use all urs for the localization of a *single* event, one can then expect to single out the fraction $2^{-N}$ of the total volume; a factor $\frac{1}{2}$ per ur. But this is erroneous; it is as if the entire universe, with galaxies, stars, planets, humans, instruments were an apparatus for the measurement of *one* position. However, we wish to know how accurate "socially feasible" position measurements can be in principle; for example, how accurately one can perform, in principle, in the real universe, how many simultaneous position measurements. "In principle" means that actually much less is measured, but that every measurement under consideration could be performed without making the simultaneous execution of the other measurements physically impossible. We make the model more precise by assuming that position measurements can only be made with ponderable matter. Actually, positions are usually measured with light but the localization will not be more accurate than relative to a reference frame which is defined with respect to a body at rest in it. The mass of ponderable matter resides mainly in the atomic nuclei, which consist of nucleons. One can assume that in principle a socially feasible position measurement cannot localize a nucleon more accurately than its Compton wavelength

$$\lambda = \frac{\hbar}{mc}. \tag{5.88}$$

Thereby an atomic *unit of length* is defined.

We call $r_c$ the radius of the universe (measured in the same arbitrary units as $\lambda$) and the volume of the universe

$$V \approx r_c^3. \tag{5.89}$$

Let us tentatively assume that $N$ denotes the number of possible events which simultaneously ("socially compatibly") can be localized in the universe up to

a volume
$$v \approx \lambda^3. \tag{5.90}$$

Thus
$$N \approx V/v = r_c^3/\lambda^3. \tag{5.91}$$

With this assumption, $N$ is not in the exponent as before; rather than $2^N$, $N$ itself will be about $V/v$. There we refer to the quantum theory of urs, i.e., the quantum theory of information. $N$ is not the number of all generally possible yes/no decisions but only the number compatible with one another, thus not proportional to the (formally infinite) number of possible states in state space but rather the number of its basis states. The simultaneous localization of $N$ events in the universe is by far not the only possible measurement in the universe but it is incompatible with the performance of the other possible measurements. In that respect the conjecture seems to make sense that $V/v$ measures the number of basis vectors of the space of states which can be built from urs.

Now above, however, we did not define $N$ as the number of basis tensors in tensor space of the actually existing urs but as the number of these urs themselves. In the space $T_N$ of all tensors of rank $N$ there are initially exactly $2 \cdot 2^N$ basis tensors, assuming that there are as many urs as there are anti-urs (s=0). Of these $2(N+1)$ are symmetric, thus contained in $T_N$. It appears as if we had defined $V/v$ by the number of symmetric basis tensors in $T_N$. The full number of physically realizable basis tensors in $T_N$ should be neither $N$ nor $2 \cdot 2^N$, but the number of basis tensors definable in the para-Bose procedure, i.e., Young's standard schemes with $S = 0$ admissible. On the other hand, without an explicit calculation, it is not clear whether the multiplicity of compatible position measurements must be equivalent to a full basis of para-Bose tensors in $T_N$. For the present model I therefore restrict attention to the estimates of my previous works, i.e., Eq.(5.91).

Now we ask a second question: How many urs are invested in a nucleon at rest? How many nucleons can one then build from $N$ urs? According to presently accepted cosmological estimates one should have about

$$r_c = \gamma \lambda, \qquad \gamma \approx 10^{40}. \tag{5.92}$$

For the localization of a nucleon, to the accuracy $\lambda$, one presumably needs $\nu = 3r_c/\lambda$ urs. This estimate is logically independent of the previous one; it defines a basis in the space of all alternatives, different from the previous one. Quantum theoretically one must superpose about $r_c/\lambda$ wave functions of wavelength $r_c$ to obtain localization down to $\lambda$; the factor 3 expresses that this must happen in 3 dimensions. Let $n$ be the number of nucleons in the universe; then it follows that if all urs were accommodated in nucleons, at $\nu$ urs per nucleon,

$$n = N/\nu \approx 10^{80}. \tag{5.93}$$

140    5 Models of particles and interaction

This number agrees very well with the empirical density of hydrogen in the universe; factors of order 10 or 100 are naturally uncertain in such rough estimates.

This discussion provides us with two empirical plausibility arguments in favor of the proposed theory. One was already known to Einstein and pertains to the order of magnitude of the gravitational constant $\kappa$: the radius of a stable Einstein space is for the empirical value of $\kappa$ just as large as one must at least assume based on current cosmological knowledge; empirically, at any rate, it cannot be smaller. The second argument follows only from ur theory: the estimate of the number $n$ of nucleons in the volume $V$ according to Eq.5.93. *This is the first testable conclusion from the ur theory which does not merely reproduce known theories.*

Generally it says for a closed universe

$$N = \gamma^3, \quad \nu = \gamma, \quad n = \gamma^2. \tag{5.94}$$

$N$ and $\nu$ are defined only ur-theoretically, but $n$ is measurable, however, and follows from the theory to the correct order of magnitude. Of course, ur-theoretically this does not demonstrate *that* the urs must arrange themselves to a first approximation all in particles of the same rest mass. It merely shows that one obtains an empirically plausible estimate of the number of particles *if* the urs arrange themselves in that way.

Furthermore, this estimate, again from earlier works, does not yet take into account the para-Bose representation of particles. In it $\nu$ must presumably be the para-Bose order of a nucleon, and thereby $N$ the para-Bose order of the universe, as far as its content is organized in nucleons. The urs in about $10^{80}$ electrons and about $10^{90}$ photons of the background radiation would qualitatively not change very much. In the context of quantum electrodynamics (Sect. 5.4e) one could argue that if $n$ massless fermions can bind $N$ massless bosons, then each fermion will bind $\nu = N/n$ bosons, hence gain the rest mass $p = \nu$. Conversely, a statistical argument ought to show that in three-dimensional space $N$ massless bosons should predominantly combine into massive particles of $N^{1/3}$ massless bosons each. The sharpness of the rest masses must then be explained statistically, like the sharpness of melting points; perhaps in the sense that the spread of $p$ of a massive fermion about the value $\nu$ is only of order $\nu^{1/2}$.

For now, all these proposals remain programmatic.

## 5.6 General theory of relativity

### 5.6.1 The problem of the structure of space[13]

In Chap. 2, Sect. 2d,e and 7–10 we sketched the history of the problem of space in classical physics. Physicists mostly followed Newton's opinion that space

---
[13] Cf. Sect. 6.1. I owe important encouragement to a conversation with Th. Görnitz.

and time are entities *in* which bodies move. Mathematicians in the nineteenth century, from Gauss over Bolyai, Lobachevsky, Riemann, Lie, Klein up to Hilbert made it clear that mathematically many different structures of space are possible and available for the description of nature. Einstein initially linked space and time and then postulated a Riemannian geometry for the spacetime continuum. Quantum theory has without any critique taken over the classical description of the spacetime continuum.

In Chap. 3 we then reconstructed abstract quantum theory from postulates about empirically decidable alternatives and the concept of probability, without any reference to position space but under the assumption of a linear time. In Chap. 3 and 4 we derived the existence of a three-dimensional position space and its linkage with time according to the special theory of relativity based on abstract quantum theory and the hypothesis of ur alternatives. There, however, we obtained uniquely only the *local* structure of the spacetime continuum: it permits everywhere a tangent Minkowski space. For the *global* structure we had the choice of an Einstein, Minkowski, de Sitter, or anti-de Sitter space, depending on the chosen symmetry group. Our task is now to overcome this arbitrariness.

It is natural to assume that the local structure of space in second approximation, i.e., the local curvature, is determined by the distribution of matter. In doing so we would be exactly at Einstein's formulation of the problem.

Einstein saw himself compelled to a dualistic theory in which the metric field of the spacetime continuum and matter interact with one another. The two laws of this action are:

1. *The law of motion.* A point mass moves, if no other forces are acting on it, along a geodesic of the spacetime continuum.
2. *The field equation.* The energy–momentum tensor $T_{ik}$ of matter determines the curvature according to

$$G_{ik} = -\kappa T_{ik}, \qquad (5.95)$$

where $G_{ik}$ is related to the quantities of the Riemannian curvature according to

$$G_{ik} = R_{ik} - \tfrac{1}{2} R g_{ik} \qquad (5.96)$$

The question is whether we can derive these laws ur-theoretically.

The *law of motion* is the generalization of the law of inertia for a curved space. For urs and free fields built from them we formulated the equations of motion in Einstein space and wave equations in the tangent Minkowski space. Both agree with the quantum theoretical inertial motion. We find here again the ur-theoretic "reversal of the arguments." The special solution which we could construct from the tensor space into a homogeneous space of the symmetry group satisfy the general laws known from present physics. It is then our task to demonstrate that these laws either hold for all states of the system, or that they hold to some other approximation. For Einstein's law of motion we can check this only after we have reconstructed its Riemannian space exactly or approximately.

Einstein's field equation posed an as yet unsolved problem to quantum theoreticians: it is nonlinear. Its quantization, however, has only been accomplished in a linear approximation. It is conceivable that a consistent quantization procedure for the nonlinear field equations might yet be found. Here, however, we heuristically assume that this will not be possible, and that it is in fact not necessary. Perhaps in this technical failure there hides a deeper problem whose meaning we should try to understand.

Between the general theory of relativity and quantum theory there is a tension in principle: the general theory of relativity is *essentially local*, quantum theory is *essentially nonlocal*. All the laws of the general theory of relativity are formulated in terms of differential geometry; it is a purely local interaction theory. Indeed, the nonlocality of quantum theory is evident in Bohr's principle of the individuality of processes. In the mathematical form of the Schrödinger equation, the nonlocality is not immediately apparent. But the Schrödinger wave is a probability wave which, by virtue of observation, instantaneously changes over the entire space (Sect. 9.2b). The most conspicuous model for this is the thought experiment of Einstein, Podolsky, and Rosen (Sect. 9.3a). The mathematical form of quantum theory is actually also nonlocal in a certain sense. The Schrödinger equation of the nonrelativistic many-body problem is formulated in configuration space with forces acting at a distance. Relativistic quantum field theory works with local field equations, yet the field strengths are not state variables but operators whose measured values are only specified by probabilities. Furthermore, also in the special-relativistic theories of interaction, the mathematical consistency of the local nonlinear operator theory has not been demonstrated. Finally, our approach is based on discrete alternatives, hence is nonlocal at its foundation. The question is how the local phenomena are to be described in it.

As an example of the problem I mention here a conversation with E. Fermi from the year 1949. He told me that in quantum theory Einstein's equivalence principle was wrong. As evidence he used the thought experiment of the diffraction of a matter wave by a diffraction grating consisting of a periodic gravitational field. The deflection angle is given by the ratio of the grating constant to the wavelength. According to the equivalence principle, according to Fermi, the diffraction should deflect all bodies with the same velocity in the same direction. Actually, however, the wavelength and thus the deflection at a given velocity depends on the inertial mass. J. Ehlers[14], to whom I told this recently, at once pointed out that the equivalence principle by necessity leads to a differential-geometric theory, in which it is then used only locally. Thus, the general theory of relativity has no problem describing the diffraction of a wave. The nonlocal character of quantum theory is revealed by the

---

[14] I am grateful to J. Ehlers for extensive conversations about the relationship between the general theory of relativity and quantum theory.

fact that the wavelength also determines the momentum of a particle at the same time. Ehlers closed the discussion with the remark that in the hoped-for future reconciliation "both theories must lose some feathers."

We will consider the problem in two steps. In subsection $b$ we forgo quantizing the metric field, thus treating general relativity in principle as a classical theory. Then in Eq. (5.95), not only $G_{ik}$ but also $T_{ik}$, i.e., matter, must be described classically. In subsection $c$ we consider $G_{ik}$ and $T_{ik}$ to be quantum variables, and discuss the presumably mutually enforced abandonment of locality. In subsection $d$ we briefly look at the consequences for cosmology.

### 5.6.2 Classical metric field

In Sect. 2.8 we discussed Leibniz's monism of matter and space. According to Leibniz, space should only be the logical class of certain relations between bodies (e.g., distance, angle between relative directions, etc.). Relations are two-valued predicates. The duality of matter and space is then only the apparently unavoidable logical duality of subject and predicate, ontologically speaking of substance and attribute. This idea, taken up again by Mach, could not be implemented by Einstein precisely *because* he had to describe space in terms of differential geometry. The metric tensor thus became a field quantity just like any other.

Quantum theory, however, offers us the opportunity to reintroduce Leibniz's interpretation. In special-relativistic quantum field theory, space and time coordinates are not observables but parameters. In our reconstruction they appear as group parameters of a symmetry group of the tensor space. They can be used to coordinate a homogeneous space of the group. Functions in this space, whose values are again (complex) numbers and not operators, serve as representation vectors of the group. If we manage to construct a theory in which these groups are used only locally, the spacetime coordinates must go over into the parameters of a classical Riemannian space. We then had a quantum field theory of matter in a classical metric field.[15] All that consists of urs would then be "substance" in the classical, abstract meaning of the word; the spacetime coordinates, however, would be, exactly as Leibniz wanted, relations between the possible states of the substance. In this interpretation it would be a misunderstanding if one wanted to quantize the general theory of relativity.

The question is now whether we can construct such a theory on the basis of the ur hypothesis. As in electrodynamics and the general theory of particles, two different problems must be solved:

---

[15] This is the difference between our approach and that of Finkelstein (see 4.9c). He does not quantize momentum but space and time. A meeting of both theories is then probably only possible in the quantum theory of the metric field (subsection $c$ below).

144     5 Models of particles and interaction

1. Finding the field equations,
2. determining the constant $\kappa$.

1) As in other theories we expect the form of the equation to be already determined to a large extent by invariance requirements. Here we need merely reiterate Einstein's own ideas within the ur-theoretic context. The expectation of a Riemannian space we have already made plausible, although naturally thus far not strictly proven. Einstein's equation (see Sect. 2.10) is the simplest field equation invariant under arbitrary coordinate transformations. In our present approach Eq.(5.95) connects the essentially classically understood quantity $G_{ik}$ with the tensor $T_{ik}$ which necessarily is built from operators. Hence the equation can anyway only hold in the limit in which the $T_{ik}$ can be approximated classically. It thus suffices to assume that Eq.(5.95) occurs as the first term in a perturbation calculation. In any case there is no more to be expected.

2) The proof of our reconstruction of the general theory of relativity would be the derivation of the empirical value of $\kappa$. In a reversal of the arguments we construct again first a model that satisfies the equation and only then ask about the general validity of the equation. The model is the Einstein space as we used it in Sect. 5.5d. Einstein derived his world model as a solution of this field equation, and the cosmological estimates pertaining to the Hubble effect gave the correct relationship between $\kappa$ and the radius of the universe $r_c$. That is, if the real universe can be approximately represented by an Einstein space, then it is a solution of Einstein's field equation, and thus a model we are looking for.

In Sect. 5.5d we tried to show that an Einstein space with $n$ urs or quasi-particles should contain just $N^{2/3}$ massive particles, whose rest mass defines a length $\lambda$ such that $\gamma = r_c/\lambda = N^{1/3}$. Choosing now atomic units defined by

$$\hbar = c = \lambda = 1, \qquad (5.97)$$

the empirical value of $\kappa$ then becomes:

$$\kappa \approx 10^{-40}. \qquad (5.98)$$

That thereby

$$\kappa \approx \lambda/r_c \qquad (5.99)$$

simply means that the empirical universe is to a good approximation a solution of Einstein's equation. We can then say, according to our interpretation of the empirical value of $\lambda$, that in ur theory we must have

$$\kappa = N^{-1/3}. \qquad (5.100)$$

Here, compared to Einstein, only the order of the causal arguments is reversed. According to Newton, the strength of gravitation is determined by the mass and location of the bodies. In more precise language (which changes from the mathematical "determine" to the physical "cause") one can say that the bodies are the cause of the particular form of the gravitational field. Analogously one usually reads Einstein's equations in such a way that matter, via the tensor field $T_{ik}$, causes the curvature of space $G_{ik}$. We on the other hand proceed from a curved compact space as representation space of finite quantum theoretical alternatives. Then we ask in what units lengths are measured in human experience, and choose for it the Compton wavelength of that elementary particle which essentially determines the rest masses of bodies. The radius of the universe, measured in these units, is, if one thinks of the universe as being held together by gravitation, the reciprocal of the gravitational constant. This is the same relationship between atomic and cosmological units as the statement that the Bohr radius of a molecule consisting of two nucleons, held together solely by gravitation, is about equal to the radius of the universe.

The stationary Einstein space is admittedly not a solution of Eq.(5.95). One either must add to the equation a "cosmological term" or let the universe expand, as in Friedmann's models. We return to this in subsection $d$ Görnitz has shown (cf. Sect. 6.1) that Einstein's equation can be exactly satisfied in expanding, ur-theoretically consistent models.

Now it would be necessary to shift from the global view in Einstein space to a local view in the general Riemann space. For this we only need a differential-geometric relationship of second order between the distribution of matter and the metric field. The procedure would be a generalization of the definition of the tangent Minkowski space to the construction of global, four-dimensional, curved world models which are locally osculating to the particular Riemann space, i.e., can be adjusted to it up to the second derivative. The procedure has yet to be carried out. The program has been tried by Görnitz with osculating de Sitter spaces. At the present stage we can say that if the procedure gives the simplest possible field equation at all, thus Einstein's, then it will, as the Einstein space satisfies this equation, also yield the correct value of $\kappa$.

### 5.6.3 Quantum theory of gravitation

The theory of the classical metric field we have just formulated can at any rate only be an approximation. This already follows from $T_{ik}$ not being a classical field. Furthermore, in the previous chapter about interacting fields, we have seen indications that it also cannot be strictly described as local operator fields. Nowadays one often uses the invariance of the laws of interaction under local gauge groups as an argument for the locality of the interaction. But in Sect. 5.5c we have seen that for this the "local" gauge invariance in tensor space is probably sufficient. If one asks *which* feather general relativity and quantum theory must lose at their reconciliation, it is now natural to expect that it is the same feather for both: that the theory can be strictly presented by means of functions in the classical spacetime continuum.

The remark that distances in space and time ought to be *measurable* points in the same direction. Einstein arrived at the special theory of relativity by studying the conditions of this measurablity. If in quantum theory every measurable quantity is described by an operator, then this should also hold for space and time. This was the starting point of the theories of Snyder (1947) and Finkelstein (1968). Of course ur theory initially led us to "rigid" coordinate spaces described by group parameters; in such spaces it would suffice to depict rulers and clocks as real instruments which determine, with finite precision, space and time coordinates. If, however, we use the rigid spaces only locally, then the manner of their continuation is dictated by the metric field. Hence one must consider it as being measurable. This ultimately leads to the requirement of quantizing Einstein's field equation, thus to a quantum theory of the gravitational field.

Gupta and Thirring (Sect. 2.10) quantized the gravitational field as a massless spin-2 field in Minkowski space. Analogously we need to study the theory of a $(s = \pm 2)$-field in tensor space. The arguments of Sect. 4.9e make it plausible to expect this theory to be connected in a natural way with the theory of fields to $(s = \pm 1)$ and $s = 0$. This would suggest a unified field theory along the lines of Einstein and Kaluza-Klein (for this see Schmutzer 1983). For now, this too remains a program.

If this theory could be carried out, would it be monistic, dualistic or pluralistic in the sense of Sect. 2.10? The monism of Leibniz would be retained. Distance and time intervals, insofar as they are uniquely measurable, remain relations between events. The performance of these measurements, however, depends then on a real field, the gravitational field. In this sense the theory is pluralistic. But if all fields are built up from urs, the theory ultimately is then unitarian.

### 5.6.4 Cosmology

It would be premature to apply the unfinished theory of this chapter to the open questions of cosmology. Therefore let us make only a few basic remarks.

If the spacetime continuum is only the medium of an approximate, thus ostensible description of reality, then this surely holds also for outer space and world history as it is used in present cosmology. In Chap. 11 we further pursue the question of whether we can theoretically advance beyond this foreground.

During the discussion of outer space, we expressed in Sect. 5.5d the strong suspicion of a *cosmic finitism*. This conjecture is also in our reconstruction of quantum theory not at all self-evident. If every decidable alternative is finite, this does not imply that the number of decidable alternatives in the universe is finite. Of course we always will decide only a finite number of alternatives. But we needed the hypothesis of the cosmic finitism for a statistical argument of thermodynamic nature to explain the sharp rest masses. Thermodynamical statistics does not depend on how much *we* can decide.

Cosmic finitism leads to ascribe a finite volume to outer space. "Outer space" thereby denotes that space which influences the rest masses of the particles we observe here. This volume need not remain constant in time. In

Sect. 4.3f we suspected that the number of urs increases with time. The radius of the universe, measured in units of $\lambda$, should then grow as well. This recalls Dirac's conjecture that $r_c/\lambda$ is the age of the universe, measured in atomic units of time. The obvious conclusion of Dirac and Jordan was then that $\kappa$ decreases inversely proportional to time. The present stage of cosmological experience is inconsistent with this conjecture.

We leave this question open, and with it the ur-theoretic description of cosmology.

# 6

# Cosmology and particle physics

*by Thomas Görnitz*

## 6.1 Quantum theory of abstract binary alternatives and cosmology

The last section of Chap. 5, which deals with general relativity, closes with the statement "We leave this question open and with it the ur-theoretic description of cosmology."

In this chapter we present results obtained since 1985 in the extension of Weizsäcker's Ansatz, and which were in part presented in my two books "Quanten sind anders"[1] and "Der kreative Kosmos."[2] The chapter begins with a critical examination of the stature of cosmology within the structual of physics. The presentation of results will be followed by a comparison with new experimental data.

### 6.1.1 Critical examination of cosmology as part of natural science

Natural science in general and physics in particular rely on experimental data and search for general laws. Something that is completely inaccessible to empiricism can therefore not be part of natural science, is thus, as the case can be, pure mathematics or can be considered a part of philosophy or theology.

For an object that as a matter of principle is unique, the thought of a general law is at least irritating. Something unique can be described, perhaps understood but—as long as it is to be understood as a unique object—it does not make much sense as an object of natural science. Even in jurisprudence a law enacted for just a single citizen is something improper, and this is even more so in natural science.

---

[1] Görnitz, Th.: *Quanten sind anders*, Heidelberg, Spektrum (1999)
[2] Görnitz, Th. u. B.: *Der kreative Kosmos*, Heidelberg, Spektrum (2002)

## 6 Cosmology and particle physics (by Th. Görnitz)

If the universe is characterized as all that about which it is in principle not impossible to obtain some knowledge then it becomes apparent that cosmology on various accounts represents a frontier of any natural science.

Firstly, universe encompasses all about which knowledge could possibly be obtained, and it is therefore a border for science, which occupies itself with what can be known and beyond which there is only room for speculation.

The multiplicity of all possible knowledge necessarily embodies a certain unity; hence the universe is also necessarily one, as expressed etymologically by the very term "UNIverse." But then one at least must ask very critically what one is doing if one understands a cosmological model in the same way as a solution of general laws, which makes good sense for objects *in* the universe. In physical cosmology nowadays the nonlinear partial differential equations of the general theory of relativity are considered to be basic. Weizsäcker asks[3] in his diplomatic way about the meaning of the other solutions of this system of equations, if only a single one of these solutions describes reality, and then about the meaning of the system of equations. I think one ought to phrase this question much more clearly. The meaning of a differential equation and of the set of all its solutions are equivalent. Therefore if of the infinitely many solutions at most one is realized and all others do not occur in reality, then the system of equations asserts far too much and essentially will not be physics but pure mathematics—which by all means is also something beautiful. Now one can object that for other theories as well, never are all infinitely many solutions of the basic equations realized. This is correct but there are no compelling reasons that sooner or later, or in other places, the other solutions might be realized as well. In the general theory of relativity, however, circumstances are different. Every exact solution is always an entire universe in its full temporal and spatial extent. For such a solution there is within physical reality no "earlier or later or somewhere else" where additional possibilities of realization might exist. Some see a way out in the image of "multiverses." But this is pure mathematics, as physics is always related to empiricism, and the only thing certain about all that could possibly exist beyond our universe is that in principle one cannot obtain empirical data about it.

Approximately—e.g., for massless planets in the field of a central star or for the emission of gravitational waves from a double pulsar (in linear approximation)—general relativity obtains very good and convincing results such that a physics in gross contradiction to it would not inspire much confidence. On the other hand, after the also publicly recognized successes of chaos theory, it has become a truism that for nonlinear systems the deviation of approximate and exact solutions can become arbitrarily large in time.

Ur theory in principle aims at an approach toward the basic questions of physics. Hence the general theory of relativity, as heuristic for cosmology, should by all means be considered but should not be elevated to an axiom. In the spirit of the reversal of the historical order of arguments I thus propose

---
[3] See p. 50.

## 6.1 Quantum theory of abstract binary alternatives and cosmology

to interpret general relativity as a theory of local deviations from what is to a first approximation smooth, unperturbed cosmology.

Cosmological research is rightfully proud of the multiplicity of empirical data made available through modern observational techniques. However, one scarcely ever reflects on the implications of the assumption that cosmology is based on experiment. One rarely considers that the use of a cosmological model with an actually infinite position space and the claim of an empirical foundation of such a model are mutually exclusive. If space is actually assumed to be infinitely large, then all empirics relates to exactly 0% of the whole, as, after all, 5 to infinity or 100000 to infinity or $10^{20}$ to infinity is always zero. In all other areas of empirical research this would be considered a slim empirical basis—in cosmology often not. Nor is one saved from this dilemma by the "cosmological principle" which asserts that the universe everywhere is essentially the way we see it from our point of view. But if it actually is infinitely extended, then the idea that the vanishingly small fraction visible to us is characteristic of the whole is in its empirical basis not better founded than the rest of cosmology.

As I consider an empirical foundation very important also for cosmology, it will be shown that a model can be developed from an ur-theoretic Ansatz which at every finite time has a position space with a finite—naturally variable—volume.

### 6.1.2 Position space in ur theory

The three-dimensionality of position space belongs to one of the empirical facts of natural science whose justification is scarcely ever even considered. A theory like ur theory which attempts to start with fundamental questions naturally cannot bypass this problem. But it also cannot allow itself to postulate that "actual" space has 11 or 26 dimensions from which, through a process analogous to Kaluza-Klein, all but the 4 dimensions of spacetime in our everyday world are compactified, and that the 11 or 26 dimensions are so self-evident that they do not need any further justification.

In *Aufbau* (here section 3.3α, p. 83) it is explained that position space is distinguished on the grounds that it can be understood as parameter space of the strength of interaction. If there is a parameter at all on which, directly or indirectly, all interactions depend, then one will be able to describe actually observable objects and their states most directly if this parameter occurs as an independent variable.

According to the ur hypothesis there exists for every object a representation in which it appears to be built up from quantized binary alternatives.

For the single ur there is a symmetry group of its states which contains as subgroups $SU(2)$, $U(1)$ and complex conjugation. $U(1)$ transforms the complex phase and is therefore seen as the group of the temporal symmetry of the ur which—as long as it remains a single ur—is in a stationary state.

In Sect. 4.1b it is explained that the relations between a set of urs remain unchanged if they all are subjected to the same transformation from $SU(2)$. These relations, however, will change between different sets if they are transformed by respectively different group elements. At this level of extreme abstraction from all hitherto assumed theoretical elements in physics, one can speak neither of particles nor of forces. Therefore such a change of relations between subsets of urs is a first indication of interaction. These changes are parametrized via the three-dimensional group $SU(2)$. Therefore the hypothesis is made that a homogeneous space of the symmetry group will furnish a mathematical model of position space. The maximal homogeneous space of $SU(2)$, i.e., the group itself as $\mathbb{S}^3$, is used as model of three-dimensional compact position space. A smaller space would point to an ineffective subgroup for which there are no indications.

It is known from the representation theory of compact groups that the group itself as homogeneous space is also distinguished by the fact that the Hilbert space of square-integrable functions on it furnishes the so-called regular representation. In the regular representation there occur all irreducible representations of the group, with a multiplicity according to their dimension.

If we accept this choice of position space we then obtain at the same time the possibility of visualizing a single ur.

In classical physics it is quite natural to break a system into parts in order to understand it more easily; it is also natural that the parts are also smaller, in a spatial sense, than the original system. Every child who takes a mechanical clock apart has this experience. In the description and verbalization of elementary particle physics one thus far still adheres to this picture although quantum theory can offer here an essential expansion of the horizon of theoretical ideas. The talk about quarks as "smallest constituents of matter," where the concept "constituent" almost invariably suggest a picture of Lego blocks, cannot be banished from the representations of elementary particle theory. This despite the fact that quantum theory in its henadic structure[4], overcomes the picture of "building blocks" in physics and that for instance the only thing certain about the theoretically very successful structure of quarks is that they do not exist "as particles" one could localize and present in space and time. Also a quark-gluon plasma cannot create a single quark.

The untenable identification of "simpler" with "smaller" is broken by quantum theory. It opens up the theoretical possibility that "simpler" may also be "larger" in the spatial sense than something more complex constructed from it. The most simple entity conceivable from which more complex structures can be built are quantized binary alternatives. The aim of ur theory is therefore not a spatial but rather a logical atomism.

An ur corresponds, along with its states, to a two-dimensional representation of $SU(2)$. Interpreting it as the two-dimensional subrepresentation of

---

[4] *"henadic"* (Greek) aiming at unity, Th. Görnitz: *Quanten sind anders*, Heidelberg, Spektrum (1999)

## 6.1 Quantum theory of abstract binary alternatives and cosmology

the regular representation we obtain thereby an important visualization. An ur can be understood as a fundamental mode of oscillation of cosmic position space with only one nodal plane. The state of an ur is represented by a function on cosmic position space $S^3$ which divides it into two halves much like the sine function divides the unit circle. Therefore an ur is to be considered primarily a cosmic entity, local objects can only be represented by means of wave packets of many urs.

Using first the circle $\mathbb{C}^1$ for visualization, then with one "qubit" merely a subdivision into two halves is possible. The corresponding wave function can be visualized as a sine function.

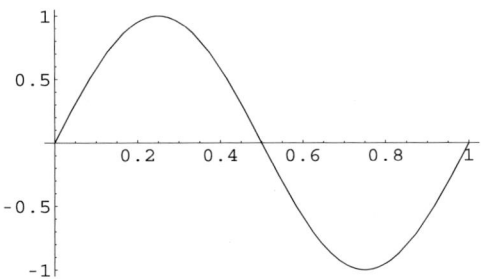

**Fig. 6.1.** The graph of $\sin(2\pi x)$

If many bits are available, in our example many sine functions which—as usual in quantum theory—are multiplied by one another, then with many bits one can localize. As an example we consider products of "qubits"—in our model powers of $\sin(2\pi x)$—and build from them quantum theoretical superpositions, i.e., we sum over possible combinations.
The more powers that are included, the sharper the localization:
to be then almost localized at a point:
A generalization to two dimensions can still be visualized in a picture, with $S^2$ as the surface of a sphere in three-dimensional space. A qubit would then subdivide the surface of the sphere into two halves, e.g., upper and lower, here represented by different shading:
A representation of $S^3$ would require an embedding in a four-dimensional space which unfortunately cannot be visualized in a picture.

Present-day physics describes something extended almost automatically as field, as a quantum field. Quantum fields are objects for which there exist representations in which they are built up from arbitrarily many quantum particles which in turn have infinite-dimensional state spaces. With this in mind it would be a basic misunderstanding to imagine an ur as a quantum field. An ur has merely a two-dimensional state space and is therefore "infinitely far away" from a quantum field and its non-denumerable many states.

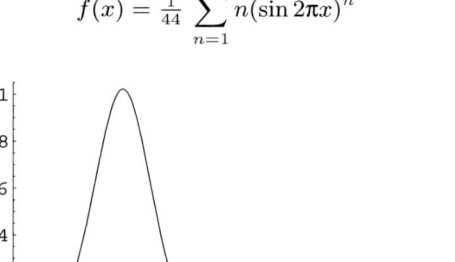

Fig. 6.2. Graphs of powers of sin(2π$x$)

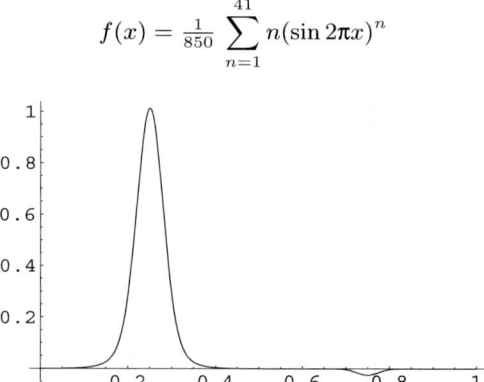

Fig. 6.3. Graphs of powers of sin(2π$x$)

Quantum fields were incorporated into physics when one attempted to bring together the special theory of relativity and quantum theory.

The special theory of relativity is valid in the approximation of a Minkowski space. In this approximation one obtains outstanding results which naturally suggest extending these successful methods to other areas. In cosmology one then runs into problems. In Minkowski space there is a strict symmetry for the reversal of time which runs there from minus to plus infinity and which is strictly homogeneous. It is well known that conservation of energy follows from this homogeneity of time. In cosmology observations suggest that time thus far was of finite duration. Cosmic expansion breaks the homogeneity of time, and thereby also the basis for conservation of energy. In

6.1 Quantum theory of abstract binary alternatives and cosmology    155

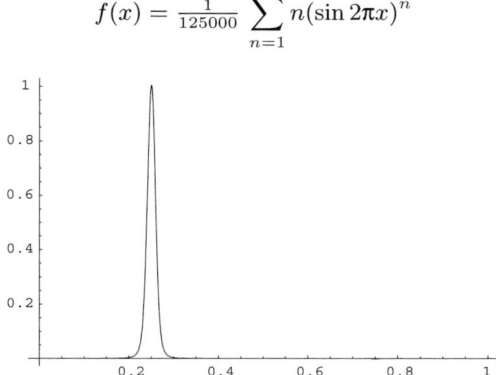

$$f(x) = \tfrac{1}{125000} \sum_{n=1}^{501} n(\sin 2\pi x)^n$$

**Fig. 6.4.** Graphs of powers of $\sin(2\pi x)$

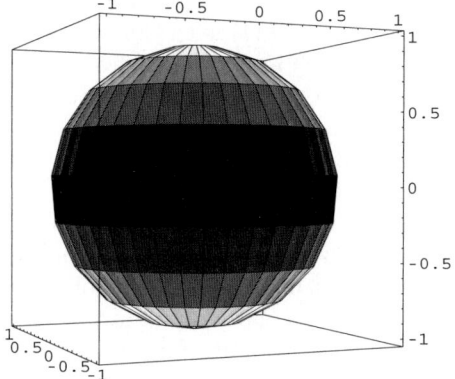

**Fig. 6.5.** Graph of a qubit on $\mathbb{S}^2$

addition, the cosmic background radiation furnishes a distinguished universal reference frame and limits the equivalence of all inertial frames, as required by special relativity, to a very reasonable local approximation which should be used as long as there are no cosmological questions involved.

If one tries to employ the concept of a quantum field in cosmology one often obtains results which differ by about 100 orders of magnitude from empirical results. If, however, we take urs as a starting point, which in quantum theory are nonlocal as a matter of principle, we obtain results which are not so grossly inconsistent with observations.

## 6 Cosmology and particle physics (by Th. Görnitz)

### 6.1.3 The cosmological model

In the usual parametrization of $SU(2)$ (see, e.g., Sect. 4.1b), the general element can be represented in the form

$$U = \begin{pmatrix} w + iz & ix + y \\ ix - y & w - iz \end{pmatrix} \tag{6.1}$$

with the subsidiary condition

$$w^2 + x^2 + y^2 + z^2 = 1. \tag{6.2}$$

The 1 on the right side of (6.2) is, however, without physical significance. To obtain from the homogeneous space of the group a physical model which can be identified with position space it is necessary to find a second length, as at most the ratio of this length to the radius of $\mathbb{S}^3$ can be of physical significance.

One reason why the ur-theoretic Ansatz has scarcely been noticed thus far was perhaps the incompatibility of the cosmological model presented in *Aufbau* (here Sect. 5.2c and 5.2d, pp. 144) with Einstein's theory. In this model a second length is postulated using an argument from the theory of elementary particles. This model requires a cosmological constant, which in the 1980s was decidedly unpopular for world models, as quantum field theoretical attempts to justify it led to values which ultimately differed by a factor of order $10^{120}$ from a realistic value compatible with observations. In addition this model from *Aufbau* led to a theory of gravitation with a variable gravitational constant. Nowadays, since the investigations of Dirac, Jordan and Ludwig, such theories have shifted somewhat outside the focus of physical interest.

We define here instead the second length via purely group theoretical arguments; this appears more appropriate for the level of abstraction desired here rather than a recourse to the not yet defined concept of a particle.

Starting with the fact that the universe contains not just one ur but a multitude of them, their quantum theoretical description is then given by the tensor product of the individual descriptions. The states of an ur are specified by a 2-dimensional (spin-1/2) representation $^2D_{1/2}$ of $SU(2)$ which is to be understood as a subrepresentation contained in the regular one and thereby represented in the Hilbert space

$$\boldsymbol{L}^2(SU(2)) = \boldsymbol{L}^2(\mathbb{S}^3). \tag{6.3}$$

Assuming $N$ urs in the universe, they define a product representation $\left(^2D_{1/2}\right)^{\otimes N}$ of $SU(2)$. It can be decomposed into irreducible components $^nD$ which in turn can be represented by functions on $\mathbb{S}^3$. These, however, now contain higher frequencies. If $R$ denotes the radius of $\mathbb{S}^3$, then the wave functions of an irreducible representation $^nD$ have wavelengths of order $R/n$. They permit—as in a Fourier series—the construction of sharper localized wave packets with increasing $n$.

## 6.1 Quantum theory of abstract binary alternatives and cosmology

Decomposing the representation $\left({}^2D_{1/2}\right)^{\otimes N}$ of $SU(2)$ into its irreducible components, we obtain a Clebsch–Gordan series of the following form

$$\left({}^2D_{1/2}\right)^{\otimes N} = \bigoplus_{j=0}^{|N/2|} \frac{N!(N+1-2j)}{(N+1-j)!j!} \left({}^{2|N/2|-2j+1}D_{|N/2|-j}\right) \tag{6.4}$$

where we defined $|N/2| = k$ for $N = 2k$ or $N = 2k+1$. To find the distribution of wavelengths of the irreducible representations, we must analyze the multiplicity factor

$$f(j) = \frac{N!(N+1-2j)}{(N+1-j)!j!}. \tag{6.5}$$

The representations with the largest wavelength, ${}^1D_0$ and ${}^2D_{1/2}$, correspond to $j = |N/2|$, and have a multiplicity of order

$$f(N/2) = \mathcal{O}\!\left(2^N N^{-3/2}\right). \tag{6.6}$$

For $N \gg 1$ the maximum of the multiplicities occurs at about

$$j_{\max} \approx \tfrac{1}{2}(N - \sqrt{N}). \tag{6.7}$$

To this belongs the representation ${}^{1+2\sqrt{N}}D_{\sqrt{N}}$.

From $D_0$ to $D_{\sqrt{N}}$ the multiplicity thereupon increases from

$$f(N/2) = \mathcal{O}\!\left(2^N N^{-3/2}\right) \tag{6.8}$$

to

$$f\!\left(\tfrac{1}{2}(N-\sqrt{N})\right) = \mathcal{O}\!\left(2^N N^{-1}\right) = \sqrt{N}\,\mathcal{O}\!\left(2^N N^{-3/2}\right). \tag{6.9}$$

After this maximum we see an exponential decrease in the multiplicity $f(j)$ between $j = \tfrac{1}{2}(N - \sqrt{N})$ and $j = 0$ from the order of magnitude $2^N$ to the value 1.

We see that the multiplicities of the Clebsch–Gordan decomposition of the tensor product are large for representations ${}^kD$ with $0 \le k \le 2\sqrt{N}$, after which they fall off exponentially so that it seems to be reasonable to neglect these few representations. In a universe containing $N$ urs localizations sharper than $R/(2\sqrt{N})$ will therefore not be found. The smallest realizable length of the order of magnitude

$$\frac{R}{\sqrt{N}} = \lambda_0 \tag{6.10}$$

will therefore be identified with the *Planck length*.

The radius of the universe thus acquires the order of magnitude

$$R = \lambda_0 \sqrt{N}. \tag{6.11}$$

Thus $R$, measured in units of $\lambda_0$, will grow with the square root of the number of urs.

To fix the rate of growth, a unit of time needs to be defined. Corresponding to the Planck length we choose the Planck time as unit; then the only universal velocity known to physics, the speed of light $c$, acquires the numerical value 1.

At this level of abstraction, making the assumption that this single universal velocity is defined in terms of the growth of the radius of the universe $R$, we then obtain the relationship between the radius of the universe $R$ and the age of the universe $T$, measured in Planck units,

$$R = T. \tag{6.12}$$

For all quantum phenomena it is known that the energy is inversely proportional to the wavelength of the particular object. If a single ur has a wavelength of order $R$, then according to this argument one must ascribe to it an energy of order $1/R$. For a universe with $N$ urs one obtains thereby a total energy $U$ of order

$$U \approx N(1/R) \approx R \approx \sqrt{N}. \tag{6.13}$$

Since the volume increases as $R^3$, the energy density $\mu$ in this model is

$$\mu \approx 1/R^2. \tag{6.14}$$

In this model there is no conservation law for total energy, reasonably enough, as there is also no homogeneity of time. What one ought to require, however, is the validity of the First Law of thermodynamics as such a closed universe probably represents the only system that is actually closed, as a closed universe does not allow any exits at all for anything. This postulate defines a cosmological pressure. From

$$dU + p\,dV = 0 \tag{6.15}$$

it follows, neglecting terms of order 1,

$$dR + p\,d(R^3) \approx dR + p\,3R^2\,dR = dR(1 + 3R^2 p) = 0. \tag{6.16}$$

As $R$ has been assumed to be variable and thus $dR \neq 0$, it follows that

$$p \approx -1/3R^2. \tag{6.17}$$

This automatically determines an equation of state for urs

$$p = -\mu/3. \tag{6.18}$$

The above arguments suggest a closed universe expanding at the speed of light. The derivation of this model is certainly unconventional, and when it was first introduced in 1987[5] it did not yet conform to the already classified

---

[5] Görnitz, Th. *International Journal of Theoretical Physics*, **27**, 527–542 (1988), **27**, 659–666 (1988), Görnitz, Th., Ruhnau, E. *Int. J. Theor. Phys.* **28**, 651–657 (1989).

## 6.1.4 Comparison with general relativity and with experiment

We now look at the present model from the standpoint of a popular standard treatise on cosmology[7]. Whereas in older discussions of cosmology pressure in the equation of state

$$p = \omega \mu c^2 \tag{6.19}$$

is restricted to the "Zeldovich-interval" $0 \le \omega \le 1$, this restrictive older standard assumption is soon dropped in Coles and Luchin. This is also justified in the light of general relativity, as its energy conditions[8] require even less: the weak energy condition is

$$\mu \ge 0 \quad \text{and} \quad \mu + p \ge 0, \tag{6.20a}$$

the dominant energy conditon

$$\mu \ge 0 \quad \text{and} \quad \mu \ge p \ge -\mu, \tag{6.20b}$$

and the so-called strong energy conditon requires

$$\mu + 3p \ge 0. \tag{6.21}$$

All three conditions are satisfied by the model presented here, whereas they are violated, e.g., by the assumption of a cosmological constant.

For such a cosmological model, its substratum being described by $\mu$ and $p$ and which behaves like an ideal liquid, it is known that Einstein's equations reduce to those of a Friedman–Robertson–Walker universe. Its metric has the form

$$ds^2 = dt^2 - R^2(t)\left[(1-r^2)^{-1} dr^2 + r^2 d\Omega^2\right] \tag{6.22}$$

and $R(t)$ satisfies the equations[9]

$$\frac{d^2R}{dt^2} = -\frac{4\pi}{3} G \left(\mu + \frac{3p}{c^2}\right) R \tag{6.23}$$

and

$$\left(\frac{dR}{dt}\right)^2 + Kc^2 = \frac{8\pi}{3} G \mu R^2 \tag{6.24}$$

---

[6] Kramer, D., Stephani, H., Herlt, E., MacCallum, M.: *Exact solutions of Einstein's field equations*, Dt. Verlag der Wissenschaften, Berlin (1980)
[7] P. Coles and F. Luchin: *Cosmology*, Wiley, Chichester et al. (1995)
[8] Hawking, S.W. and Ellis, G.F.R.: *The large scale structure of spacetime*, Cambridge University Press (1973), p. 90
[9] Coles and Luchin, p. 11.

where we have included the constants in (6.23) and (6.24) ($G$ is the gravitational constant). The constant $K$ has the value $K = 1$ for a closed, $K = 0$ for a flat, and $K = -1$ for a hyperbolic position space.

For comparison with experiment one introduces the Hubble parameter

$$H(t) = [\mathrm{d}R(t)/\mathrm{d}t]/R(t) \tag{6.25}$$

and the deceleration parameter $q_0$

$$q_0 = [R(t)\,\mathrm{d}^2 R(t)/\mathrm{d}t^2]/[\mathrm{d}R(t)/\mathrm{d}t]^2 \quad \text{at } t = t_0, \tag{6.26}$$

where $t_0$ denotes the present age of the universe. Setting

$$H_0 = H(t_0) \quad \text{and} \quad R_0 = R(t_0), \tag{6.27}$$

one expands $R(t)$ in a series

$$R(t) = R_0\left[1 + H_0(t - t_0) - \tfrac{1}{2} q_0 H_0^2 (t - t_0)^2 + \ldots\right]. \tag{6.28}$$

With the critical density

$$\mu_{0c} = 3H_0^2/(8\pi G) \tag{6.29}$$

and the density parameter

$$\Omega = \mu_0/\mu_{0c} \tag{6.30}$$

(6.24) can be transformed into

$$\left[\frac{\mathrm{d}R(t)/\mathrm{d}t}{R_0}\right]^2 - \frac{8\pi}{3} G\mu(t) \left[\frac{R(t)}{R_0}\right]^2 = -\frac{Kc^2}{R_0^2}. \tag{6.31}$$

At $t = t_0$ we have

$$\left[\frac{\mathrm{d}R(t_0)/\mathrm{d}t}{R_0}\right]^2 - \frac{8\pi}{3} G\mu_0 \left[\frac{R(t_0)}{R_0}\right]^2 = -\frac{Kc^2}{R_0^2}. \tag{6.32}$$

or

$$H_0^2 - H_0^2 \frac{8\pi G}{3H_0^2} = -\frac{Kc^2}{R_0^2}, \tag{6.33}$$

$$H_0^2 \left(1 - \frac{\mu_0}{\mu_{0c}}\right) = -\frac{Kc^2}{R_0^2}, \tag{6.34}$$

$$H_0^2 (1 - \Omega) = -\frac{Kc^2}{R_0^2}. \tag{6.35}$$

We see that $\Omega = 1$ implies a flat universe, i.e., $K = 0$.

Similarly a universe without content, i.e., $\mu = p = 0$ and thus $\Omega = 0$, has a solution with $K = 0$, for which $\mathrm{d}^2 R/\mathrm{d}t^2 = 0$ and thus $q_0 = 0$ and $H_0 = 0$. In addition, for $\Omega = 0$ there is one further solution with $K = -1$, known as the

## 6.1 Quantum theory of abstract binary alternatives and cosmology

"empty universe," which serves as a reference for the comparison of different models.

A closed model like the present one means $K = 1$ and therefore $\Omega > 1$.

If the expansion velocity is to be $dR(t)/dt = c$, it follows from (6.35)

$$\Omega = 1 + \left[\frac{c}{H_0 R_0}\right]^2 = 1 + \left[\frac{c}{dR(t_0)/dt}\right]^2 = 1 + \left[\frac{c}{c}\right] = 2. \tag{6.36}$$

According to the concepts of ur theory one should be able to build "normal quantum particles and -fields" from urs. This means that some of the urs in the universe will make their appearance as nonrelativistic particles, i.e., with $p = 0$, or also as relativistic quantum objects, i.e., massless objects with $p = +\mu/3$. But this implies that it becomes necessary to introduce a third type of cosmic object, which behaves like the vacuum or a cosmological term, i.e., which satisfies an equation of state $p = -\mu$ and thereby $p < 0$.

We cannot rule out the possibility that an additional fraction of urs belongs to a possible "dark remainder" not organized in any way as particles, and thus immediately satisfies the equation of state for urs $p = -\mu/3$. For the remaining part which appears as matter (massive particles), light (massless particles) and vacuum, we decompose the energy–momentum tensor into three corresponding terms

$$_{\text{ur}}T^k{}_i = {}_{\text{matter}}T^k{}_i + {}_{\text{light}}T^k{}_i + {}_{\text{vacuum}}T^k{}_i, \tag{6.37}$$

or

$$\begin{pmatrix} \mu & & & \\ & \frac{\mu}{3} & & \\ & & \frac{\mu}{3} & \\ & & & \frac{\mu}{3} \end{pmatrix} = \begin{pmatrix} \mu_{\text{matter}} & & & \\ & 0 & & \\ & & 0 & \\ & & & 0 \end{pmatrix} + \begin{pmatrix} \mu_{\text{light}} & & & \\ & -\frac{\mu_{\text{light}}}{3} & & \\ & & -\frac{\mu_{\text{light}}}{3} & \\ & & & -\frac{\mu_{\text{light}}}{3} \end{pmatrix} + \begin{pmatrix} \lambda & & & \\ & \lambda & & \\ & & \lambda & \\ & & & \lambda \end{pmatrix}. \tag{6.38}$$

As the ground state energy of any system depends on its spatial extent, the idea of a cosmological constant which retains its initial value unchanged, despite the increasing spatial extent of the universe, is from the point of view of quantum theory completely incomprehensible. In our model we obtain with $_{\text{vacuum}}T^k{}_i$ a quantity which, at a given time, effectively acts like a cosmological term but nevertheless can react to an expansion of the universe.

The ratio $\omega$ between the energy densities of matter and light on the one hand and the vacuum energy on the other is in our model a free parameter which, however, can only have a limited range of variation. With

$$\omega = \frac{\mu_{\text{matter}} + \mu_{\text{light}}}{\lambda} \tag{6.39}$$

we obtain

$$\lambda = \frac{\mu}{\omega+1},$$

$$\mu_{\text{light}} = \mu \frac{2-\omega}{\omega+1}, \qquad (6.40)$$

$$\mu_{\text{matter}} = 2\mu \frac{\omega-1}{\omega+1}.$$

Then it follows from $\mu \geq 0$ as well as $\mu_{\text{matter}} \geq 0$ that

$$2 \geq \omega \geq 1. \qquad (6.41)$$

Defining an effective cosmological constant at time $t_0$

$$_{\text{eff}}\Lambda = -\lambda(t_0),$$

we obtain in the vicinity of this time an effective Einstein equation of the usual form

$$_{\text{eff}}G^k{}_i + {}_{\text{eff}}\Lambda \delta^k{}_i = -\kappa\big({}_{\text{matter}}T^k{}_i + {}_{\text{light}}T^k{}_i\big). \qquad (6.42)$$

The present model therefore solves several of the so-called cosmological problems.

The flatness problem searches for an explanation of why the radius of curvature of the universe is presently so large. Here the almost flat space presents itself as the consequence of a sufficiently long expansion at the speed of light.

The horizon problem poses for many models the difficult question of why the background radiation is identical, to an accuracy of $10^{-5}$, from regions of the universe which according to these models could never have been in causal contact. In our case this poses no problem as the constant expansion with $c$ does not allow any additional horizons besides the finiteness of the universe. The entire content of the universe is thereby causally connected.

The size of the cosmological term can also easily be explained by the present model, as it must be of the same order of magnitude as the energy density of matter plus light. An initial "fine tuning," accurate to 100 orders of magnitude as in other models, is superfluous.

Now there follows a comparison with present cosmological data.

The red shift $z$ is defined via the change of the wavelength due to the cosmic expansion,

$$z = \frac{\lambda_0 - \lambda_e}{\lambda_e}. \qquad (6.43)$$

There $\lambda_0$ is the wavelength we measure for the same emission process here and now, and $\lambda_e$ the wavelength measured from a distant emission. The connection to cosmological quantities is obtained from

$$1 + z = \frac{R_0}{R}. \qquad (6.44)$$

## 6.1 Quantum theory of abstract binary alternatives and cosmology

In astronomical observations one uses the luminosity distance $D_L$, which is obtained from the absolute luminosity $L$ and the apparent luminosity $\ell$:

$$L = 4\pi D_L^2 \ell. \tag{6.45}$$

$D_L$ is a function of the cosmic conditions. For this one defines with (6.29) relative densities

$$\Omega_i = \frac{\mu_i}{\mu_{0c}} \tag{6.46}$$

and equations of state for $\mu_i$

$$\mu_i = \text{const.}\ V^{-1-\alpha_i} \tag{6.47}$$

For normal matter $\Omega_M$ we have $\alpha_M = 0$

$$\mu_M = \text{const.}\ V^{-1} = \text{const.}/R^3$$

For a cosmological term $\Omega_\Lambda$ there is $\alpha_\Lambda = -1$

$$\mu_\Lambda = \text{const.}\ V^{-1+1} = \text{const.}/R^0$$

For relativistic particles and radiation $\Omega_{rad}$ we have

$$\mu_{rad} = \text{const.}\ V^{-4/3} = \text{const.}/R^4$$

and for abstract quantum information, urs, $\Omega_u$ we have $\alpha_u = -1/3$ and

$$\mu_u = \text{const.}\ V^{-1/3} = \text{const.}/R^2$$

Incidentally, the equation $\alpha = -1/3$ also holds for noncommuting strings. In addition one defines a quantity $\kappa_0 = \sum_i \Omega_i - 1$.

For different cosmological models the luminosity distance is determined according to the formula[10]

$$D_L H_0 = (1+z)|\kappa_0|^{-\frac{1}{2}} S\left\{ |\kappa_0|^{\frac{1}{2}} \int_0^z dz' \left[ \sum_i \Omega_i (1+z')^{3+3\alpha_i} - \kappa_0 (1+z')^2 \right]^{-\frac{1}{2}} \right\} \tag{6.48a}$$

where $S\{x\} = \sin x$ for $\kappa_0 > 0$ (closed universe), $S\{x\} = \sinh x$ for $\kappa_0 < 0$ (open universe) and $S\{x\} = x$ for $\kappa_0 = 0$ (flat universe).

In particular, for a flat universe with $\kappa_0 = 0$ we have

$$D_L H_0 = (1+z) \int_0^z dz' \left[ \sum_i \Omega_i (1+z')^{3+3\alpha_i} \right]^{-\frac{1}{2}} \tag{6.48b}$$

and for the empty one with $\Omega_i = 0$

---
[10] Schmidt, B. P. et al, Astrophysical J. (1998) 507, 46

$$D_L H_0 = (1+z)\sinh\left\{\int_0^z \frac{dz'}{1+z'}\right\} \tag{6.48c}$$

Astronomers employ the so-called distance modulus, which is obtained from the apparent magnitude $m$ and the absolute magnitude $M$:

$$m - M = 5\log_{10}\left(\frac{D_L}{10\,\text{parsec}}\right).$$

In the article by Schmid et al., (1998) Fig.1, the difference $\delta$ for $m - M$ between an empty universe and various other models is given as a function of the red shift:

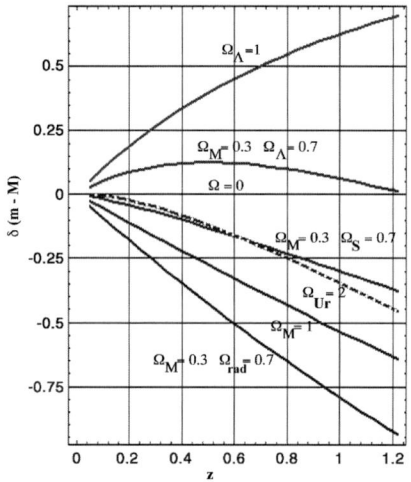

**Fig. 6.6.** Difference $\delta(m - M)$ between an empty universe and various other models ($\Omega_\Lambda$ cosmological constant, $\Omega_M$ matter, $\Omega_{rad}$ radiation, $\Omega_S$ strings, $\Omega_{Ur}$ urs) as a function of the redshift $z$.

In a recently published article by Tonry et al.[11] results derived from supernova Ia data are presented. The authors use a diagram logarithmic in $z$:

Comparing a corresponding representation with our model, we see that the above measurements are compatible with it, and become optimal for values of $z$ close to 1.

In the same article one can find an estimate of the ratio of $H_0$ to $t_0$, which is almost 1. This also yields a very good fit to the present model, which requires $H_0 t_0 = 1$. It would then no longer be necessary to postulate an accelerated expansion of the universe.

---

[11] Tonry, J.L. et al., Astrophysical J. **594** (2003), 1–24

6.1 Quantum theory of abstract binary alternatives and cosmology    165

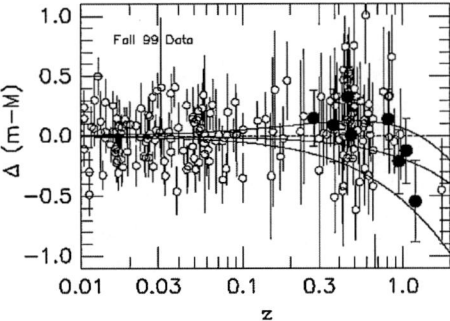

**Fig. 6.7.** The autumn 1999 and other data points are shown in a residual Hubble diagram with respect to an empty universe. From top to bottom the curves show $(\Omega_M, \Omega_\Lambda) = (0.3, 0.7), (0.3, 0.0), (1.0, 0.0)$, respectively, and the autumn 1999 points are highlighted. (From Tonry et al, 2003, Fig. 8.)

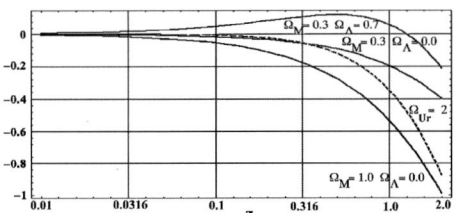

**Fig. 6.8.** Comparison of the three cosmological models from Tonry et al., 2003 with the ur-theoretic model.

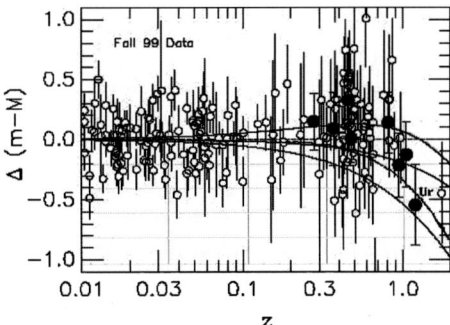

**Fig. 6.9.** Comparison of the three cosmological models from Tonry et al., 2003 with the ur-theoretic model.

# 6 Cosmology and particle physics (by Th. Görnitz)

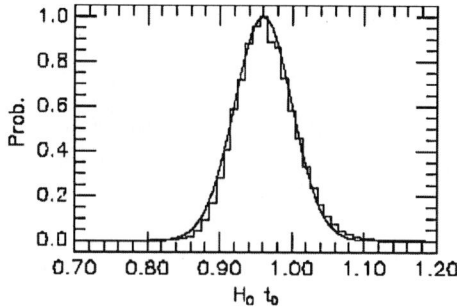

**Fig. 6.10.** The probability distribution for $H_0 t_0$ given the SN Ia observations is tightly constrained to 0.96±0.04, and an approximating Gaussian curve (from Tonry et al., 2003, Fig. 15).

At present there exists a preference for models of a flat universe, which from our point of view, however, ought to be reconsidered due to the problems mentioned in the introduction. The main argument for a flat universe comes nowadays from the investigation of the background radiation with the WMAP-satellite.[12] This radiation decoupled from matter at a temperature of about 3000 K to 10000 K. At that time this radiation was so extremely isotropic that it had only deviations of the order of 0.1 K. However, as remarked by Kegel,[13] the interpretation of these fluctuations is by no means unique. Its interpretation as density fluctuations which then led to fluctuations in the gravitational potential assumes that by the time of the decoupling of matter and radiation, significant deviations from thermodynamic equilibrium must already have been present. On the other hand, one must consider that in thermodynamic equilibrium, fluctuations in the gravitational potential are always related to temperature fluctuations, which compensate this Sachse–Wolf effect. The interpretation as density fluctuation also assumes that one can specify temperature fluctuations at the time of the decoupling to better than 0.1 K. And one must at least bear in mind that the observed density fluctuations might also be due to fluctuations in velocity of a few km/sec.

A recently published article by Schwarz et al.[14] shows that neither the dipole moment of the background radiation, which registers the relative motion of our galaxy with respect to this radiation, nor its quadrupole and octopole moments are of cosmological origin.

As long as there are no other insights available we do not want to give up our Ansatz $\Omega_{total} = 2$. But we leave it open that the part of the energy density $\mu$ of the urs in (6.38) which shows up in known media ($\mu_{light}, \mu_{matter}, \lambda$) need

---

[12] Spergel, D.N. et al, Astrophysical J. Suppl. 148 (2003), 175
[13] Kegel, W. H: Extragalaktische Systeme und Aufbau des Kosmos, Goethe-Universität Frankfurt, Vorlesungsskript 2000/2001
[14] Schwarz, D.J., Starkman, G.D., Huterer, D., Copi, C.J.: PRL **93**, 221301 (2004)

## 6.1 Quantum theory of abstract binary alternatives and cosmology

not be identified with the total energy density.

$$\Omega_{\text{total}} = 2 = \Omega_{\text{matter}} + \Omega_{\text{light}} + \Omega_{\text{vacuum}} + \Omega_{\text{rest-urs}}.$$

The part $\Omega_{\text{rest-urs}}$ of the "remaining urs" could then explain what in other models is called "dark matter" or "quintessence"—but now with the advantage that an interpretation as abstract, cosmologically defined quantum information is possible.

### 6.1.5 The path to elementary particles

The foregoing discussion has shown how the concept of abstract quantum information can be related to the structure of the universe at large. How does it look at the other end of the scale, usually occupied by elementary particles?

In his early investigations[15] Weizsäcker had estimated the number of quantum bits which form a nucleon as $10^{40}$, with a total number of $10^{120}$ urs in the universe. Thereby one obtained for the number of nucleons in the universe a value of order $10^{80}$, which corresponds to an acceptable estimate of the actual baryon density.

By and large, Weizsäcker's estimate went unnoticed. Later I could show in four papers[16] that these arguments could be connected, via the entropy of black holes, to other established areas of physics. If one hypothetically constructed one single black hole from the entire matter of the universe then one obtained upon the addition of a single nucleon an increase of its Bekenstein-Hawking entropy of said order of magnitude.

In addition, it was shown[17] that already from the state of the theory at that time one could estimate the ratio of photons to baryons to be of order $10^{10}$. Order of magnitude estimates for the temperature of the background radiation and the rest masses of the neutrinos gave at that time values of $10^2$ K and $10^{-2}$ eV.

### 6.1.6 Black holes

From the ur-theoretic Ansatz, the Bekenstein-Hawking entropy of black holes follows quite naturally from the number of quantum states hidden behind the horizon, which, however, are to be related to quantum bits and not particles or fields.

---

[15] C.F.v. Weizsäcker: *Vorlesungen*; Hamburg 1965, Engl.: *The Unity of Physics*, in: T. Bastin (Ed.) *Quantum theory and beyond*, University Press, Cambridge 1971, *Einheit der Natur*, Hanser 1971, and *Aufbau der Physik*.

[16] Görnitz, Th. *International Journal of Theoretical Physics*, **25**, 897–903 (1986), **27**, 527–542 (1988), **27**, 659–666 (1988); Görnitz, Th., Ruhnau, E.: *Int. J. Theor. Phys.* **28**, 651–657 (1989).

[17] Görnitz, Th. *International Journal of Theoretical Physics*, **27**, 527–542 (1988)

It is also important that from this quantum theoretically based model there follows an alternative model for the physically meaningless interior solution of a black hole. The idea that all matter in its interior disappears at a mathematical point follows from the nowadays strange-looking decision to use only general relativity for this area of physics and to ignore quantum theory. In all classical areas—like the atom—neglecting the validity of quantum theory leads to singularities which can be avoided upon taking it into account. If one applies the idea of a negative cosmological pressure, as follows from the ur-theoretic model, to the interior of the Schwarzschild horizon, the Schwarzschild singularity is transformed into a Friedmann singularity.[18] The interior is then to be described as a complete FRW-universe.

The Schwarzschild metric is a special case of a spherically symmetric metric which in general is of the form

$$ds^2 = e^{2A}dt^2 - e^{2B}dr^2 - r^2(d\theta^2 + \sin^2\theta \, d\phi^2) \tag{6.49}$$

where the coefficients $A$ and $B$ may not depend on the angles:

$$A = A(r,t), \quad B = B(r,t).$$

According to Birkhoff's theorem, a spherically symmetric vacuum solution must be static, i.e., is the Schwarzschild solution, and $A$ and $B$ do not depend on time. We choose

$$A(r,t) = A(r) = \tfrac{1}{2}\ln(1 - R_s/r), \tag{6.50a}$$

$$B(r,t) = B(r) = -\tfrac{1}{2}\ln(1 - R_s/r), \tag{6.50b}$$

with a constant $R_s$ which will turn out to be the Schwarzschild radius, and obtain in the usual way the Schwarzschild metric

$$ds^2 = (1 - R_s/r)dt^2 - \left[(1 - R_s/r)^{-1}dr^2 + r^2(d\theta^2 + \sin^2\theta \, d\phi^2)\right] \tag{6.51}$$
$$\text{for } R_s < r < \infty.$$

For the interior solution one usually requires the pressure to vanish at the Schwarzschild radius and that there exists a smooth transition from the exterior to the interior solution. From a quantum physics point of view this seems to be at least inconsistent, as the existence of a boundary demands a positive ground state energy of the enclosed volume (in the sense of quantum mechanics, the horizon of a black hole forms perhaps the only truly inescapable "box" in the universe). Assuming the same ground state or vacuum for interior and exterior solution thus can only be justified by an ad hoc suspension of the laws of quantum physics.

As it is not possible at all to learn something concrete about the state of matter in the interior, it appears to be plausible to use the equation of

---

[18] Görnitz, Th., Ruhnau, E. Int. J. Theor. Phys. **28**, 651–657 (1989).

state (6.18) of quantum bits for the interior. For according to Bekenstein and Hawking the amount of information in the interior of a black hole can be estimated which, as indicated, in its concrete form is not known outside and therefore referred to as entropy. It can be shown[19] that the Ansatz

$$A(r,t) = 0 \tag{6.52a}$$

$$B(r,t) = B(r) = -\tfrac{1}{2}\ln\bigl(1 - r^2/R_s^2\bigr) \tag{6.52b}$$

which leads to the line element

$$ds^2 = dt^2 - \left[\bigl(1 - r^2/R_{s^2}\bigr)^{-1}dr^2 + r^2\bigl(d\theta^2 + \sin^2\theta\,d\phi^2\bigr)\right] \tag{6.53}$$
$$\text{for } 0 < r < R_s$$

belongs exactly to the postulated energy–momentum tensor

$$T^0{}_0 = \mu \quad \text{and} \quad T^\alpha{}_\beta = -p\,\delta^\alpha{}_\beta \quad (\alpha, \beta = 1, 2, 3)$$

We find for the total mass

$$2M = R_s$$

whereby $R_s$ turns out to be the Schwarzschild radius. Under the coordinate transformation

$$r' = r/R_s$$

we obtain from the line element (6.53) for the interior of the black hole the form

$$ds^2 = dt^2 - R_s^2\left[\bigl(1 - r'^2\bigr)^{-1}dr'^2 + r'^2\bigl(d\theta^2 + \sin^2\theta\,d\phi^2\bigr)\right] \tag{6.54}$$

i.e., a stationary Robertson–Walker solution, more precisely an Einstein space with radius of curvature $R_s$.

Summarizing once more the result of these calculations, we obtain—if quantum bits are defined as here via the entropy of black holes and cosmology—from the equation of state $p = -\mu/3$ that for such a closed space the Schwarzschild and Friedmann radius become equal.

## 6.2 Ur-theoretic vacuum and particle states

All states of arbitrary elementary particles are obtained in relativistic quantum mechanics and quantum field theory via irreducible unitary representations of the Poincaré group. Strictly speaking, these representations exist only in the physical approximation where spacetime can be identified with Minkowski space. In the cosmological model presented here, spacetime approaches locally a Minkowski space in the limit of an infinite number of

---

[19] Görnitz, Th., Ruhnau, E. *Int. J. Theor. Phys.* **28**, 651–657 (1989).

urs. This is consistent with the fact that potentially infinitely many urs are also needed to construct the above mentioned unitary representations of the non-compact Poincaré group. The construction of elementary particles from urs has been explicitly demonstrated for the massless representations of the Poincaré group for arbitrary spin[20]. For massive particles, states with spin 0 and spin 1/2 have been explicitly constructed[21]. Here we present a short summary.

### 6.2.1 Multiple quantization

In quantum theory, operators that always increase or decrease the number of objects present by one unit are referred to as creation and annihilation operators. Such mathematical forms permit therefore the description of a variable number of these objects. It turns out that in their totality they can deal then with physical objects of a higher degree of physical complexity. The first example for this, based on Einstein's idea of photons, was the description of an electromagnetic field in terms of an undetermined number of these field quanta. For this one defines a so-called vacuum state $|\Omega\rangle$ corresponding to zero number of such objects. By repeated application of the creation operator $a^\dagger$ to the vacuum, one creates more and more of the objects. The annihilation operator $a$ diminishes the number of objects always by one and gives the null result when applied to the vacuum:

$$a|\Omega\rangle = 0.$$

For an arbitrary state with $N$ objects one chooses the normalization of $a^\dagger$ and $a$ such that

$$a^\dagger a |N\rangle = N |N\rangle.$$

Since the early days of quantum mechanics, one postulated for these operators the so-called canonical commutation relations

$$[a, a^\dagger] := aa^\dagger - a^\dagger a = 1, \tag{6.55}$$

which lead to the desired properties. These commutation relations are also known as Bose commutators, in contrast to the Fermi commutation relations

$$\{a, a^\dagger\} := aa^\dagger + a^\dagger a = 1.$$

Whereas an arbitrary number of Bose objects can occupy a state, at most one Fermi object can be in one state.

---

[20] Görnitz, Th., Graudenz, D., Weizsäcker, C. F. v.: *Intern. J. Theoret. Phys.* **31**, (1992) 1929–1959

[21] Görnitz, Th. and Schomäcker, U.: *Group theoretical aspects of a charge operator in an ur-theoretic framework*, talk given at: GROUP 21, Applications and Mathematical Aspects of Geometry, Groups, and Algebras, Goslar, (1997)

Görnitz, Th. u. B.: *Der kreative Kosmos*, Heidelberg, Spektrum (2002)

The construction of a quantum field from its field quanta has been called "second quantization." In ur theory it turns out that the relativistic particles themselves, which in physics occur as field quanta, can be created by an analogous process of a "second quantization" from quantum bits.

### 6.2.2 Elementary objects

As "elementary" one denotes an object for which no internal changes can be defined. In physics one often uses for this the concept of a "particle." A change of its state can then be only accomplished by means of external changes, such as varying the position or velocity, for example. An electron might serve as an example of an elementary object, having according to our present knowledge no internal structure. Another possibility would be an atom in its ground state. There internal changes are ruled out by definition, otherwise it would no longer be in the ground state. As in general it takes a relatively large amount of energy to lift an atom out of its ground state, it is sometimes useful to describe an entire atom as "elementary." Thus if elementary objects do not possess any internal differences then one must also expect that objects of the same sort cannot be distinguished from one another.

By *motion* one means a transformation that merely alters the state of an object, but not the object itself. For elementary objects one therefore obtains—as already mentioned—all changes of state through motion. In a (3+1)-dimensional spacetime with constant curvature there is a 10-parameter group describing motions of an elementary object, namely 4 translations in the 4 directions of the coordinates, 3 rotations of the three spacelike coordinates among themselves, and 3 "hyperbolic" rotations of the spacelike coordinates and the timelike coordinate. The corresponding transformation groups are the de Sitter group SO(4,1) and the anti-de Sitter group SO(3,2), and in the flat case the Poincaré group which is the semi-direct product of the Lorentz group SO(3,1) with the Euclidean translations T(4). In the case of the anti-de Sitter universe we deal with a cyclical time coordinate, which is not very physical. The de Sitter universe can be represented as a universe with a compact position space, first contracting and then expanding again.

In elementary particle theory, where the curvature of space is not felt at all, it is reasonable to work almost always in Minkowski space such that the elementary objects can be classified by irreducible representations of the Poincaré group. Such a representation, and thus the particle it describes, is uniquely determined by invariant quantities. The invariants which characterize a representation are called Casimir operators. They determine the mass as the square of the four-momentum and—via the Pauli–Lubanski vector—as well as the spin of the object. In this way mass and spin prove to be fundamental properties of all elementary particles, following simply from their stability under motions.

We wish to show now that every elementary object can be described by a certain combination of urs, i.e., quantum bits. Motion then implies that certain combinations of urs are added and others removed from the object. As an arbitrary state of the elementary object can be obtained this way it is thereby completely understood and described. In the previous section we have made it clear that a quantum bit is something quite nonlocal. It is therefore understandable that very many urs are needed to describe a particle with sharp position or also one with sharp momentum, and that the number of urs in the particle has only a probability distribution. Conversely it also holds that a fixed number of urs may lead, with a certain probability, to completely different particles.

For non-physicists the idea might appear somewhat strange that the mathematical description is also linked to an ontological statement, yet physics can only be understood if one accepts that the concepts and mathematical structures represent a mirror image of the objects and their interactions described by these concepts. The range of applicability of a theory then extends as far as this correspondence is valid. If—as is assumed here—the fundamental substance of the universe is given as protyposis, i.e., that it essentially has the character of information, then it is not surprising that good theories are distinguished by being able to offer a good description of the behavior and properties of this information.

### 6.2.3 Creation and annihilation operators for urs

By means of generalized commutation relations for the creation and annihilation operators of urs one can obtain the elementary objects in de Sitter space.[22] It turns out that in the limit of a flat space these representations go over into two representations of the Poincaré group corresponding to a particle as well as its antiparticle. If one wants to describe these two separately then one must define urs as well as anti-urs. The set of indices comprises then the values $\{1,2\}$ for urs and $\{3,4\}$ for anti-urs. The introduction of such "four-urs" was justified in Sect. 4.1. As long as one postulates the Bose commutation relations (6.55) for them, one can thereby obtain only massless particles.

To also describe objects with positive rest mass, one must go to "para-Bose commutation relations." This can be interpreted as the introduction of a hidden property which one can imagine as different "sorts" of urs. The para-Bose order $p$ then registers how many "sorts of urs" there are in a particle ($p = 1$ yields Bose commutation). The para-Bose operators obey trilinear commutation relations

---

[22] Görnitz, Th., Weizsäcker, C. F. v.: *De Sitter representations and the particle concept in an ur-theoretic cosmological model*, in Barut, A. O., Doebner, H.-D. (eds.): *Conformal Groups and Related Symmetries, Physical Results and Mathematical Background*, Lect. Notes in Physics 261, Springer, Berlin etc. (1986)

$$[a_t^\dagger, \{a_r^\dagger, a_s^\dagger\}] = 0 \qquad [a_t, \{a_r, a_s\}] = 0 \qquad [a_t^\dagger, \{a_r, a_s\}] = -2\delta_{rt}a_s - 2\delta_{st}a_r$$
$$[a_t^\dagger, \{a_r^\dagger, a_s\}] = -2\delta_{st}a_r^\dagger \qquad \qquad [a_t, \{a_r^\dagger, a_s^\dagger\}] = +2\delta_{rt}a_s^\dagger + 2\delta_{st}a_r^\dagger$$

where $r, s \in \{1, 2, 3, 4\}$. The effect on the vacuum state $|\Omega\rangle$ of urs is given by

$$a_s a_r^\dagger |\Omega\rangle = \delta_{rs} p |\Omega\rangle,$$

with the para-Bose order $p \in \{1, 2, 3, 4, \ldots\}$. It turns out, however, that values greater than 4 do not generate any new representations of the Poincaré group.

These symbols originated historically and are not particularly convenient for computer-aided investigations. We therefore introduce the following new notation: we write $e[r]$ ("Erzeuger") for the creation operators, $v[s]$ ("Vernichter") for the annihilation operators, and $p[0]$ for the para-Bose order.

For the generators of the group representations, i.e., for the so-called Lie algebra, one needs bilinear compositions of the creation and annihilation operators of urs. We now define the notation $f[r, s]$ for a bi-creator, $w[r, s]$ for a bi-annihilator, and $d[r, s]$ for the operator which gives rise to a "rotation" between urs:

$$f[r, s] = f[s, r] = \tfrac{1}{2}\bigl(a_r^\dagger a_s^\dagger + a_s^\dagger a_r^\dagger\bigr),$$
$$w[r, s] = w[s, r] = \tfrac{1}{2}\bigl(a_r a_s + a_s a_r\bigr),$$
$$d[r, s] = \tfrac{1}{2}\bigl(a_r^\dagger a_s + a_s a_r^\dagger\bigr).$$

The $n$-th power of such a bilinear operator, e.g., $f[r, s]$, will be written in the form

$$\bigl(f[r, s]\bigr)^n = f\bigl[r, s, p[n]\bigr],$$

analogously for the other operators. We denote the noncommutative product of operators by the symbol **.

### 6.2.4 Representations of the Poincaré group

The representations of a Lie group, for example the Poincaré group, can be constructed from the representations of the Lie algebra that contains an infinitesimal neighborhood of the unit element of the group. The elements of the group are obtained from those of the Lie algebra through exponentiation. With $\exp(0) = 1$ one indicates that the unit element of the group, which does not induce any change under group multiplication ($ex = xe = x$), corresponds to the null element of the Lie algebra ($0 + x = x$).

Given these preliminaries, we can now construct the generators of the Poincaré group, i.e., the Lie algebra, from ur-creation and -annihilation operators (the coordinates $\{0, 1, 2, 3\}$ in the generators $poinc[\ldots,\ldots]$ correspond to time and space in Minkowski space, and are unrelated to the indices $\{1, 2, 3, 4\}$ of urs):

The four translations turn out to be

$$P^1 = \text{poinc}[t,1] = (-\tfrac{1}{2})\{d[2,1] + d[1,2] + d[3,4] + d[4,3]$$
$$+ w[2,3] + f[3,2] + w[1,4] + f[4,1]\},$$

$$P^2 = \text{poinc}[t,2] = (-\tfrac{i}{2})\{d[1,2] - d[2,1] + d[4,3] - d[3,4]$$
$$+ w[2,3] - f[3,2] - w[1,4] + f[4,1]\},$$

$$P^3 = \text{poinc}[t,3] = (-\tfrac{1}{2})\{d[1,1] + w[1,3] - d[2,2] - w[2,4]$$
$$+ f[3,1] + d[3,3] - f[4,2] - d[4,4]\},$$

$$P^0 = \text{poinc}[t,0] = (-\tfrac{1}{2})\{d[1,1] + d[2,2] + d[3,3] + d[4,4]$$
$$+ w[1,3] + f[3,1] + w[2,4] + f[4,2]\};$$

(6.56a)

the rotations are

$$J^1 = \text{poinc}[2,3] = (-\tfrac{1}{2})\{d[2,1] + d[1,2] - d[3,4] - d[4,3]\},$$
$$J^2 = \text{poinc}[1,3] = (-\tfrac{i}{2})\{d[2,1] - d[1,2] - d[3,4] + d[4,3]\},$$
$$J^3 = \text{poinc}[1,2] = (-\tfrac{1}{2})\{d[1,1] - d[2,2] - d[3,3] + d[4,4]\};$$

(6.56b)

and the so-called boosts rotate between the timelike and one spacelike coordinate

$$\text{poinc}[0,3] = (-\tfrac{i}{2})\{w[1,3] - f[3,1] - w[2,4] + f[4,2]\},$$
$$\text{poinc}[0,2] = (-\tfrac{1}{2})\{w[1,4] + f[4,1] - w[2,3] - f[3,2]\},$$
$$\text{poinc}[0,1] = (-\tfrac{i}{2})\{w[1,4] - f[4,1] + w[2,3] - f[3,2]\}.$$

(6.56c)

We observe that there are no rotations between urs and anti-urs, thus the difference between the number of urs and anti-urs remains unchanged under the action of the group.

### 6.2.5 The Lorentz vacuum

A state in which there are no particles present is usually referred to as a *vacuum*. As we have already defined the vacuum $|\Omega\rangle$ of urs, we will for clarity denote the vacuum of particles by *lvac*. It is characterized as having mass and spin 0, and being invariant under the action of the Poincaré group, which means that it must be annihilated by every generator:

$$\text{poinc}[a,b]\, lvac = 0.$$

After a simple but somewhat lengthy calculation one finds

$$lvac = \sum_{p[1]} \sum_{p[2]} \frac{(-1)^{p[1]+p[2]}}{p[1]!p[2]!} f[1,3,p[1]] * *f[2,4,p[2]] * *|\Omega\rangle \quad (6.57)$$

where the sums over $p[1]$ and $p[2]$ extend in each case from zero to infinity.

Clearly the statement that there are no particles in Minkowski space already contains a potentially infinite amount of information. For urs, the Lorentz vacuum has the properties of a Dirac sea, the annihilation of one ur being equivalent to the creation of an anti-ur. For example, we have

$$e[1] \ lvac = -v[3] \ lvac \quad \text{or} \quad v[4] \ lvac = -e[2] \ lvac.$$

Furthermore

$$w[1,3] \ lvac = (-p[0] + f[1,3]) \ lvac,$$
$$w[2,4] \ lvac = (-p[0] + f[2,4]) \ lvac,$$

and on the other hand

$$w[1,4] \ lvac = f[2,3] \ lvac \quad \text{as well as} \quad w[2,3] \ lvac = f[1,4] \ lvac,$$

and finally

$$d[1,3] \ lvac = -f[1,1] \ lvac \quad \text{etc.}$$

## 6.3 Relativistic particles

### 6.3.1 State of a massless boson with any spin

The explicit expression of the Lorentz vacuum we just found allows us to describe states of particles in simpler form. As all generators $poinc[a,b]$ annihilate the vacuum $lvac$, only the differences between the respective particle states from $lvac$ need to be considered.

To begin with, let us describe a massless boson, i.e., a quantum particle constituting a force field of infinite range (like, e.g., the electromagnetic field), with arbitrary integral spin ($p[2] = 0, 1, 2, \dots$) in a concrete state. Massless particles must move with the speed of light, with no loss of generality in the direction of the third axis. Then both components of the momentum in the direction of the first and second direction vanish,

$$P^1 = P^2 = 0$$

and energy and the third component of momentum coincide (as usual we set the speed of light to 1):

$$P^0 = P^3 = m_+/2.$$

Denoting the para-Bose order with $p[0]$, the particular state has then the following form

$$\Phi(m_+, p[2]) = \sum_{p[1]} (-m_+)^{p[1]} \frac{(2p[2] + p[0] - 1)!}{p[1]!(p[1] + 2p[2] + p[0] - 1)!}$$

$$f[3, 1, p[1]] * * f[1, 1, p[2]] * * lvac.$$

For massless objects spin can only have a component parallel or anti-parallel to the direction of motion. This quantity is called helicity, and its magnitude is equal to the total spin. From

$$J^3 \, \Phi(m_+, p[2]) = -p[2] \, \Phi(m_+, p[2])$$

one can see that the state $\Phi(m_+, p[2])$ has the helicity $-p[2]$. This integral value of spin is carried by the pairs of urs which are added to the Lorentz vacuum by powers of $f[1,1]$. Replacing $f[1,1]$ by $f[3,3]$ one obtains a helicity $+p[2]$.

Whereas momenta and energies are due to infinite sums of ur-anti-ur pairs, spin is generated by only a few urs. This may also be the reason why thus far in all experiments about quantum information the spin of the particles is used to represent qubits. This property becomes even more conspicuous if the state is written in the following form (all $f[a,b]$ commute):

$$\Phi(m_+, p[2]) = f[1, 1, p[2]] * *$$

$$\sum_{p[1]} (-m_+)^{p[1]} \frac{(2p[2] + p[0] - 1)!}{p[1]!(p[1] + 2p[2] + p[0] - 1)!} f[3, 1, p[1]] * * lvac$$

### 6.3.2 State of a massive boson with spin 0

States for particles with rest mass have a much more complicated structure. Here we specify the state of a spinless boson at rest. It satisfies the conditions (as mentioned before, we employ here units where the speed of light has the value 1, thus $E = mc^2$ reads $E = m$)

$$P^1 = P^2 = P^3 = 0 \quad \text{and} \quad P^0 = m$$

as well as

$$\left(J^1\right)^2 + \left(J^2\right)^2 + \left(J^3\right)^2 = 0.$$

The corresponding state turns out to have the following form

$$\Phi(m, 0) = \sum_{p[1]} \sum_{p[2]} \sum_{p[3]} (p[0] - 2 + p[1] + p[2] + p[3])! \, (-1)^{p[1] + p[2] + p[3]}$$

$$m^{2p[1] + p[2] + p[3]} \frac{(p[0] - 2 + p[1] + p[2] + p[3] + p[1] + 1)!}{(p[0] - 2 + p[1] + p[2])!}$$

$$\frac{(p[0] - 2 + p[1] + p[2] + p[3])!}{p[1]! p[2]! p[3]!}$$

$$f[4, 1, p[1]] * * f[3, 2, p[1]] * * f[3, 1, p[2]] * * f[4, 2, p[3]] * * lvac$$

Obviously the para-Bose order $p[0]$ must be greater than 1, Bose commutators only give rise to massless objects.

### 6.3.3 State of a massive fermion with spin 1/2

The particles which represent "matter" with its resistance to compression have half-integral spin and obey the Pauli principle; they are called fermions. Here we also want to present the case of such a particle with rest mass $m$. Such an object can be at rest, then the three momenta vanish, $P^1 = P^2 = P^3 = 0$, and the energy is equal to the rest mass, $P^0 = m$. Let the spin point into the direction of the negative third axis ($J^3 = -1/2$):

$$\Phi(m,1/2) = \sum_{p[3]}\sum_{p[2]}\sum_{p[1]} (-1)^{p[2]+p[3]+p[1]}\, m^{2p[1]+p[2]+p[3]}$$

$$(p[0] - 1 + p[2] + p[3] + p[1])!$$
$$\{(p[0] + p[2] + p[3] + 2p[1])!\, (p[0] - 2 + p[1] + p[3])!\}^{-1}$$
$$\{(p[0] - 1 + p[1] + p[2])!\, p[1]!\, p[2]!\, p[3]!\}^{-1}$$

$$\left\{ e[1] + \frac{m\, e[2] * *f[4,1]}{(p[0] + 1 + p[2] + p[3] + 2p[1])(p[0] - 1 + p[1] + p[3])} \right\}$$

$$* *f\big[4,1,p[1]\big] * *f\big[3,2,p[1]\big] * *f\big[3,1,p[2]\big] * *f\big[4,2,p[3]\big] * *lvac$$

## 6.4 Outlook

The results which were only briefly outlined here are presented in more detail in the book "Der kreative Kosmos."[23] There ideas from the ur hypothesis were further developed. Through a concrete link to cosmology and the quantum theory of black holes we presented a generalization of the Einsteinian equivalence of matter and energy to an equivalence of these two quantities and a cosmologically defined abstract, i.e., also devoid of meaning, quantum information. This, however, is only possible if, as we do here, one goes beyond Weizsäcker's dictum[24] of "an 'absolute' concept of information has no meaning" and interprets this concept in such an abstract way that neither sender nor receiver and especially no meaning is attached to it. For this level of abstraction to be attainable we propose for this absolute, abstract, and cosmologically founded quantum information the term "protyposis." The Greek root of "to imprint" will indicate that form, information, or even meaning can be imprinted on it.

---

[23] Görnitz, Th. u. B.: *Der kreative Kosmos*, Heidelberg, Spektrum (2002)
[24] See p. 217 below.

To think of quantum information in such an abstract manner has scarcely been considered in physics thus far. But this is now on the agenda, as it on the one hand emphasizes the structure of physics in its entirety and the natural necessity of a quantum physics description of all domains of experience, from universe to elementary particles. Furthermore it may help to overcome the still often heard allegations of the conceptual "incomprehensibility" of quantum theory.

It is moreover particularly important, as it makes possible a description of man, including his thoughts and experiences, by means of natural science, which is not feasible without such a further conceptual development of physics. By means of it the science of man can free itself from prejudices which are particularly prevalent nowadays and which stem from the world view of the nineteenth century. But with the new insights[25] the picture of man can be related to those empirical findings every healthy adult constantly experiences for himself, namely to think reflectively, to be able to make free decisions, and to possess a conscious mind which to him is no less real than his body.

---

[25] Görnitz, Th. u. B.: *Der kreative Kosmos*

# Part II

# Time and information

# 7
# Irreversibility and entropy[1]

## 7.1 Irreversibility as problem

At every instant there exists an unmistakable difference between past and future. A factual event $A$ which I have experienced I will not with a clear mind mistake for a possible event $B$ which I am expecting or dreading. But when later on $B$ actually happens, then $B$ is just as factual as $A$. Is there then still a qualitative difference between the two? The answer appears to be easy: $A$ is and remains *earlier* than $B$. That in itself is a perfectic statement and the question is how it can be documented. The documentary proof is easy if the objective time of the occurrence of $A$ and $B$ has been substantiated by documents like a laboratory log book listing the day and the hour. But is this arrangement of all past events according to an objectively attestable time scale the only means of distinguishing earlier from later ones? In that case one would not speak of a qualitative difference between earlier and later.

We indeed find such a lack of distinction in temporal sequence for events that are not causally related to one another. Whether Fritz shaved in Hamburg earlier this morning than his brother Peter did in Munich, if neither remembers the time of the day, is subsequently scarcely decidable on the basis of documents or laws of nature. In general there will also be scarcely an observer for whom at any point in time Fritz's shaving was known as a past event and Peter's shaving as an event not yet happened. But whether Fritz has first lathered himself and then shaved or vice versa, one knows afterward with great reliability, as the lathering is the natural prerequisite for a successful shave and not conversely. *Causality* defines in very many cases unambiguously which sequences of events are possible and which are impossible. One must run a movie backwards to see how totally absurd a world appears in which these sequences are reversed. Arranging past events along possible causal sequences is indeed the remnant of the qualitative difference between the past and future

---

[1] This chapter stems from the 1965 lecture *Zeit und Wahrscheinlichkeit*. It is an elaboration of my essay (1939).

which one must expect if all events of the particular sequence have happened. For a current event, that which may or must happen in the immediate future is in general different from what happened in the immediate past. A car continues to move in the direction in which it just moved, a roof tile breaking off falls to the ground and does not rise, the coffee pot on the dining table cools down and does not become hotter, etc. Thus, from the point of view taken in this book, there is initially no problem at all. The objective ordering of past events into earlier and later ones is the natural consequence of the objective causal ordering according to which in the respective present the future follows the past. We also expect, of course, the same order for the distant future.

A problem occurs, however, in classical physics, as this physics, in its basic equations, no longer encounters the irreversibility of events which is so self-evident in everyday life. Physics finds itself confronted with the completely baffling fact, for an uninitiated mind, of the *reversibility* of the elementary occurrences. For all further considerations we must scrutinize this fact very carefully. In preparation for this analysis of the concept of reversibility, we initially consider the above three examples of causally determined temporal sequences, as each of them demonstrates another aspect of the problem.

A car that continues to move illustrates the *law of inertia*. Historically there the phenomenon of reversibility first became visible. According to pre-Galilean thinking, a cause is required for a body's position to change; such a cause is called a force. The motion occurs in the direction the force acts; the reverse motion can only happen if in the environment or the nature of the body itself there is something else, such that a force acts in the opposite direction. According to the law of inertia, however, the body persists, precisely in the absence of a force, in a state of constant velocity; the same body, under the same external conditions, moves in the opposite direction if from the outset it had the velocity reversed. In the simplest case of only *one* spatial coordinate $x$, the equation of force-free motion

$$\ddot{x} = 0 \tag{7.1}$$

is a law that admits reversibility, in short, a "reversible law," in the sense that to every solution $x(t)$ there exists another solution $x'(t) = x(-t)$. Somewhat casually this is often expressed by saying the equation admits time reversal. Actually, of course, the time is not reversed in the solution $x'(t)$ (which does not have any comprehensible empirical meaning) but the direction of motion; we have

$$\dot{x}'(t) = -\dot{x}(-t). \tag{7.2}$$

Of course for this "reversible" law our previously asserted connection between causality and temporal order of events also persists in a certain sense. The law of inertia is a "deterministic" law: the state of an object at one time determines its state at a later time. In order to be able to say this, one must satisfy two conditions. On the one hand, one must ensure that in the time interval considered the environment of the object do not have any influence on the object which is not considered in the equation, which is what is meant by that just *this* equation governs the motion. On the other hand, one has to *define* the "state" of the object in a way that includes all conditional properties of the object on which the changes of the conditional properties under consideration depend. We initially had only considered its *position x*. Its change, however, does not solely depend on it, the position, but on the *velocity* $\dot{x}$. The state thus must be characterized by position *and* velocity. The law now says that this is sufficient as no change of the velocity occurs. The state which obeys the principle of determinism is thus the *phase* in the sense one uses this word when one refers to the totality of all possible positions and velocities (or momenta) as *phase space*. Introducing the mass $m$ of the object (which is superfluous for its inertial motion but necessary for later examples) and defining its momentum $p$ to be

$$p = m\dot{x}, \qquad (7.3)$$

then the phase is a vector with the two components $x$ and $p$, satisfying the pair of equations

$$\begin{aligned}\dot{x} &= p/m \\ \dot{p} &= 0.\end{aligned} \qquad (7.4)$$

This equation does not admit "time reversal," i.e., the vector

$$\begin{aligned}x'(t) &= x(-t) \\ p'(t) &= p(-t)\end{aligned} \qquad (7.5)$$

is not a solution of (7.4) but of the equations

$$\begin{aligned}\dot{x}'(t) &= -\dot{x}(-t) = -p(-t)/m = -p'(t)/m, \\ \dot{p}' &= 0.\end{aligned} \qquad (7.6)$$

A solution of (7.4) is only

$$x'(t) = x(-t), \quad p'(t) = -p(-t). \qquad (7.7)$$

If the state is completely characterized, e.g., by its phase, then our previous assertion stands that laws of nature determine which state is earlier and which is later. This is contained in the manifestly evident sentence: The car continued to move in the same direction.

Nevertheless, it makes good sense to call inertial motion reversible. But this does not mean that the sequence of its phases could be traversed in the reverse order, only the sequence of its *positions*. Here we come across a phenomenon we have already considered in Chap. 4: the distinction of the concept of position before other physical concepts, the fact, to put it differently, that all physical objects, whatever else their properties might be, have not only time but also space in common. For a discussion of this fact, our present conceptual apparatus is not sufficient. We restrict ourselves therefore to a more abstract terminology: the state parameters fall into two classes such that those of one class are sufficient to formulate the laws of nature if one accepts that the laws are then not described by differential equations of first but of second order. In our case, reversibility is a property of (7.1) where only $x$ occurs as independent variable. Formulated in $x$ alone, the laws of nature do not distinguish an objective succession of earlier and later; a body can move along a straight line, consistent with the law of inertia, in each of the two directions. Precisely because of this the specification of position alone is not sufficient to determine the subsequent development but also the velocity must be given, which, as regards position, could be called a "development trend."[2] Also for more complicated cases one can in general consider in this way the state parameters of the second class to be an expression of the development trend of the parameters of the first class, if the reversal of the temporal succession of the parameters of the first class is accomplished through a reversal of sign of the parameters of the second class, and if with every system of values of the parameters of the first class always values of both signs of the parameters of the second class are compatible.

Thereby our definition of reversibility can be tied in with one customary in thermodynamics, initially formally completely different. There one calls a process leading from a state $P$ of the object to a state $Q$ of the same object reversible if there exists a process leading from $Q$ back to $P$ without a permanent change of the parameters of the object not contained in the definition of the state or in the environment. Here is initially a different version of the problem, insofar as in thermodynamics one in general considers processes which represent a temporal succession of equilibrium states, thus of states that do not change at all by themselves. The changes in the state of the object are enforced by changes in the state of the environment (supply or removal of heat or work). In the thermodynamic case the development trend is thus not in the conditional state of the object but in the environment (mostly envisioned as being arbitrarily changeable). Another difference is that the cited definition admits the case that the return path from $Q$ to $P$ runs through a different succession of states of the object than the direct path. If, however, the process leading from $P$ to $Q$ is reversible in each of its steps then it can be performed in exactly the reverse order. Then both of our definitions can be identified in the following manner: a process is reversible if there is a class of attributes of the state whose temporal succession can be traversed in one or the reverse direction, depending on the values of certain other quantities (the so-called "development trends"), without leaving behind a permanent change of attributes of the object not taken into account, or in the environment. Naturally this definition is now so abstractly formulated that it will

---

[2] See also the discussion of the concept of velocity in Gernot Böhme: *Über die Zeitmodi*, Göttingen 1966.

attain a precise meaning only after an accurate specification of what must be understood under *object, environment, state, quantity, change,*

A sufficient mathematical condition for reversibility is that there are, as state parameters of the first class, certain real quantities $q_k$ ($k = 1, 2 \ldots f$) whose change is governed by the extremum principle

$$\delta \int_{t_1}^{t_2} L(q_k, \dot{q}_k) \, dt = 0. \tag{7.8}$$

The extremum principle has the Euler equations

$$\frac{\partial L}{\partial q_k} - \frac{d}{dt} \frac{\partial L}{\partial \dot{q}_k} = 0, \tag{7.9}$$

which are invariant under the transformations

$$\begin{aligned} q'_k &= q_k(-t), \\ \dot{q}'_k &= -\dot{q}_k(-t). \end{aligned} \tag{7.10}$$

Now we can turn to both of our other examples. The free fall of a roof tile serves as an example of a *motion under an external force*. We know from mechanics that it satisfies a law of the form (7.8) or (7.9), and is therefore reversible. The reverse motion is that the tile rises from the ground and reaches zero velocity at the roof so that it could remain there, given a suitable support. Here the apparent paradox is that we just have mentioned above the non-occurrence of the reverse motion as an example of the succession of earlier and later states as described by laws of nature. Apparently this non-occurrence signifies something other than irreversibility in the just defined sense of the word reversibility. The tile falls reversibly. Also for reversible motion the laws of nature determine their temporal sequence, for the tile as well as inertial motion; like the car that continues to move in the same direction, the tile continues to fall once it begins falling, but rises once it begins rising (with the sole exception of the highest point of the parabolic motion). That we in fact often observe falling but not rising roof tiles is thus a completely different distinction of earlier and later than the one considered thus far.

In the language of mechanics the reason for this is easily found: certain initial conditions happen much more often than others. If I know that right now there is a tile freely in the air, below the edge of a roof, then I can conclude with overwhelming plausibility that it is falling from the roof and not being thrown up by men or some apparatus; one usually puts tiles on a roof and leaves them there, but one does not usually throw tiles from below onto roofs. (When there are just roofers at work, one is aware that they themselves throw the tiles up from below, and can therefore conclude that the opposite holds.) If one is content with the allusion to such well-known facts of life, one will no longer see a problem here. Yet one can ask why life just proceeds in this way, that it favors certain directions of reversible processes. To get away from the complications of human life, one can consider, e.g., a block of ice cleaving from the face of a glacier high above a rock face. Here again, the pure falling motion, from the breaking-off to impact on the ground, is reversible. Such chunks of ice fall with a certain regularity, but never rise and reattach to the glacier. Apparently the initial conditions are created by processes which themselves are irreversible. Hence one will only be able to understand the distinctiveness of certain initial conditions when one also understands the occurrence of irreversible processes.

This leads us to the third example, the coffeepot, cooling off on the dining table. As known from thermodynamics, this process is actually irreversible. The equation of heat conduction, for one coordinate,

$$\dot{T} = \alpha \frac{\partial^2 T}{\partial x^2}, \tag{7.11}$$

is of first order in time and therefore fixes the value and in particular the sign of $\dot{T}$. The development trend is here not an independent variable of state, and on the other hand the second law of thermodynamics teaches that the equalization of temperature cannot be reversed by any process which leaves object and environment otherwise unchanged. Thus there exist irreversible processes, and they are the ones which determine though laws of nature the development trend of the process and which permit, also retroactively, an objective distinction between earlier and later events.

Only now does there arise the problem which is described, in the manner of presenting physics that has become traditional since the nineteenth century, as the problem of the irreversibility. Physics has found irreversible processes only in thermodynamics. The kinetic theory of heat has taught us to understand heat as a hidden motion of atoms. About this motion we can assume that it satisfies, as any other motion, the laws of mechanics. Mechanical motions, however, as we have learned, are reversible. Thus also heat processes must be reversible. What is the origin of their factual or apparent irreversibility?

Statistical mechanics shows that the reversal of a process, which in phenomenological thermodynamics is described as irreversible, may indeed occur. The successful description of fluctuation phenomena like, e.g., Brownian motion provides ample support for the way of thinking of statistical mechanics. According to it, irreversible processes are merely more frequent or more probable, their reversals less frequent or less probable processes.

But why do certain events occur more often than their reversals? The answer of statistical mechanics is that the required atomic initial states occur more often than the states required for the reversal. But why do certain initial states occur more often than others? In the second example we have already come across this question. There we have without any detailed discussion referred to genuinely irreversible processes which preferentially produce the required initial states. Now we seem to be caught in a logical circle if we reduce the relative frequency of irreversible processes, compared to their reversals, to the frequency of the occurrence of certain initial states. Why doesn't everything we experience occur in a reversed sequence of events if what can scarcely be questioned is also a solution of the equations of atomic mechanics?

This difficulty has in fact often been discussed in the past; Boltzmann and Gibbs were fully aware of it, and a classic article by P. and T. Ehrenfest (1906) is devoted to it. But to my knowledge a satisfactory answer was neither given then nor in the textbooks at any time since (which rather hush up the problem). Boltzmann proposed a solution, refutable in my opinion, Gibbs restricted attention to an appropriate but difficult remark to understand, P. and T. Ehrenfest were aware that they had not solved the problem. Most of the newer attempts (e.g., Reichenbach) take up again, in different form, Boltzmann's erroneous attempted solution. In the following we will describe the problem in three successive steps, namely as the problem of human experiments, objective documents, and cosmic occurrences.

## 7.2 A model of irreversible processes

All theories of irreversible processes contain the following parts:

a) a model of elementary processes (e.g., free motion and collisions of spherical gas molecules,
b) a system of reversible fundamental laws (e.g., the equations of classical mechanics of point masses or elastic spheres),
c) an approach to the statistical treatment of the processes under consideration,
d) the derivation of the thermodynamic quantities and their laws from this approach, in particular their irreversible change.

We wish to avoid here those complications that have nothing to do with our main problem of irreversibility, in particular questions related to the special form of the equations of motion b); we are looking therefore for a model, as simple as possible, that is just capable of illustrating the transition from reversible basic laws b), via the statistical approach c), to irreversible thermodynamic laws d), and to make it transparent for the discussion. For this we choose a game with balls, first discussed by P. and T. Ehrenfest, which we will call the *entropy game*.

In each of two urns $A$ and $B$ there will be $N$ balls; to be specific we will assume $N = 100$. At the beginning of the game there will only be white balls in $A$, only black balls in $B$. A "move" of the game consists of picking randomly one ball from each of the two urns and putting them into the other urn. After each move each of both urns will be thoroughly mixed again such that the probability to pick in the next move a particular individual ball is the same for all $N$ balls contained in it. One asks about the number $n_k$ of white balls in urn $A$ after the $k$-th move. Through $n_k$ all other numbers are fixed after the $k$-th move; the number of black balls in $B$ is also $n_k$, and the number of black balls in $A$ and white ones in $B$ is $N - n_k$.

Our model in fact assumes the elementary laws *b*) to be statistical (equal probability for each ball to be picked). Therefore it is unsuitable for discussing the problem of whether one can derive statistical laws from basic deterministic laws at all. The problem will not occupy us any further as quantum theory is a theory of fundamental statistical laws. Here we only remark that difficulties may occur if one were to reconcile the assumption of the *strict* validity of the frequency predictions of probability theory with deterministic basic laws (ergodic problem, etc.). Yet I would suspect that the *meaning* of probability predictions explained in Chap. 3, above all their necessary relationship to the approximate validity of all predictions as implied by assumption $B$, would prove all these difficulties to be unfounded. At any rate we are not concerned here with the compatibility of deterministic basic laws with statistical thermodynamics, but rather of reversible basic laws with irreversible thermodynamics.

The basic statistical law of our model is now indeed reversible. Given a state $P$ characterized by specifying which individual balls are in $A$ and which in $B$ (one can imagine them being individually labeled by engraving numbers, like $w1$ to $w100$ for the white, $b1$ to $b100$ for the black balls). By exchanging two specific balls, whose numbers we indicate by the letters $\alpha$ and $\beta$, $P$ goes over into another state $Q$. Then $Q$ is transformed into $P$ by exchanging $\beta$ and $\alpha$ (we always specify first the ball moved from, e.g., $A$ to $B$, then the other). The probability of finding in the state $P$ the ball $\alpha$ in $A$, and in $B$ the ball $\beta$ is $P^{PQ}_{\alpha\beta} = 1/N^2$; the probability of finding in state $Q$ the ball $\beta$ in $A$ and in $B$ the ball $\alpha$ is also $P^{QP}_{\beta\alpha} = 1/N^2$. Hence

$$P^{PQ}_{\alpha\beta} = P^{QP}_{\beta\alpha}. \tag{7.12}$$

This is what is *meant* by saying that the basic law is reversible: the probability of every transition between two states is equal to the probability of the reverse process. The same also holds if $Q$ is reached from $P$ not by a single exchange but via several intermediate steps; one must then separately compute the probability that the transition is accomplished in a fixed number $\Delta k$ of moves. There is to every sequence of states $P_1 P_2 \ldots P_k \ldots P_K$ with fixed $K$ also a possible sequence $P'_1 P'_2 \ldots P'_k \ldots P'_K$ with $P'_{K-k} = P_k$, and the sequence of the $P'$ has the same probability of occurrence, given $P'_1 (= P_K)$, as the sequence of the $P$, where $P_1 (= P'_K)$ is given. This is the analog of the equation (7.2). Naturally it does not follow as in the deterministic case that the two sequences $P_k$ and $P'_k$, once they are in the specified mirror image-like relation, must also retain it under a continuation to indices less than 1 and greater than $K$. Being mirror image-like is not a *character indelebilis* of two sequences.

The statistical approach *c*) is now based on the idea that we are not interested in states characterized by the specification of the individual distributions of the balls but only in how many balls of a given color are in one urn, thus only the number $n_k$. We therefore collect all states belonging to the same number $n_k$ in one class. We will also call the individually characterized states *microstates*, the classes *macrostates*. The number of microstates belonging to the macrostate $n_k$ is

$$W(n_k) = \left(\frac{N!}{(N-n_k)!\,n_k!}\right)^2. \tag{7.13}$$

According to Stirling's formula one can, for large $n_k$ and $N$, derive in the usual manner the approximation for the logarithm of $W$:

$$H = \ln\frac{W}{(N!)^2} = -2\left[n_k \ln n_k + (N-n_k)\ln(N-n_k)\right]. \tag{7.14}$$

In the language of information theory, $-H$ is the information gained by specifying that the microstate belongs to the class $n_k$, if initially nothing was known about the microstate. We denote this specification itself by $n_k$. Then the information of $n_k$ is the logarithm of the probability to find $n_k$, if $n_k$ is known, divided by the probability of finding $n_k$, if $n_k$ is unknown. The former probability is 1, the latter is $W/(N!)^2$. One can also call

$$w(n_k) = \frac{W(n_k)}{(N!)^2} \tag{7.15}$$

the *thermodynamic probability* of the macrostate $n_k$. It is a fixed attribute of $n_k$ and to be sharply distinguished from the probability that in a certain situation, for instance with a certain prior knowledge, one can predict to find $n_k$ at a certain time.

We investigate now the laws d) for the time development of the macrostate. Having obtained in the $k$-th move $n_k = n$, then $n_{k+1}$ can only assume one of the three values $n+1, n$ and $n-1$. From the for all microstates common probability that a certain ball $\alpha$ from $A$ is exchanged with a certain ball $\beta$ from $B$, which we call

$$P = \frac{1}{N^2}, \tag{7.16}$$

we obtain for the probabilities

$$w_+ = w(n_{k+1} = n+1) = \frac{(N-n_k)^2}{N^2},$$

$$w_c = w(n_{k+1} = n) = \frac{2n_k(N-n_k)}{N^2}, \tag{7.17}$$

$$w_- = w(n_{k+1} = n-1) = \frac{n_k^2}{N^2}.$$

Therefore

$$\frac{w_+}{w_-} = \frac{(N-n_k)^2}{n_k^2}. \tag{7.18}$$

Consequently $w_+$ is greater than $w_-$ for $n < N/2$, and $w_+$ is less than $w_-$ for $n > N/2$. In other words, it is likely that the value of $n$ approaches $N/2$ with every step. As $H$ and $w(n)$ have a maximum at $n = N/2$, one can also express this by saying that the entropy (the thermodynamic probability) is likely to grow with every move as long as is has not attained its maximum possible value. This is Boltzmann's $H$-theorem, tailored to our special case. It demonstrates the irreversibility of the dynamical law for the macrostates.

Although the result is provable from the given premises, it must still appear paradoxical. The basic laws are reversible, subsuming the microstates into classes does not distinguish any direction of time, and an irreversible dynamical law is still obtained for these classes. Where has the irreversibility been smuggled into the proof?

First one can show with P. and T. Ehrenfest that the formation of classes does indeed not distinguish a direction of evolution. For this purpose we consider many sequences of states starting with an arbitrary microstate, itself randomly picked. As almost all microstates belong to those classes where $n$ differs only little from $N/2$, one mostly will start with an $n$ close to the "equilibrium value" $N/2$. Thus in general the entropy will not increase at all but remain constant, with small fluctuations. It can only grow if it has assumed a value different from the maximum value. Generally this it will only have because it has decreased before from the equilibrium to this non-maximal value. Quantitatively, these relationships can be easily derived from the formulas given above. The probability with which an arbitrarily chosen state of the sequence belongs to $n$ is given by

$$w(n) = \left(\frac{1}{(N-n)!\,n!}\right)^2. \tag{7.19}$$

To simplify the argument we assume $n < N/2$; for $n > N/2$ one would obtain *mutatis mutandis* the same. This state can—if one thinks of a sequence of directly consecutive states with $n_k = n$ as one, longer lasting state—occur in four different ways:

α. before and after $n_k = n - 1$ (maximum of $n$)
β. before $n_k = n - 1$, after $n_k = n + 1$ (rise)
γ. before $n_k = n + 1$, after $n_k = n - 1$ (fall)
δ. before and after $n_k = n + 1$ (minimum of $n$)

The probability that afterwards $n_k = n+1$ is, according to (7.17), (by omitting $w_0$, the normalization has changed)

$$w'_+ = \frac{(N-n)^2}{N^2 - 2n(N-n)}. \tag{7.20}$$

The probability that $n_k = n+1$ beforehand is $w'_+$, *because* on the average the number of rises equals the number of falls. The probabilities that $n_k = n - 1$ before and after are

$$w'_- = \frac{n^2}{N^2 - 2n(N-n)}. \tag{7.21}$$

The derived probabilities of the four indicated cases are products, where we denote the denominator by $N'$:

$$w_\alpha = w'_- w'_- = \frac{n^4}{N'^2}, \quad w_\delta = w'_+ w'_+ = \frac{(N-n)^4}{N'^2},$$

$$w_\beta = w_\gamma = w'_- w'_+ = \frac{n^2(N-n)^2}{N'^2}. \tag{7.22}$$

Now we assumed $n < N/2$, thus

$$N - n > n; \tag{7.23}$$

and therefore

$$w_\delta > w_\beta. \tag{7.24}$$

Even

$$\frac{w_\delta}{w_\beta + w_\gamma} = \frac{(N-n)^2}{2n^2} \tag{7.25}$$

is $> 1$, as well as

$$n < \frac{N}{1 + \sqrt{2}}. \tag{7.26}$$

In words: a state with $n < N/2$ is with the largest relative frequency a minimum of $n$, thus the extremum of a fluctuation. The surplus of $w_+$ over $w_-$ in (7.17), on which our proof of the $H$-theorem depended, is thus merely a consequence of $w_\delta > w_\alpha$ and therefore $w_\beta + w_\delta > w_\gamma + w_\alpha$, i.e., that a state with $n < N/2$ more often represents a minimum than a maximum of $n$. That is, it is true that *after* such a state $n_k$ will mostly be greater than $n$, but it is equally true that $n_k$ *before* the same state was also mostly greater than $n$.

Thus the $H$-theorem does not prove at all any asymmetry of the time-development of events, but on the contrary the full symmetry under the assumptions made thus far. The erroneous appearance of irreversibility arose only because we naturally and spontaneously related the concept of the probability for the transition from $n_k$ to $n_{k+1}$ to a step from the present into the future, and not to a step from the present into the past. The $H$-theorem provides irreversibility only if the deduction about the *future* by means of the probabilities (7.17) is *allowed*, but the one relating to the *past* is *forbidden*. Precisely this is suggested in a sentence of Gibbs which P. and T. Ehrenfest quote with the remark that they were unable to understand it: "Now it is rarely the case that probabilities of prior events can be determined from those of subsequent events." The basic idea of the present book developed from the attempt to make just this sentence understandable.

We first consider its application to the future. From the outset, in Chap. 3 we developed the concept of probability such that it applies to statements about the future. In our model there is nothing in the way of this application. If I now find $n$ white balls in urn $A$, then I can predict only with probability the subsequent development of the game. The theoretical calculation of this probability can only be based upon the presently stated ideas and calculations. Hence I will predict that $n$ is likely to increase. Someone who wishes to actually perform this experiment will find the frequency predictions of this section confirmed, practically with certainty, if he always thoroughly mixes the balls in the urns in sufficiently many experiments.

Now an application to the past. According to Chap. 3, perfectic statements refer to facts. For these it is only important whether or not we know them. We consider the two cases separately.

If we know the facts about the previous course of the game, namely all $n_k$ from the beginning of the game thus far, then there is indeed absolutely no reason to deduce them by means of probabilities. In our present theoretical analysis we consider only two typical, extreme cases.

*Case 1:* Among the past $n_k$ the "equilibrium value" $N/2$ has already occurred once. (That can happen if, contrary to our special assumption made above, one started with $n_1 = N/2$ or if the game has already lasted that long that equilibrium had been reached already once.) In this case one *knows* that $n_k$ has decreased from that value $N/2$ to the present value $n < N/2$, perhaps with fluctuations between. Thus one *knows* that the entropy has on the average decreased in this part of the game. According to the statistical theory this can occasionally happen, and we know that it happened in just this case. Hence we rightly "epignosticate" (the mirror-image of the word "prognosticate") a higher entropy for the past than for the present. If we do not know all $n_k$ but only that once $n_k = N/2$, then we epignosticate correctly by means of the probability formulas (7.17).

*Case 2:* The game started with $n_1 = 0$ and has not yet lasted for $N/2$ moves. Then $n < N/2$, as it did not have any opportunity at all to grow to $N/2$. Now we correctly epignosticate a smaller entropy for the past than for the present. If we do not know the values of $n_k$ between $n_1 = 0$ and the present $n_k = n$, then we will "epi-predict" (retroactively predict as being then in the future ) the first ones from $n = 0$ and (7.17); to approach the present state a more complicated probability argument is necessary, utilizing the knowledge $n_1 = 0$ and $n_k(\text{now}) = n$.

We see from these examples that where we have actual knowledge of past facts, this knowledge completely supersedes the probability statements about the respective facts and it also determines the way probability arguments are to be applied to unknown but with the know facts causally or statistically connected facts. Thus we are certainly not entitled, as we are as regards the future, to conclusions about the past from the probabilities used in the $H$-theorem.

## 7.2 A model of irreversible processes

But this is not yet sufficient. Actually, the second law of thermodynamics has been found *empirically*. In other words, we *know* nowadays that in the past, wherever humans produced a reliable verification, the entropy of a closed system on the average increased or at most remained constant. The epignosis with the probabilities used in the $H$-theorem is thus always false (up to fluctuation phenomena), the epiprediction according to just these probabilities always correct. To proceed carefully enough we will also consider this knowledge separately for two cases: first for cases which today are actually known through remembrance or documents, and second for cases whose factual course is not documented for us in this way but only made available.

The known cases teach us that in the past (with the exception of the observation of "probable" fluctuations like the Brownian motion), it was practically never fluctuations, but always an increase in entropy that was observed, i.e., that the past known to us was practically always of the type given by case 2 above, and not case 1. The actually observed courses of events used to begin with a state which was not in thermodynamic equilibrium. This is not surprising for experiments performed by humans themselves. In an experiment one usually creates an initial state that deviates from equilibrium (e.g., 100 balls of the same color in one urn) and then watches what happens; generally only then will something of interest to us happen at all. But also the cases which we have not prepared ourselves are of this kind. This is not merely a consequence of our selection like that only those cases are of interest to us. Nature never presented us with decreases of entropy in physically decidable cases, but constantly with improbable initial states with their implied increase in entropy. Only situations with constant entropy (for instance, established temperature equilibria) do we often leave out of consideration for lack of interest.

From here the transition to processes which factually are not observed presents itself. As with all other empirically discovered laws of nature we unhesitatingly also apply the second law of thermodynamics to the profusion of unobserved processes in the past. Also, non-observed temperature differences even themselves out in time *after* the cause responsible them no longer acts, and so on. The Second Law i s nothing short of being *the* tool to arrange past events in temporal sequence. It is not only a consequence of our means of observation, but also an objective law of all natural phenomena, insofar as we can dare at all to draw conclusions from experience.

Can we make the general validity of this law comprehensible from simpler or more plausible premises? Precisely this is promised by its derivation from statistical mechanics. But thus far we have only seen that it can justify the law for all natural phenomena which at present are still in the future. For events which have now already happened, the facticity of the past was only sufficient to also demonstrate the theoretical lack of justification of the empirically wrong conclusion from the $H$-theorem that in the past the entropy decreased on the average. However, we arrive at the full Second Law if we utilize the full structure of time, namely that every past event was once a present event. If *then* the conclusions about the future according to the laws of probability theory were justified, then at that time it was legitimate to predict an average increase in entropy for events which were then in the future. If afterwards this increase in entropy had not happened on the average, then one had an empirical proof that the application of probability theory to the future was unjustified. In other words, the empirical success of the probability predictions for the respective future, in the time now past, substantiates at the same time the general validity of the Second Law for just this time, in the sense of the usual generalization of empirical results to factually unobserved processes.

We may summarize this as follows. The structure of time itself is necessary and sufficient for the justification of the Second Law. Of course here it is also meant that additionally the known and here utilized but not yet fully analyzed premises of physics hold, e.g., the applicability of the concept of an object and presumably also the reversibility of the basic laws. Our assertion will thus be made more precise through our subsequent discussion. At any rate "necessary" now means that merely using the parameter of time without taking into account the difference between the factual past and the possible future leads to the wrong result that entropy decreased in the past. And "sufficient" means that the description of the past as bygone future is enough to conclude that entropy has increased in the past as well.

This result, however, needs to be supplemented by a *consistency argument*. We must show that with the Second Law as a universally valid law of nature time *may* indeed have the structure we have described. That this is not trivial we will demonstrate by the example of the concept of a document. This consistency argument may also contribute to the illumination of the ideas presented thus far. Physicists in particular are often inclined to consider the structure of time presented here to be "merely subjective". This often leads them to search for an "objective reason" for the "distinction of one direction of time." The phenomena which are specified as such objective reasons are essentially just those which we will present in the following two sections as conditions for the consistency of our arguments. Therefore only after they have been discussed may we talk about the possibility and desirability of this "objective justification."

## 7.3 Documents

The consistency problem mentioned at the end of the previous paragraph presents itself as follows. The Second Law can be derived from the structure of time. It contains among others the facticity of the past, as opposed to the possibility of the future. About the facts of the past we can be certain in each individual case only because there are at present documents about these facts. But there are currently no documents about future events. Without this asymmetry of the concept of a document in regards to temporal modes, our description of the difference between the two non-present tenses by means of the concepts "factual" and "possible" would be without any possibility of verification; without them the concept of "experience" would have no sensible meaning. Now the existence of documents for the past but not for the future appears to be a physical fact for which one will look for a reason in the laws of physics. It is reasonable and, as we will show later, with an appropriate interpretation also correct to look for precisely this reason in the validity of the Second Law. Hereby our argumentation seems to be caught in a circle. We justified the Second Law in terms of the structure of time and the structure of time via the Second Law. We will try to prove that this is not an erroneous circle, but merely evidence for the consistency of our assumptions. Perhaps we can draw upon an admittedly simplistic example from logic. If two statements $A$ and $B$ are logically equivalent, then one can deduce $A$ from $B$ and $B$ from $A$. This statement is not a *circulus vitiosus*. This would only be the case if $A$ claims to have derived $B$ from it, from $B$ again $A$, and one then alleged to have proved the truth of $A$; the only thing that has been proved is that $A$ and $B$ are either both true or both false. In the present case, however, the structure of time and the Second Law are not logically equivalent. We cannot assert such an equivalence, inasmuch as we have not formulated the structure of time in a logically precise manner. This would indeed be quite difficult if our conjecture that the structure of time is itself a prerequisite for logic were true. Actually we wish to make clear that the structure of time cannot be derived from the Second Law; it is a necessary but not sufficient condition for the structure of time as we describe it.

We now go step by step through the line of the reasoning just sketched.

First, is it true that there are documents of the past but not of the future? Or, to put it differently, what do we mean by such a sentence? In the eighteenth century one occasionally used the following as an example for a causal deduction. An explorer arrives in the South Seas on an island uninhabited by humans and finds the figure of the Pythagorean theorem sketched on the beach. He will reliably conclude that a short time before, say at most a month, humans (presumably Europeans) had been on this island. I now ask whether he can, in the same vein, conclude from this figure that in an equally near future humans (even Europeans) will be on this island? Obviously not. Cause and effect are not interchangeable. The presence of geometrically versed humans can become the cause for a figure in the sand but the figure in the sand is not cause for the presence of humans versed in geometry.

It is useful here to discuss a possible objection. The figure in the sand might after all become the cause of the presence of geometrically adept humans. For example, the explorer who discovers the figure will presume that the persons who drew the figure might return and he could, in order to meet them, remain longer than planned on the island. Nevertheless, one will say that the figure in the sand is a document about the past presence of humans but not a document about their future presence. With documents we only mean the effects of what they document. This asymmetry in the terminology is related to a difference in structure.

## 7 Irreversibility and entropy

One notices this difference already in the reliability of the conclusions about the past and the future. If the figure in the sand is the reason for a longer stay of the explorer, whose first landfall occurred without being caused by the figure (which was then still unknown to him), then one will find this causal connection rather loose. If one knows that there is this figure on the island, one will predict only with a small probability that afterwards there will be humans on it. Conversely, however, one concludes almost with certainty that in the near past humans were here: "It was the devil's work if no humans were here." Here the devil obviously represents only the possibility that the laws of nature known to us had been breached in a highly visible way or must be amended. Excluding from the discussion the ever-present possibility of an erroneous deduction, as one always does when one tests the consequences of a theory (and not the possible flaws), we can say that documents are often suitable to give us *certainty* about a past fact.

This certainty obtained from documents is initially sharply to be distinguished from the certainty we obtain from reversible laws of nature. The latter actually permit the same kind of deductions about the past and the future. From the present positions and velocities of the stars one can calculate with the same accuracy a solar eclipse which must have happened 2500 years ago (at the time of Thales) as one which must occur in 2500 years. In both cases the reliability of the deduction depends on three things, namely the accuracy to which we know the present data; the accuracy of the assumed laws of nature; and the accuracy to which one can treat the system of bodies involved as a closed system, i.e., to which one can disregard external perturbations not considered in the calculation.

Assuming the first two conditions to be fully satisfied, there always remains uncertainty about the third. A dark object might, in 2400 years, cross paths with the solar system, and by its gravitational interaction deflect the moon by a few arcminutes from its path. The solar eclipse would then not occur at the calculated time. At present we cannot know whether this will happen. Whether the same happened 2400 years ago we know just as little from our present astronomical data. But if we have a report of historians that Thales had predicted the solar eclipse of the year 585 and this eclipse coincides with our calculations, then we immediately know quite reliably that since 585 BC there has been no perturbation of the moon's orbit of that magnitude. This much certainty we acquire from documents.

Nor is the difference between the certainty from documents and the certainty from laws of nature restricted to reversible laws of nature. If I apply an irreversible law, like that of heat conduction, to a closed system, then I have at any rate a comparable reliability of the deductions about the near past and the nearer future. As long as the coffee pot stood undisturbed on the dining table, it must have cooled down; as long as it stands there undisturbed, it will further cool down. Now here precisely the irreversible character of the process implies time limits for a meaningful epiprediction. If the coffee has reached room temperature, still remaining undisturbed, then nothing new will happen; the detailed prediction depends then only on the presently neglected influence of the environment. On the other hand, the assumption of the pot standing there undisturbed cannot be extrapolated without limits into the past. At every time the undisturbed coffee must be hotter than at the subsequent time, and one can specify the point in time before which it must be above the boiling point, thus must be not liquid coffee at all if it had already been then and from then on always been undisturbed. I.e. the coffee pot standing there proves in turn to be a document, exactly due to the irreversibility of the process, from which one can conclude with certainty that the pot with hot coffee could have come to this place at the earliest before a certain time that can be specified (e.g., a quarter of an hour), or perhaps had been robbed of a thermally insulating cover.

Finally let us remark that by no means a thing or a state must be human made to be a document. The bone of an ichthyosaurus in the Swabian Alps is a document that here (as we can estimate today, about 100 million years ago) lived an ichthyosaurus, but certainly not that after a similarly long time will an ichthyosaurus live here. The lead content in a uranium-bearing mineral furnishes documentary proof that this mineral has been lying in the ground chemically undissolved for, e.g., $2 \cdot 10^9$ years, but certainly not that it will continue to lie there for a similar length of t ime. The light from a new star in the Andromeda nebula proves to us that $2 \cdot 10^6$ years ago, a star started to shine there, but obviously not that it will still be shining in $2 \cdot 10^6$ years. The phenomenon of a document is thus an objective, physically demonstrable fact existing completely independent of human memory; the latter we will address below (see p. 200).

Thereby we come to the second question: what physical laws lie behind the fact that there are documents of the past, but not of the future? We suspect at once that reversible laws cannot be responsible for something like this. This leads us to suspect the Second Law. But we wish to proceed more slowly, and first formulate the phenomenon itself in more abstract terms.

The present result we can express by saying that a document, like the Pythagorean figure in the sand of a South Sea island, offers us much information about the past and little information about the future. The concept of a document can thus be associated with the concept of *information*. Verbally we can introduce the concept of information by saying that information contained in a statement (an event) $A$ is the suitably measured increase in the probability with which certain statements $B, C \ldots$ can be made if $A$ is known, compared to the case that $A$ is not known. We restrict attention to one single statement $B$ and introduce the logarithm base 2 of the ratio of both probabilities as what has become now usual measure of information; the information of $A$ relative to $B$ would then be

$$H_A(B) = \log_2 \frac{w_A(B)}{w(B)}. \tag{7.27}$$

Here $w(B)$ is the probability of $B$ to begin with, $w_A(B)$ is the conditional probability for $B$ under the assumption of $A$; one could also write $w(B)$ as the conditional probability $w_{A\cup\overline{A}}(B)$. To specify the full content of information of $A$, one must sum over all possible statements $B$:

$$H_A = \sum_B H_A(B). \tag{7.28}$$

For all those $B$ which are independent of $A$, $w_A(B) = w(B)$, thus the logarithm vanishes, and they do not contribute to the sum. Here, however, we are interested in the relative information $H_A(B)$, namely what a document $A$ permits us to deduce about a past event $B$ or a future event $C$. Our present finding says

$$H_A(B) \gg H_A(C) \quad \begin{array}{l}(A \text{ present}) \\ (B \text{ past}) \\ (C \text{ future})\end{array} \tag{7.29}$$

So that we can compare this finding with the Second Law, let us now assume $B, A$, and $C$ to be formally possible macrostates of the same object, which remains isolated from its environment during the entire time interval in question. For the unconditional probabilities $w(B)$ and $w(C)$, and equally also for $w(A)$, one will choose the thermodynamic probabilities according to (7.15), precisely because they are meant as probabilities without any further information. Thus we thereby assign also a probability to past and present events as well as to future ones. This is meant in the sense explained in Sect. 3.1; it is the probability of finding, in investigations, the particular event as factually having happened.

First one can see that

$$w(A) < 1, \tag{7.30}$$

if fact, if $A$ is to be a good document, it must be much less than 1. Namely suppose $w(A) = 1$; then the conditional probabilities $w_A(B)$ and $w_A(C)$ must equal the unconditional ones. The occurrence of an event that is certain beforehand does not increase our knowledge. Now there is the question of how the conditional probabilities are to be computed subject to the requirement (7.30). If one knows special causal relations between $B$ and $A$ or $A$ and $C$, then one will be able to draw conclusions from them. These are then implications from laws of nature of the usual kind, which on the average will teach us as much about the past as well as the future event. They are not a topic of our present theory; deductions from documents we have just distinguished from deductions from (special) laws of nature. Hence we assume that there are no special laws of nature known to connect $A, B$, and $C$. On the other hand, let the one general law of nature we are just dealing with be known, the Second Law. It says that the thermodynamic probability of a state, as long as no equilibrium is reached, is on the statistical average constantly increasing. For the cases which are of interest to us one can disregard fluctuation phenomena. Thus we can say that the thermodynamic probability of the states were always increasing from $B$ over $A$ to $C$; due to (7.30) the case of the equilibrium can only happen after $A$. If we therefore know $A$, then $B$ must be one of the states with

$$w(B) < w(A). \tag{7.31}$$

Without any special causal information we can only conclude that $B$ is one of the states which satisfy condition (7.31). But these are only a fraction of all formally possible states, and as the sum of the thermodynamic probabilities extended over *all* formally possible states is 1, it is certainly still admissible for the moment to take

$$\sum_B w(B) \ll 1, \tag{7.32}$$

and in fact is generally very small compared to 1. On the other hand, for all admissible $B$, the sum of their probabilities, subject to the condition $A$, clearly must be 1 (after all, one of them must have occurred):

$$\sum_B w_A(B) = 1. \tag{7.33}$$

Hence we will also have

$$\sum_B H_A(B) > 1; \tag{7.34}$$

$A$ contains much information about $B$. For the future event $C$ there follows from

$$w(A) < w(C) \tag{7.35}$$

only a much weaker constraint, as according to the law of large numbers the variance of the distribution will be very small, i.e.,

$$1 - \sum_C w(C) \ll 1. \tag{7.36}$$

This implies

$$\sum_C H_A(C) \approx 0. \tag{7.37}$$

In other words, if the state $A$ is significantly out of equilibrium, then almost all formally possible microstates lie in macrostates which are closer to equilibrium than $A$. Therefore the statement that $B$ is farther from equilibrium contains much information, while the statement that $C$ is closer to equilibrium than $A$ is practically meaningless.

Thus, we have deduced from the Second Law that from documents one can only draw conclusions about the past. Of course we have imposed restrictive conditions which are not satisfied in general: isolated system, lack of any special causal knowledge. But the deviations from these conditions do on the average not single out any temporal modality. An external influence on the system could make the inference from documents more dubious, the knowledge of special causal connections can make them more precise. But in the end, the validity of our argument is not affected, on the average over all cases. Further details can then only be supplied if the nature of the external influences and the causal chains is specified. It may suffice to point out that in all previously discussed examples a certain absence of perturbations was guaranteed and a certain causal information was present, but that the conclusion about the past was based on demonstrable irreversible processes. The drawing in the sand, the petrified bone, the uranium mineral containing lead are all relatively unchanging objects that could only be created by means of irreversible processes. As a counterexample, compare the attempt to sketch the Pythagorean figure on an elastic surface like rubber or into the sea. Light emerging from the nova in the Andromeda galaxy, while not a fixed object, is instead a process with entropy still steadily increasing; the reversal would be a concentric spherical wave, contracting into the star, and "such things do not happen."

If one believes that corresponding to human thought there is a material structure obeying the laws of physics, then the fact that we have a *memory* for the past but not for the future fits without effort into our arguments. *Storage* in the sense of cybernetics is a reservoir for documents and, if they are to be material, they must contain information about the past and not about the future. There the assumption of the material nature or the validity of all physical laws for the structure assigned or attributed to human thought is not necessary at all; only those assumptions need be made which suffice for the derivation of the Second Law. For this it is only necessary that there actually exist in thought alternatives bridging time which permit a meaningful arrangement of the formally possible states into classes according to the distinction of micro- and macrostates.

In this way, one has demonstrated the required consistency proof. The structure of time implies the Second Law, and this implies the distinctiveness of the past in terms of the existence of documents. It should, however, be easily seen that the Second Law, if one introduces there only parameter-time, does not imply the full structure of time. From it there does not follow at all the distinction of always one point in time as present, nor that the past has happened or the future not yet. Just as little does it contain any distinction at all between facticity and possibility; the thermodynamic probability is in the parameter-time merely a measure of the size of certain classes of microstates. Only the meaning of probability we described in Chap. 3 brings into the theory our antecedent understanding of the structure of time; it does not belong to the mathematical formalism but to semantics.

## 7.4 Cosmology and the theory of relativity

Physicists who consider the structure of time we have placed uppermost to be "merely subjective" (whatever the meaning of the words "merely" and "subjective") see in general two ways out or two objections to the analysis given here. On the one hand they consider it possible to reduce the irreversibility of events to cosmological assumptions; on the other hand they think that the theory of relativity had abolished the distinction between space and time and thus pulled the rug out from under theories like the one presented here. We must show that on the one hand, these loopholes do not exist and that these objections are wrong, and on the other hand outline at least the main features of how to talk about cosmology and the spacetime continuum consistent with the structure of space and time.

We consider two typical attempts to explain irreversibility cosmologically. The first is due to Boltzmann, the second is widespread among present-day physicists. The first we call the *fluctuation hypothesis*, the second the *initial hypothesis*. I would presume that no other attempt at such an explanation can be found which could not be discussed by means of a suitable combination of arguments that we must advance against these two hypotheses.

The *fluctuation hypothesis*, introduced by Boltzmann in the final chapter of his *Lectures on Gas Theory*, says that the universe is enormous in space and time and, on the average, over sufficiently large regions in space and time, it is everywhere and always in thermodynamic equilibrium. To equilibrium belong fluctuations, and now and then there occur very large fluctuations. Separated from one another by billions of light years and eons, here and there, now and then, there occur fluctuations which are so much spread out in space and time that they must be interpreted as the appearance of an entire universe. The universe we live in is such a fluctuation. Although such an enormous fluctuation is extremely improbable, in the infinite reaches of space and time practically it does occur once with certainty (strictly speaking even infinitely often). That we live precisely in it is, however, not improbable at all, as here the question is about the conditional probability that a human, if he is living at all, lives in such a universe. This conditional probability is large as only such a universe offers him the conditions he needs for life. One still could argue that thereby the Second Law is not really explained, as the fluctuation in entropy without doubt will have an extremum value (minimum of entropy) and from it entropy would increase in both directions of time; thereby the probability would be the same that we are living in a phase of decreasing or a phase of increasing entropy, and the validity of the $H$-theorem would at best be a mere coincidence. Here a supporter of the fluctuation hypothesis would object that there is no objectively distinguished direction of time, rather humans would measure time in each of both branches "in the direction of increasing entropy" (so Boltzmann).

The *initial hypothesis*, on the other hand, fits the notion closer to present-day astronomy that the universe is a one-time occurrence, which had its origin at an approximately specifiable time. This assumption makes it possible to carry over the arguments of the previous paragraph from human experiments to the entire universe. A human experiment usually starts with a specially chosen state of the object which, just because it is specially chosen, in general will have characteristics that imply a smaller entropy than that of equilibrium. Thereby it is then clear, as discussed above, that its entropy *after* this initial state will probably increase. In the same manner every cosmological model which represents the universe as a unique, one-time occurrence with a beginning in time implies the specification of certain attributes of the initial state. These attributes, just because they are specified (e.g., a homogeneous distribution of hydrogen and the like), already imply a value of the entropy different from equilibrium. From this it follows already that the entropy of the universe (or, if one wants to avoid this concept, the entropy of every sufficiently isolated finite part of the universe) will increase from the initial state until one has reached equilibrium. Now one can further conclude like Boltzmann that humans in the universe would measure time "in the direction of increasing entropy." Thereby one also avoids the allegation of a *petitio principii* of having arbitrarily put the simple state at the temporal beginning of the universe instead of choosing the equally possible assumption for it to be at the end of time, whereby the increase in entropy and direction of time would apparently point in the opposite direction. Defining the direction of time as the direction of increasing entropy one can incorporate the chosen model of the universe into a speculative larger model, having, e.g., periodicity in time or mirror symmetry; thereby the two hypotheses we discuss would converge to just one.

We wish to show now that the fluctuation hypothesis is in all probability demonstrably false and that the initial hypothesis does not lead beyond the consistency arguments of the previous paragraph.

We state more precisely the *goal of the proof* of the fluctuation hypothesis. It starts with the assumption, fundamental indeed for Boltzmann's statistical explanation of the Second Law, that the measure he calls the "thermodynamic probability" of a macrostate can be used to estimate the probability for the occurrence of this macrostate. Now the state of the universe in which we find ourselves is thermodynamically extremely improbable; hence, by assumption, it is also extremely unlikely that it occurred. If something that improbable actually happened, and it is even the basic fact of our entire existence, then Boltzmann asks himself *with good reason* whether he can ever trust his assumption at all. Hence he must show that estimating the probability of the present state of the universe from a thermodynamic probability is a different problem from the corresponding estimate for individual physical systems in the universe. He hopes to show this by the remark that for individual physical systems the existence of human observers may already be taken for granted but not for the universe as a whole. One cannot reasonably ask: "How probable would a human being find the universe such as we know it?" without asking about the conditions that humans can exist at all. If the universe such as we find it adheres to these conditions then the *conditional* probability that a universe like ours exists is 1 *if* humans who can ask about its probability are present.

Now, however, according to Boltzmann's own premise, it is not clear at all why an entire universe like ours should be the condition for the existence of one human being. The strict proof to the contrary cannot be given for the sole reason that one cannot strictly argue for or against Boltzmann's assumption without causal knowledge of physiology, which of course we do not possess in this form. We restrict attention solely to what Boltzmann uses, namely the estimate of thermodynamic probabilities. We can now easily specify two conceivable states in the hypothetical infinite universe of Boltzmann which certainly have significantly higher entropy than the universe known to us in its fluctuation minimum, as postulated by Boltzmann. The first one is a human that came into being through fluctuation, all by himself without any environment, in an environment approximately in equilibrium whose dimensions we choose such that it has the same number of formally possible microstates as the part of the universe we know about.[3] Of course to our mind it is absurd to assume a human appearing all by itself as a result of a fluctuation. But according to Boltzmann's own initial assumption it is much more likely than that an entire universe such as our own sprang into being by itself through a fluctuation. If Boltzmann drops this assumption of his, then his problem disappears, but apparently so also does his entire statistical interpretation of the Second Law. If he retains it, then our objection shows that he has not solved his problem. A second example is our present universe, as we know it, with the assumption that the minimum of entropy occurs precisely now, or at any rate shortly before the childhood of the oldest living person today. Precisely from the Second Law, the part of the universe known to us now has, without doubt, a higher entropy than 1000 years ago. According to Boltzmann's own assumption it must then be much more likely that the present universe originated directly from a fluctuation than that there belonged to its prehistory the state which we physicists assume to be its state 1000 years ago. Of course the present universe contains many documents of events 1000 years ago. But (see Sect. 7.3) these documents are only documents about past events, if we can indeed assume the Second Law for the past. According to this assumption one must conclude that a universe containing countless "documents" of events which did not happen is much more likely to arise from a fluctuation than a universe in which initially all these events occur, and thereby create the documents.

As we have remarked, these arguments are not fully rigorous as they disregard the causal connections of special events with other special events. But they do not disregard them to a greater degree than *any* statistical justifica-

---

[3] This example is due to Landau.

tion of the increase in entropy. If it is, e.g., mechanically possible at all that the wind erases the figure in the sand, then for reversible laws of mechanics it is then *certainly* also possible that it engraves it into a previously smooth patch of sand; one must merely reverse all atomic motions of wind and sand precisely. If it is also possible that an isolated human, in an environment in thermal equilibrium, dissolves in time through death, decay, and dispersal of the remains, then for reversible basic laws it is *therefore* also possible that he arises from the reverse process. The same holds for the second counterexample. The absurdity of our counterexamples is thereby not diminished; it only reveals the absurdity of Boltzmann's opinion that a universe like ours could arise through fluctuations. One avoids this absurdity and still saves the statistical interpretation of the Second Law if one restricts Boltzmann's assumption about the connection between "thermodynamic probability" and probability in the usual sense to the respective future, as we do here.

This restriction now appears to be furnished by the *initial hypothesis*. It seems to imply the growth of entropy in the direction of time which leads from the origin to the equilibrium state. It implies, as in Sect. 7.3, that a present state of an object that deviates from the maximum value of the entropy allows strong conclusions about the past, but only weak conclusions about the future. It implies that the respective past can be considered factual, but not the future, and this seems to be the structure of time we are using. As already remarked at the end of Sect. 7.3, this, however, does not contain the distinction of one point in time as present; the "flow" of time is one component of the structure of time which in any case remains unexplained. In this sense the initial hypothesis remains arguably only a sharpening of our consistency arguments through specialization of the cosmological assumptions. But in addition the structure of time has also been used in the elements of the initial hypothesis which we just referred to without any critique. It is important to recognize this, as it is related to classically unsolved problems of statistical mechanics.

According to classical statistical mechanics the microstate is the actual state of an object. The macrostate is a class of microstates; its specification is thus an incomplete characterization of the actual state. What is actually known about an object is not even the macrostate in the sense considered thus far but the Gibbs *canonical ensemble* to which the object belongs. Accurately one can only observe states which, to a sufficient approximation, are in equilibrium. In just this approximation they have a definite temperature but not definite values of other quantities like, e.g., energy. For canonical ensembles one can then prove the laws of phenomenological thermodynamics; there the growth of entropy is probable but not certain. One can legitimately speak in this way if one chooses the path described in this book. However, according to the initial hypothesis the use of the concept of probability is to be eliminated from the foundations of the theory and to be justified only secondarily, as approximation. Thus we must ask how events present themselves if initially one completely avoids the concept of probability.

If one does this, i.e., if one only speaks about the causal development of the microstate, then one loses all the concepts of the theory by which one wanted to derive the "distinction of the direction of time." The reversible mechanics of atoms does not really distinguish a direction of time. Assuming the initial state of the universe to be uniquely characterized as a definite microstate, this only implies a unique microstate for every later point of time, thus no decrease in information at all with time. According to the laws of mechanics, the same sequence of microstates could be traversed in the reverse order (Poincaré's reversibility objection). Furthermore, it can be a periodic or almost periodic function of time (Poincaré's recurrence objection), and it is difficult or impossible to prove that the time-average of a physical quantity over such a course is equal to the (ensemble) average assumed in statistical mechanics (ergodic problem). All these half or completely unsolved problems of statistical mechanics do not even occur if one introduces the concept of probability in the way we have done. Then the only empirically verifiable meaning of the statistical assertions is one which can only be studied on a large number of systems of the same kind, i.e., the only averages which can be considered at all are ensemble averages, and the reversal and recurrence objections disappear, as demonstrated in Sect. 7.2. Thereby one must allow that probability predictions themselves can be empirically verified only with a certain probability (see Chap. 3). The three problems exist only for a theory with the ambition of further justifying the good empirical accord of the probability calculus by means of arguments from atomic mechanics. Our present objection to this ambition is that it fails immediately due to the fact that a rigorous (classical) theory of atomic mechanics does not contain the concept of probability, hence also cannot justify its success. Naturally one can specify for every microstate the macrostate to which it belongs. But micro-mechanically it simply does not follow that a microstate must be succeeded by another one which belongs to a macrostate of higher entropy. It follows just as little that this happens *in the majority of the cases*, so long as it is not specified how the "number of cases" is to be measured. This can only be accomplished by making an assumption about the a priori probability of the individual case; and this a priori probability only then has the empirically required meaning if it is understood as the probability for the occurrence of an event, i.e., if the concept of probability is already on hand.

Under these circumstances, the only thing the initial hypothesis could attempt to accomplish would be to make comprehensible that humans, if their bodies consist of atoms, "can only measure time in the direction of increasing entropy." One might perhaps say that physiological processes can only take place if their direction is determined thermodynamically. Fluctuations do in fact happen but if they are large they kill the human in whose body they are occurring. Living humans have engrams in the brain which according to the Second Law are documents of the past but not of the future. In this way the Second Law ensures that the human understanding of time, and the law itself, is guaranteed by the initial state of the universe. But (apart from the fact, as repeatedly emphasized, that thereby the distinction of the respective present is not explained) here lies the circle that the initial state of the universe guarantees the Second Law only for those physicists who already have the concept of entropy at their disposal, and who describe the initial state as a macrostate or a Gibbs canonical ensemble. If the initial state is determined microphysically, then this does not imply anything rigorous about the entropy of the later state, and the growth of entropy follows only "with some probability." This is not a *circulus vitiosus* if we understand the initial hypothesis merely to be an affirmation of our approach in the sense of a consistency argument. If it is assumed to *prove* the distinctiveness of one direction of time, then it is a *circulus vitiosus*.

Thus there is no discernible prospect of justifying the structure of time cosmologically, only the possibility of proposing cosmological hypotheses compatible with the structure of time.

Now there appears to be one obvious objection from the theory of relativity. Have not, since Minkowski's famous address,[4] "space by itself, and time by itself" faded "away into mere shadows" such that "only a kind of union of the two" is preserved as an independent reality? From this sentence one can see the disadvantage of a rhetorical gift for scientists. As Minkowski very well knew, the difference between the timelike and the spacelike separation of two events is Lorentz invariant. It has relativistically invariant meaning to say of two events that happen on the same material object which one is earlier and which one is later. Thereby, as long known, the Second Law is compatible with the special theory of relativity. Equally compatible with it is the distinction of a point in time as respective present and with it our entire reasoning up to now.

However, one could formulate the objection more subtly: what primarily appears to many physicists as "only subjective" in the structure of time is precisely this distinction of one point in time as respective present. They could say now that if this distinction were objective, then it must be defined for the entire universe which point in time is the respective "now." But that contradicts Einstein's realization that simultaneity of spatially separated events is only defined relative to a particular reference frame; it would objectively distinguish one reference frame. Also for this the answer is simple nothing in our analysis of the structure of time compels us to extend the concept of the present over large spaces, even the entire universe. This is even indicated by linguistic usage. For instance, if one speaks of a person who is present, one means somebody who is *here now*. That for events happening one after another the concept "here" (same location) depends on the reference frame has been known for a long time; Einstein also recognized that for events happening side by side the concept "now" (simultaneity) is dependent on the reference frame. The concept of relativistic causality, which plays such an important

---

[4] H. Minkowski, Lecture 1908, reprinted in: H. Lorentz, A. Einstein, H. Minkowski: *Das Relativitätsprinzip*, 1958.

role in present quantum field theory, precisely formulates certain necessary conditions for the Lorentz invariance of the structure of time. There one denotes as the future of an event $A$ all those events that can still be acted upon from $A$, as the past for $A$ consists of all events that can act upon $A$. In this concept of the action the difference between factual and possible is tacitly assumed, in the same way as we reduced in Sect. 7.1 the temporal order of two events to the difference between cause and effect. The past for $A$ are events which, if $A$ is present, can be known as facts, the future of $A$ consists of those upon which, if $A$ is the present, one can still exert an influence. Yet Einstein's discovery indeed represents an *extension* of the structure of time which phenomenologically could scarcely have been anticipated. To the three kinds of events or propositions which can be distinguished—the presentic, perfectic, and futuric—he adds a fourth group of events with spacelike separation. In the prerelativistic mode of thinking, one unhesitatingly would have subsumed them under presentic events, the corresponding proposition under presentic propositions. According to the theory of relativity, however, such propositions are not presentic, as their content is presently not phenomenologically demonstrable; they are not perfectic, as their content is not given as fact, and they are not futuric in the full sense, because their possibility signifies only uncertainty, but not being subject to influence. A full logic of temporal propositions must from the outset be so designed as to leave room for this fourth class of propositions; here, however, we do not further pursue this topic (for this see Mittelstaedt 1979).

That it is possible to operate in the Minkowski spacetime continuum like in a four-dimensional space is due to the fact that according to the special theory of relativity one thing is certain for events of all four classes, no matter what else can be known or unknown about them: the formally possible values of their space and time coordinates, relative to arbitrary Lorentz systems. The spacetime continuum as the totality of all formally possible positions and times forms, to quote Einstein, a finished tenement, with events moving in. *That* at the point $x, y, z$ at time $t$ something has happened, is happening, or will happen (for the fourth class our language does not have a separate expression), that is clear from the outset; only *what* has happened there, is happening, or will happen is conditional. That we can know such a thing a priori (before any individual experience) is not self-evident at all. In *Zeit und Wissen* II 4.3 we remark that from our knowledge of the past and the present it does not even follow logically that there must be a future at all; but this is presumed in all of physics. Likewise, from the outset, prerelativistic physics attributed Newtonian space to all events as the totality of their possible positions. The special theory of relativity (which was not created without Hume's influence) was the first to reject the naïveté of anticipating the subsumption of all conceivable events by replacing the old a priori space and time metric by a new one.

The general theory of relativity goes even farther. It shows that even the formally possible measurement results for positions and times, as measured with rulers and clocks, are not fixed a priori but depend on the conditional distribution of matter. A priori it only provides the small-scale topology and the general laws of the metric (Riemannian geometry). The contingency of the fundamental metric tensors implies also a contingency of the large-scale topology, and thereby creates cosmological possibilities which were previously unknown. It also appears to me that the general theory of relativity is only one step on the path of a continuing analysis of the formally possible space and time structure, and in particular that nowadays it is too early for an adequate appraisal of its cosmological models. As these models have been brought into connection with the structure of time on various occasions, we can say in conclusion a few words about them.

In these models irreversible processes are mostly disregarded. For this reason there arise in some of them structures which scarcely appear compatible with the structure of time described here. For example, there can be strictly periodic pulsating models of the universe which might create the impression that the events in the universe could be altogether strictly periodic, thus the Second Law, e.g., would only be restricted to one phase of the pulsation. Stranger still is that in some models, like the one of Gödel, there exist timelike world lines which represent a rapid motion of an object relative to the substratum of the universe and which close in themselves in time. An astronaut taking off on such a world line from the Earth could, in a contiguous flight, return to the starting point in space *and* time and thus travel along the same world line infinitely many times. To us, he would be one who came from space and departed into it again; to himself he would be one who in the future would be the one he had already been. Gödel deduced from this result, mathematically demonstrable in his model, the "ideality" (i.e., unreality) of time.

One can, however, assert in good conscience that all these results are merely the consequence of inadmissible approximations. Without calculation one can qualitatively understand how the result will change if one no longer neglects the irreversibility of processes in the universe. A periodically pulsating model of the universe can be treated formally as a pulsating sphere of gas. The total energy of such a sphere is constant (neglecting the radiation into the exterior space which can be omitted in the model of the universe). Its terms kinetic energy, potential energy of gravitation, gas and radiation pressure vary periodically. If one now introduces irreversible processes like friction, then the process is no longer periodic, and eventually approaches equilibrium. Exactly this will happen with a pulsating universe; no wonder one obtained without friction a result incompatible with the assumption of the existence of an irreversible process, for what one does not put in one cannot get out. It is the same with Gödel's model. If the astronaut is alive or if only a clock is flying with him which registers the flow of time in documents (punch cards, tear-off-calendar), then something irreversible is happening on the flight. Then the second flight cannot be identical with the first flight, the third not with the second, etc. If the astronaut flies, e.g., one million times and then stops,

## 7.4 Cosmology and the theory of relativity

then for us one million astronauts must arrive nearly simultaneously from the universe and depart again. Thereby initially for the astronaut himself the reality of the temporal sequence is established. To us, however, there still remain paradoxes. For example it must be impossible to shoot one of the "middle" astronauts as thereby all "later" astronauts retroactively must disappear; the facticity of the past would otherwise be violated. I would like to conjecture that an exact analysis of the irreversible processes associated with such a flight would demonstrate the impossibility of the entire process.

According to a theorem of Hawking and Ellis (1973) world lines must intersect at a finite time in every model of the universe without a cosmological constant and everywhere positive energy density. The natural interpretation of this result, from our point of view, is that they have intersected in the past, i.e., that the universe has a beginning in time; presumably only this case is compatible with the Second Law. Evidently the problems occurring in a unification of cosmological and thermodynamic questions are still far from solution.

# 8

# Information and evolution

## 8.1 The systematic place of the chapter

If one defines physics as one of several natural sciences, then this chapter does not belong to the reconstruction of physics. Information is a reflexive concept, pertaining to all sciences; evolution is a basic phenomenon of organic life, thus conceptually appears to belong to biology. Our book, however, builds up theoretical physics as the core theory of all natural sciences, at about the same level of generality as the concept of information. De facto we use this concept in the reconstruction of statistical thermodynamics as well as quantum theory. In any case, examination of the meaning of the concept of information belongs to the interpretation of the physics thus reconstructed. Furthermore, our reconstruction claims to justify physics also as the fundamental theory for biology. Therefore the relationship of information and evolution offer one of the most important tests of the feasibility of this claim.

The present chapter does not belong to the actual reconstruction of physics but rather to the interpretation of physics. It is connected to Chaps. 3 and 7 via probability and entropy, and prepares the conceptual material for the interpretation of nature as a stream of information in Chap. 10. The titles of this and the previous chapter are, in a way, reversed. Evolution and irreversibility are two basic phenomena of nature. Entropy and information are two concepts by means of which we attempt to quantitatively describe and ultimately explain these phenomena. In the previous chapter we started with the phenomenon of irreversibility and introduced the concept of entropy for its description and explanation. There entropy historically was at first a descriptive basic concept of phenomenological thermodynamics; via its probability-theoretic interpretation it became the means for an explanation of irreversibility. Phenomenologically, irreversibility is described as increase of entropy; statistically this increase turns out to be the overwhelmingly probable phenomenon. Thus far, we reported here merely the classical theory of the late nineteenth and early twentieth century. What is new in our presentation is merely the clarification of the meaning of the then naïvely-correctly employed concept of probability as expression of the openness of the future, or, as we can say, as "futuric modality."

In the present chapter, on the other hand, we begin with the introduction of the essential concept: information. It was created in the mid-twentieth century on the basis of the theory of probability (Shannon and Weaver 1949). We reinterpret it now as a temporal concept. Then we proceed to the basic phenomenon of evolution. We describe evolution as growth of information and show again that this growth is the overwhelmingly probable phenomenon. In contrast to the statistical explanation of the Second Law, this statistical explanation of the tendency of evolution cannot yet be regarded as acknowledged conviction of present science. In fact, since Kant-Laplace and Darwin one uses probability arguments "naïvely-correctly" in concrete models of evolutionary processes. But the attempt to trace back the success of these models to an abstract rule similar to the increase in entropy has led to difficulties which, at least in the general awareness of scientists, leaves an unsettled feeling . The claim of the arguments presented here is not to increase the number of successful models by another one but rather to completely elucidate the remaining abstract conceptual problems.

The starting point is the identity of the definitions of entropy and syntactic information. The ambiguity which exists in the usual parlance about the sign of information can easily be resolved by means of a temporal interpretation: entropy is potential information, negative entropy is actual information. One can then show that evolution can be explained as an increase in a suitably defined potential information, hence indeed as increase in entropy. The much-discussed difficulty of reconciling increasing entropy with evolution turns out to be merely a consequence of imprecisely defined concepts. The general interpretation of entropy as a measure of disorder is no more than linguistic and logical sloppiness.

The main idea of this discussion stems from the end of the sixth lecture on *Geschichte der Natur* (Weizsäcker 1948), and its execution from the essay *Evolution und Entropiewachstum* (Weizsäcker 1972). To this we add further investigations about the concept of information. Section 8.6 sketches the definition of a pragmatic concept of information, due to E. and C.v. Weizsäcker, which might be fundamental to biological and general system-theoretical considerations. Section 8.7 suggests the beginning of a philosophical *Kreisgang*[1] from logic through physics and the theory of evolution to the biological prerequisites for logic.

## 8.2 What is information?

What do we mean when we ask "What is information?" What answer can we hope to get?

---

[1] Cf. p. XXII, fn. 14.

## 8.2 What is information?

Information is one of the fundamental concepts of modern science. Formally we ask for an explicit definition of this concept, factually about the essence of what it represents. Giving a precise definition of a fundamental concept cannot be easy. It would thereby have to be reduced to even more fundamental concepts; this query ends in undefinables. For instance, if we ask: "What is matter?" then the initial answer can only be "Tell me to what philosophy you subscribe and I will tell you how you must define matter." To such questions we return in the third part of this book.

It might appear, however, that the problem is simpler in the case of information. This concept was introduced into science by explicit definition only a few decades ago. We will discuss this definition more closely in Sect. 4 of this chapter. Here we simply recall its main idea. It explains the concept of information in terms of the concept of probability. One can denote the information content of an event as a quantitative (logarithmic) measure of the improbability of its occurrence. This definition, however, leads to two further questions.

1. What is probability in the sense of this definition?
2. Does this definition correspond to the use we make of the concept of information in practice?

*To 1.:* We have dealt with the question "What is probability?" in Chap. 3. The philosophical debate teaches us to distinguish at least three essentially different interpretations of probability: the logical, the empirical, and the subjective. We have defined probability, starting with temporal logic, as the prediction of relative frequency. We have postponed the question of how this definition is related to the three cited interpretations to a later book. Now we can only say that if probability measures the relative frequency of a type of event, then high information content of this type means that it rarely occurs. If one encounters it, one experiences something which is not self-evident, precisely "much information." In this way we used in Sect. 7.3 the concept of information to explain the concept of document.

*To 2.:* As the concept of probability is philosophically unsettled, there naturally also originated an inconclusive philosophical debate about the concept of information. Initially, it was relatively easy to say what information is *not*. An amount of information is evidently neither an amount of matter nor an amount of energy; otherwise tiny chips in a computer could not be the carriers of very large amounts of information. Nor is information simply all that we subjectively know. The chips in a computer, DNA in a chromosome contain their information objectively, independently of what a human being knows about it. In the spirit of the Cartesian dualism prevalent in natural science one asked whether information is matter or mind, and obtained the appropriate answer: neither. Some authors then referred to it as a "third kind of reality."

We will choose the positive answer: information is a quantitative measure of *form* (*Gestalt*). Form is neither matter nor mind, but a property of material objects, and we can know about it in our mind. We can say that matter *has* form, mind *knows* form. What this brief formula means in practice we will discuss in this chapter by going through a set of problems. We do not demand of the reader to immediately accept our explanation of information as an amount of form in the sense of a "philosophical truth" but appeal initially only to his appreciation of a convenient expression. The more decisions can be made about an object, the more "form" one can recognize in it, in a general, not necessarily spatial meaning of the word. The amount of form is, as just said, a property of the object and knowable to us. In Chap. 10 we return to what form means philosophically.

In the present chapter we must clarify, among other things, the relationship between information and four other concepts: entropy, meaning, utility, and evolution.

Section 4 discusses the relationship to *entropy*. The concept later called "information" was originally defined by Shannon, who called it "entropy." We will justify this identification, and elucidate in particular the relationship between the sign of entropy and information. Positive entropy is *potential* (or virtual) information. The entropy of a macrostate measures the amount of Gestalt which somebody must know who wanted to specify the corresponding microstate. Whether one wants to define entropy as a measure of the amount of form or disorder is thus merely a distinction of different degrees of knowledge. The amount of written and printed paper stacked on my desk is, if I know what is on the paper, an extraordinary amount of Gestalt; if I (or the maid) don't know it then it is a mess.

The relationship to *meaning* leads to a much-discussed question in applications, in particular biology. Let us begin with the historical origin of the concept of information, the theory of communication. A certain telegram, e.g., in English contains an amount of information which is determined by the statistical relative frequency of all letters occurring in it in the written English language. By means of these letters the telegram transmits a message from the sender to the receiver, such as "coming tomorrow." It has, as we say, a meaning. Shuffling the same letters differently, to produce, say, "cgimmn oooorrtw," the information computed according to Shannon, which only depends on the probabilities of the letters, is the same but the meaning is lost. The *syntactic* information has remained, the *semantic* information is gone. The semantic information, however, is obviously the communicative purpose of the information. Can we define it?

This leads to the relationship of information to utility and *pragmatic information* (Sect. 8.6). We rate information by the effect it has. Semantic information is only measurable as pragmatic information.

The biological application of the concept of information must be in this light, in particular its relationship to *evolution* (Sect. 8.5).

## 8.3 What is evolution?

*Evolution* mainly refers to the emergence of the multiplicity of forms of organic life in the course of Earth's history. The emergence of a multiplicity of forms is of course not restricted to the subject matter of biology. On the one hand there is a rich spontaneous emergence of Gestalt in the inorganic realm, incorporated nowadays under the general name of synergetics (Haken 1978). On the other, human culture as well is always devising new forms. Evolution as process involves the entire reality we know. Thus it also demands a comprehensive explanation.

The discovery of evolution and Darwin's causal, statistical approach to its elucidation led in the nineteenth century to the greatest shock of the traditional world view through science. The most common early experience teaches us the usefulness of the forms and behavior of living creatures. The word "organism" denotes just that: "organon" means tool. Aristotelian biology described this usefulness with empirical fidelity. Christian theology saw in it the work of a designing God. Darwin, however, maintained that all these forms and modes of behavior originated "by themselves," through chance and selection. Nowadays, as the victory of the theory of evolution has been decided for a long time, one avoids in the description of these forms and behaviors the word "purpose" which suggest the idea of a planning mind. One calls them "functional." Objectively, however, this is exactly the same: the forms facilitate the preservation and subsequent development of life.

A certain uneasiness about the possible causal explanation of functional forms, however, remained for some scientists. We discuss this uneasiness in the popular confrontation of irreversibility and evolution. It is usual, as mentioned above, to interpret entropy as a measure of disorder, and thereby thermodynamic irreversibility as an increase in disorder. Evolution, however, is understood an increase in possible forms, and in that sense as order. Under these premises evolution must be perceived as a process proceeding against thermodynamic irreversibility. Here, exactly the opposite thesis is to be presented: Under suitable circumstances, an increase in entropy is *identical* to the growth of forms; evolution is a special case of the irreversibility of events.

Section 8.5 illustrates this for the simplest concept, that of syntactic information. Section 8.6 takes up again the question from the point of view of pragmatic information. Finally, Sect. 8.7 classifies human knowledge in the context of evolution from the standpoint of accumulation of information.

## 8.4 Information and probability

For the time being, we adopt the usual definition of information in terms of probability. We are given a $K$-fold experimental alternative, i.e., $K$ mutually exclusive possible events $x_k$ ($k = 1, 2 \ldots K$). In the case of a decision for the alternative with probability $p_k$, we expect $x_k$ to occur. The "one-time information" $I_k$ measures the "newsworthiness" of the event $x_k$ if a decision produces just this value $x_k$. An event that has occurred is the less newsworthy, the more likely it was before; if it occurred with certainty, one will consider its newsworthiness to be zero. $I_k$ should therefore be a monotonically decreasing function of $p_k$. One usually demands the newsworthiness of a combined event consisting of two independent events to be the sum of their news values. This leads to the ansatz $I_k = -\log p_k$. The usual definition assigns to an event of probability 1/2 the newsworthiness 1 (one bit). For this one must set

$$I_k = -\log_2 p_k \tag{8.1}$$

One is interested now in the expectation value of $I_k$, i.e., the average value, over many events, of the newsworthiness expected of a one-time decision of the alternative. It is

$$H = \sum_k p_k I_k = -\sum_k p_k \log_2 p_k. \tag{8.2}$$

This is the quantity introduced by Shannon as the measure of information, justifiably called entropy.

First a word about the sign of this quantity. One has correlated information with knowledge, entropy with ignorance and consequently called information negentropy. But this is a conceptual or verbal vagueness. Shannon's $H$, including its sign, is equal to entropy. It is the expectation value of an event that has not yet occurred, thus a measure of what I could know, but right now do not know. It is a measure of potential knowledge, and in that sense a measure of a limited kind of ignorance. Exactly the same also holds for thermodynamic entropy. It is a measure of the number of microstates in a macrostate. Thus it measures how much someone who knows the macrostate could know if he also learned about the microstate. For a constant total number of microstates of a system, the increase in entropy implies indeed an increase in that amount of knowledge which someone who only knows the macrostate does not have but which he in principle could gain from the determination of the respective microstate.

The transition from "information" to entropy follows if one sets, for a given macrostate, the probabilities of all its compatible microstates equal to one another, and those of all other microstates equal to zero.

As a model we consider Ehrenfest's game of urns, Sect. 7.2. It follows that

$$H = K \cdot \frac{1}{K} \log_2 K = \log_2 K. \tag{8.3}$$

The entropy of a macrostate is the logarithm of the number $K$ of possible microstates contained in it. This we call the *potential information* contained in the microstate. It is largest for the state of thermodynamic equilibrium. Suppose it contains $K_{\max}$ microstates. In it the *actual information* about the microstates is smallest. If we arbitrarily set its value to zero, then for every other macrostate the actual information would be

$$I = \log_2 K_{\max} - H = \log_2(K_{\max}/K). \tag{8.4}$$

Thus, up to the term $\log_2 K_{\max}$, the actual information is equal to the negative entropy. It is information about the microstate which one *already possesses through* the knowledge of the macrostate.

We have defined thermodynamic entropy in terms of its relationship to two *classes of states*, macro- and microstates. In general logical terminology there corresponds one *concept* to one *class*. Entropy is thus defined as a *relationship between two concepts*. The same also holds in general for information. Here the two concepts denote *classes of events*. To the macrostate corresponds the event "decision of the $K$-fold alternative" $x_k$ ($k = 1 \ldots k$); to the microstate the event $x_k$. Instead of two concepts we will in the following also speak of two *semantic levels*.

In this way we have assumed that one already knows how the concepts "microstate" and "macrostate" are defined in the concrete case. For Ehrenfest's game we have defined explicitly both types of states ("how many balls in urn 1" and "what individual balls in urn 1"). For channels of communication (for instance the receiver of telegrams) the macrostate might be "this device will soon print out a letter of the Latin alphabet," the microstate "the device prints the letter X." In classical statistical mechanics of atoms the macrostate is defined by specifying the thermodynamic variables of state of a system (pressure, volume, temperature), and the microstate by specifying the phase point (position and momentum) of each atom of the system. These examples explain what is meant if we say that the macro- and microstates respectively are specified through a concept or a "semantic level."

It follows then that the measure of information is defined relative to two semantic levels, the two underlying macro- and microstates. An "absolute" concept of information has no meaning; information exists only "under one concept," more accurately, "relative on two semantic levels." For instance it is not absolutely defined how large the information content of a set of chromosomes of drosophila is. For a molecular geneticist the "set of chromosomes" would be a macrostate, the sequence of letter of the DNS as microstate would make sense; for a chemist the "chain of molecules as macrostate," the specification of each atom occurring in the molecule with its bindings as a microstate; for a particle physicist "material system" as macrostate, the specification of all occurring elementary particles as microstate. In this manner we give the molecular geneticist the most prior knowledge: "I am dealing here with a set of chromosomes of a living creature" and just because of this the smallest potential information in the then observed macrostate "set of chromosomes of drosophila."

Since according to (8.2) information is uniquely defined in terms of a "probability vector" $p_k$, we are also reminded that probability determines a relationship between two classes of events. One usually calls them *possible* events (the class of all $x_k$ for all $k$) and *favorable* events (the class of all events of type $x_{k'}$, with a chosen fixed $k'$ whose probability $p_{k'}$, the expectation value of the relative frequency of the $x_{k'}$, is among all the occurring $x_k$). Related to this is that every probability that can be specified by a rule is actually a *conditional* probability $p(y, x_k)$. There the condition $y$ is just the event that an experiment is done which according to some law has all the $x_k$, and only those, as possible results and which has been designed such that each $x_k$ can be expected exactly with the probability $p_k = p(y, x_k)$. If only the possible results $x_k$ are initially known but not a function $p_k$, then Bayes' procedure serves for their approximate determination, i.e., for the statistical assessment of *which* $y$ best describes the conditions of the experiment.

## 8.5 Evolution as growth of potential information[2]

### 8.5.1 Basic idea

Organic evolution is the development of *functional* forms. The responsible mechanisms have been repeatedly studied by theorists of biological evolution. These investigations go far beyond the scope of a book about the reconstruction of physics. The present section has a more modest aim. It merely studies the growth of potential *syntactic* information, thus not of special functional forms but of countable forms in general. This growth is relatively easy to describe in a mathematical model. In this model one can then demonstrate that under suitable conditions states richer in form are at the same time the more probable ones. Under these conditions the growth of the richness of forms is not opposed to thermodynamic irreversibility but a special case of it.

I might be permitted to discuss the problem at a certain breadth by beginning with some quotations from a book by Glansdorff and Prigogine (1971). Prigogine discussed this problem in detail, starting with the thermodynamics of irreversible processes. Naturally I follow his special models without any reservation yet believe that my approach, to be described again in the following, permits an even simpler description of the abstract principles of irreversibility and evolution, where here also the simplifying principle is the choice of the phenomenologically already given temporal modalities of past and future as a point of departure.

Glansdorff and Prigogine write:

"It is a rather remarkable coincidence that the idea of evolution emerged in the nineteenth century with two conflicting aspects: in thermodynamics, the Second Law is formulated as the Carnot-Clausius principle. It appears essentially as the evolution law of continuous disorganization, i.e., of disappearance of structure, introduced by the initial conditions.

---

[2] This section is an abbreviated version of my essay *Evolution und Entropiewachstum*, in: J.-H. Scharf, *Informatik*, Nova Acta Leopoldina, NF 206, vol. 37/1, Barth, Leipzig, 1972.

## 8.5 Evolution as growth of potential information

In biology or in sociology, the idea of evolution is, on the contrary, closely associated with an increase in organization giving rise to the creation of more and more complex structures." (p. 287)

"Are there consequently two different irreducible types of physical laws?" (p. 288)

The authors do not opt for this conclusion but for an opposite solution:

"The point of view considered in this monograph suggests that there is only one type of physical law, but different thermodynamic situations: near and far from equilibrium. Broadly speaking *destruction of structures* is the situation which occurs in the neighborhood of thermodynamic equilibrium. On the contrary, *creation of structures* may occur, with specific nonlinear kinetic laws beyond the stability limit of the thermodynamic branch [namely the entropy production function, C.F.W.]. This remark justifies Spencer's point of view (1862): '*Evolution is integration of matter and concomitant dissipation of motion*'.

For all these situations, the second law of thermodynamics still remains valid." (p. 288)

From the standpoint of the history and philosophy of science, this problem can be described as one of retrospective reflection. Wherever scientists have attempted to explain causally the empirically observed or suspected emergence of structures, they have proposed a direct hypothesis about the respective mechanism which made it more or less obvious that new forms could or even ought to emerge. The classic example is Darwin's theory of selection. But also for the emergence of inorganic forms (e.g., growth of crystals, creation of the planetary system, etc.) direct, more or less plausible hypotheses have been proposed. Afterwards, however, the inventors of such evolutionary models must ask themselves, or be asked, how their explanation of a quasi-irreversible tendency of formation of forms is compatible with the Second Law which after all asserts the destruction of forms and the growth of disorder. To this further inquiry four different types of answers were given (if I have not overlooked any), which can be illustrated using the examples just mentioned:

*1.* In the given phenomenon entropy actually decreases and thereby the phenomenon turns out not to be subject to the Second Law. For instance, vitalists have in general thought in this way about the development of life. As it was assumed that Darwin's theory of selection was intended to be compatible with the Second Law, this theory was discarded as well together with the Second Law. One will not be able to claim that this view has been refuted by modern biological experience. Nevertheless, I do not want to consider it any further. Here I am concerned with the analysis of the conceptual structure of the theory of selection, and thus wish to examine how it is related to the Second Law *insofar as it is true*. I hope to make plausible that there are no difficulties for it arising from the Second Law.

*2.* For the given phenomenon the concept of entropy and consequently the Second Law cannot be applied or at least not to such an extent that there would be a problem. As it is indeed difficult to estimate quantitatively the entropy of living systems, this was sometimes proposed as a way out of the dilemma between evolution and growth of entropy. I mention it here only to attest that it had not escaped my attention. But I hope to show that it is superfluous, quite apart from the fact that in my opinion it would be hard to defend at a closer inspection of the meaning of a thermodynamic approach.

3. In the given phenomenon one term of the entropy decreases due to the emergence of form, but this will be overcompensated by an increase in other terms, such that the Second Law is never violated. This is perhaps the prevailing opinion about the problem of biological development. The entropy production of the metabolism of the organisms, under the constant throughput of energy from the Sun, quantitatively surpasses by far the changes of entropy due to the development of form. The formulation of Glansdorff and Prigogine arguably must also be interpreted in this sense.[3]

4. In the given phenomenon, the development of form itself represents an increase in entropy, and is thus a direct consequence of the Second Law. This for instance is the case in the only one of the examples mentioned above which thermodynamically can be calculated explicitly, that of the growth of crystals. The thermodynamics of melting shows that at sufficiently low temperature the thermodynamic equilibrium is on the side of the crystal and not the liquid.[4] As on the other hand one is unlikely to deny that the crystal exhibits higher structure than the liquid, this example gives reason to doubt the thesis that

---

[3] Bernd-Olaf Küppers pointed out to me that Prigogine understands his arguments rather in the sense of the second answer (editors' note: cf. B.-O. Küppers, *Information and the Origin of Life*, MIT Press: Cambridge, 1990). Living systems are open systems, and one can claim that for them the Second Law is not applicable at all. For my own proposal of a solution this question of classification is not essential and I am not sufficiently familiar with the literature to express definite opinions about the subjective meaning of the authors. About the factual issue, however, one has to say that with regard to open systems one also argues in the language of thermodynamics. One ascribes to them, within margins of error that can be estimated, definite values of thermodynamic quantities like energy, pressure, volume, temperature, entropy. This is what I meant by saying that with the sentence that the second way out is hard to defend upon closer inspection of the meaning of a thermodynamic approach. That, however, for an open system the current entropy, although defined to a good approximation, does not increase is the third point of view. In this very special sense "the Second Law does not apply to them."

[4] Here I am indebted to B.-O. Küppers for pointing out an ambiguity in my expression. For isothermal growth of crystals the relevant quantity is not entropy but the free energy. My argument, however, compares the entropy of an isolated crystal with the isolated liquid at the same energy content and below the melting point, thus approximately corresponds to the adiabatic crystallization of a supercooled liquid. Here no doubt the entropy of the crystal is higher than that of the liquid.

## 8.5 Evolution as growth of potential information

growth of entropy necessarily corresponds to a reduction of structure.[5] The present account does express the opinion that the solution of the problem is perhaps always to be sought in the fourth answer. Naturally thereby it is not to be denied that the processes which usually are in interpreted in the sense of the third answer are actually occurring. The positive description of the emergence of structure according to Glansdorff and Prigogine through instabilities of entropy creating processes far from equilibrium is accepted as being convincing, including the thereby occurring decreases of the rate of entropy production. The model of evolution of Eigen (1971) was even the trigger for this account, by questioning how entropy is to be defined in it. The thesis is only that where development of form is actually occurring, with an accurate definition of the corresponding entropy, there corresponds to the growth of multiplicity and complexity of the forms an increase and not a decrease in that term of the entropy which is assigned to the information of form. If the thesis is correct then the appearance of a conflict between development of form and the Second Law is merely a consequence of an in general unfounded identification of entropy with a measure of uniformity lacking in structure, as generalized from a few examples. The heat death would be, assuming a sufficiently low temperature, not a mush but an assemblage of complicated skeletons.[6]

This thesis is a correction of an earlier thought.[7] Starting with the development of cosmic forms, in particular the planetary system, I had discussed the question of their relationship to the Second Law. Even if we are not certain about the correct model of planetary evolution, no modern astrophysicist will doubt that this process was compatible with the Second Law; on the other hand, the form of the system is so special and "ornate" that its conjectured impossibility of a mechanical explanation was once for Newton the basis of a proof of God—the proof of the existence of an engineer-like God working according to a plan. At that time I expressed the opinion that the development of differentiated forms is quite generally the consequence of the same "structure of time" as the Second Law. In short one can say that both developmental laws express that the probable will happen. This circumstance can be called structure of time ("historicity of time"), as the probable is expected

---

[5] As will be seen in the closing remarks (see p. 228) that a formulation is possible which combines what is correct in answer 3 with answer 4.

[6] To take a model from kitchen practice, when one makes fruit salad, perhaps as an addition to a Bircher muesli, consisting of apples, peaches, and bananas, one will first slice each fruit separately into the bowl and then stir to mix the various sorts of fruits. Thin slices of bananas, however, have a tendency to stick to one another (but not to apples or peaches). Accordingly, one therefore spreads the banana slices individually between the slices of the other fruit, as much as possible. If one *then* mixes with a spoon, clumps of bananas will form again bit by bit. As the mixing clearly increases the entropy, clumps of bananas have a higher entropy than a uniform distribution of single slices of bananas throughout the other fruit.

[7] *Die Geschichte der Natur*, 1948. End of the sixth lecture ($2^{\text{nd}}$ ed.), pp. 62–65.

for the future but not asserted about the past. For the Second Law the interpretation is familiar that increasing entropy is the occurrence of the probable. For the development of forms one must consider that a multiplicity of forms is a priori probable, a state completely devoid of form a priori improbable. This was discussed then only qualitatively; here it will be elaborated conceptually and carried out in more detail in a model.

In that earlier discussion I nevertheless adopted the point of view of the third answer. I was then of the prevailing opinion that the emergence of forms indeed signified a decrease in entropy, which was, however, overcompensated by the production of entropy in the accompanying irreversible processes. That was, however, as I see now, an inconsistency. The concept of entropy is so general and abstract that the specification of a high a priori probability for a state rich in form also amounts to assigning a high entropy to it. Shannon's concept of information was then not yet known, by means of which the problem will be described in the following.

Restricting attention, as in the previous examples, to *two* semantic levels and *one* thereby defined concept of information, there follows from the structure of time only the Second Law: as time progresses, the actual information of the state present at that time will decrease with overwhelming probability, its potential information ( entropy) increase. If one wants to express the development of form by means of the concept of information at all, one must introduce (at least) *three* semantic levels, with three different measures of information then defined among them. Let us denote the three levels by the three letters $A, B, C$ such that $A$ only occurs as a microstate, $C$ only as macrostate, but $B$ as a microstate relative to $C$ yet as macrostate relative to $A$. Denote by $j_{BC}$ the number of states of level $B$ which are contained in a state of level $C$; this will in general be a function of the special state $C$. Analogously one can define $j_{AC}$ and $j_{AB}$.

We now call $C$ the morphological, $B$ the molecular, $A$ the atomic level and thereby fix the idea of a model. The entire system whose states we consider consist of "atoms" whose states are completely described at the level $A$. The specification of an $A$-state is thus (according to our model) the maximally possible knowledge about the system. The atoms are supposedly capable of combining into definite forms, different kinds of "molecules." A $B$-state indicates which molecules are present, i.e., which kinds of molecules and how many of each. Whether for each molecule its position and momentum is to be specified depends on the definition of the levels; we remark on this later. Every $B$-state contains of course many different $A$-states, at least insofar as one applies classical statistical mechanics to the atoms (or if one does not specify the position and momentum of the molecule in the $B$-state). A $C$-state yields only "morphological" information about what kinds of molecules are present, not how many molecules there are of each kind. Each $C$-state contains in general again many $B$-states. This means that $j_{AB}, j_{BC}$, and $j_{AC}$ are in general large numbers.

## 8.5 Evolution as growth of potential information

The Second Law now implies that the morphological state will develop in time to ever larger values of $j_{AC}$. In general, then, $j_{AB}$ and $j_{BC}$ will also grow at the same time. $\log_2 j_{BC}$ now stands for the information one can gain by asking, for a given morphological state, how many molecules of each kind are present. $j_{BC}$ thus measures the many different ways the morphological state can be realized and thereby also the multiplicity of forms contained in it. A growth of $j_{BC}$ can thus be interpreted as a growth of the multiplicity of forms. *If* $j_{AC}$ *and* $j_{BC}$ grow at the same time, the growth of the multiplicity of forms is *in this sense* directly associated with the growth of entropy. That both are growing at the same time cannot be proved in general but only for certain states far from equilibrium (as alleged by Glansdorff and Prigogine). We will find examples for this in a model.

Another possible measure of the multiplicity of forms is the number of different *kinds* of molecules occurring in the $C$-state. This number is a characteristic of the $C$-state, which for instance will then grow with overwhelming probability if in the initial state only isolated atoms are present, whereas the equilibrium state contains different kinds of molecules in finite relative concentration.

### 8.5.2 Condensation model

We illustrate these circumstances by means of explicit calculations in the following very simplified model. So that we deal only with discrete numbers, we completely disregard the spatial coordinates of the atoms. On the other hand we wanted to illustrate that atoms have other degrees of freedom apart from the freedom to associate into molecules; this is accomplished by introducing a quantized "excitation energy." Thereby one obtains four "semantic" levels, depending on the way the excitation energy is taken into account. One could, however, perform exactly the same calculations also in the manner of the kinetics of chemical reactions by taking into account the position and momentum degrees of freedom of atoms and molecules. Our model, in which we introduce only *one* kind of atom and where we distinguish the types of molecules only by the *number* $k$ of atoms contained in the "molecule" would then become the description of a simplified condensation process in which the "types of molecules" correspond to the size of droplets. We call it therefore a condensation model.

We consider an isolated system of $n$ atoms. These atoms are combined into molecules; one molecule is characterized by the number $k$ of atoms of which it consists. Any number $k$ ($1 \leq k \leq n$) may occur. We call a free atom a molecule with $k = 1$. Molecules with large $k$ one could also call liquid droplets; hence the name "condensation model." In addition, each molecule may contain different amounts of energy. To simplify the calculation I assume energy to be quantized. There is a universal quantum of energy $E$, and each molecule can have any number $q$ of such quanta of energy. There $q$ consists of a portion $q_B$ of "binding energy" and a portion $q_E$ of "excitation energy" (that can also be interpreted as "kinetic energy" if so desired). The excitation energy of a molecule can be any nonnegative integer ($q_E = 0, 1, 2, \ldots$). The binding energy is negative, and its absolute value is equal to the number of atoms bound to the first atom in the molecule, i.e., $q_B = 1 - k$; a free atom thus has $q_B = 0$, a two-atom molecule $q_B = -1$, etc.

We distinguish now not only the "microstate" and "macrostate" but also four kinds of states at four "semantic levels":

*1. Atomic states.* The atoms are thought to be individually known, for instance numbered. An atomic state is specified by indicating for each atom with what atoms it is combined to a molecule and giving the energy of this molecule.

*2. Molecular states.* The molecules are known according to their number, type, and energy. For every molecule the number $k$ of its atoms and its energy $q$ is given.

*3. Population states.* The population of molecules are individually known. That is, for each type $k$ of molecule it is known how many molecules of this type there are (possibly zero). Population and molecular states differ only in that for molecular states the energy of each molecule is also known. We will, however, stipulate that in a population state the total energy of the entire system is known. Let it be described by the integer $Q$.

*4. Morphological states.* It is only known what types of molecules occur. Here also, let the total energy $Q$ be known.

We denote the four semantic levels by their numbers 1 through 4 and define the number $j_{xy}$ of states of level $x$ per state of level $y$; it is the generalization of the concept "number of microstates per macrostate."

Here we only sketch the calculation. A molecular state is completely characterized by a function $(k, q)$, which indicates how many molecules of type $k$ and energy $q$ occur in it, and a population state in terms of the number function $(k)$, which indicates how many molecules of type $k$ there are, as well as through $Q$. In Table 8.1 there are listed in the first column (under $K$) symbolically all possible population states under the assumption that there is a total of 6 atoms. $K$ is the list of all molecules occurring in the respective state; e.g., $K = 3, 1, 1, 1$ denotes the state where there is one molecule with $k = 3$, thus consisting of 3 atoms, and three monatomic molecules.

The second column indicates the number $j_{12}$ of atomic states which may occur in a lowest molecular state, i.e., in a molecular state without any excitation energy; in such a state $q_A = 0$ and $q = q_B = 1 - k$. For example, in $K = 5, 1$ there are 6 atomic states, as each of the 6 atoms can be the one located in the molecule $k = 1$. For a prescribed total excitation energy $Q_E$ one defines thereby a population state with energy $Q = Q_E + Q_B$; $Q_B$ is easily computed for each $K$. The energy $Q_E$ can now be distributed in various ways over the molecules. In this way one obtains for each $K$ a number $j_{12}$ of molecular states which belong to the same population state. The number

## 8.5 Evolution as growth of potential information

**Table 8.1.** Values of $j_{13}$

| | | \multicolumn{9}{c}{$Q$} | | | | | | | | |
|---|---|---|---|---|---|---|---|---|---|---|
| $K$ | $j_{12}$ | $-5$ | $-4$ | $-3$ | $-2$ | $-1$ | 0 | 1 | 2 | 3 |
| 6 | 1 | 1 | 1 | 1 | 1 | 1 | 1 | 1 | 1 | 1 |
| 5,1 | 6 | – | 6 | 12 | 18 | 24 | 30 | 36 | 42 | 48 |
| 4,2 | 15 | – | 15 | 30 | 45 | 60 | 75 | 90 | 105 | 120 |
| 3,3 | 20 | – | 20 | 40 | 60 | 80 | 100 | 120 | 140 | 160 |
| 4,1,1 | 15 | – | – | 15 | 45 | 90 | 150 | 225 | 315 | 420 |
| 3,2,1 | 60 | – | – | 60 | 180 | 360 | 600 | 900 | 1260 | 1440 |
| 2,2,2 | 15 | – | – | 15 | 45 | 90 | 150 | 225 | 315 | 420 |
| 3,1,1,1 | 20 | – | – | – | 20 | 80 | 200 | 400 | 700 | 1120 |
| 2,2,1,1 | 45 | – | – | – | 45 | 180 | 450 | 900 | 1575 | 2520 |
| 2,1,1,1,1 | 15 | – | – | – | – | 15 | 75 | 225 | 525 | 1050 |
| 1,1,1,1,1,1 | 1 | – | – | – | – | – | 1 | 6 | 21 | 56 |

of atomic states per population state follows from $j_{13} = j_{12} \cdot j_{23}$. These $j_{23}$ are given in the remaining columns of the table. There are different columns according to the given possible values of the total energy $Q$.

We now discuss the values of the $j_{13}$, i.e., the number of possible realizations of the same population state. Because of the binding energy, the lowest possible total energy is $Q = -5$, which can only be realized in the "droplet" of 6 atoms. For low values of $Q$ large molecules are favored, for high $Q$ small molecules. For example, if we consider the column $Q = 0$, we find that having just free atoms (1,1,1,1,1,1) can only be realized in *one* way, similarly a "droplet" which contains all atoms (6). In 600 possible ways one can construct 3,2,1, in 450 ways 2,2,1,1. Let us assume that every atomic state can change into every other atomic state, directly or indirectly, and that these transition probabilities are symmetrical (i.e., that the probability for a transition from $A$ to $B$ is the same as that for $B$ to $A$). Then in statistical equilibrium the probability for finding a particular population state will be proportional to the number of atomic states contained in it, and away from equilibrium the entropy defined by this number will grow on the statistical average. One immediately sees that in our example the states with complicated forms, like the one with the three types of molecules 3,2,1, are statistically strongly favored compared to the simple forms, like the mere lump of 6, or free atoms only 1,1,1,1,1,1. At sufficiently low energy the state of maximum entropy is rich in form.

The relationship among three levels $A, B, C$ described in Sect. 8.3 can be demonstrated in our model most easily between the levels "population" ($C = 3$), "molecule" ($B = 2$), and "atomic" ($A = 1$). From Table 8.1 for $n = 6$, I extract for three values of $Q$ the values of $j_{12}$ and $j_{23}$:

**Table 8.2.** $j_{12}$ and $j_{23}$ for $n = 6$

|         |          | $j_{23}$ |         |         |
|---------|----------|----------|---------|---------|
| $K$     | $j_{12}$ | $Q = -3$ | $Q = 0$ | $Q = +3$ |
| 6       | 1        | 1        | 1       | 1       |
| 5,1     | 6        | 2        | 5       | 8       |
| 4,2     | 15       | 2        | 5       | 8       |
| 3,3     | 20       | 2        | 5       | 8       |
| 4,1,1   | 15       | 1        | 10      | 28      |
| 3,2,1   | 60       | 1        | 10      | 28      |
| 2,2,2   | 15       | 1        | 10      | 28      |
| 3,1,1,1 | 20       | 0        | 10      | 56      |
| 2,2,1,1 | 45       | 0        | 10      | 56      |
| 2,1,1,1,1 | 15     | 0        | 5       | 70      |
| 1,1,1,1,1,1 | 1    | 0        | 1       | 56      |

One sees that the two quantities by no means progress in parallel, but at least for $Q = 0$ reasonably well. Even better is the parallelism of $j_{23}$, which acts here as a measure of the forms, with $j_{13}$, measuring the thermodynamic probability; here $j_{12}$ is the factor of proportionality, independent of $Q$.

But this comparison is quite formal, as one normally cannot measure the distribution of energy over the molecules. A simple measure of the amount of form would be the number $\varphi$ of different species in a morphological state. For $n = 6$ there are 4 morphological states with $\varphi = 1$, which at $Q = 0$ together comprehend 252 atomic states, in addition 5 morphological (population) states with $\varphi = 2$ and 980 atomic, finally one with $\varphi = 3$ and 600 atomic. In a statistical distribution over the population states with probability $p(\varphi)$, the "information about $\varphi$" is then

$$H\varphi = -\sum_{\varphi} p(\varphi) \log_2 p(\varphi). \tag{8.5}$$

Starting with a certain population state, e.g., $1, 1, 1, 1, 1, 1$, then $H = 0$. Letting that state develop statistically, $H$ then increases to the value that corresponds to the equilibrium distribution; it is about 1.32, whereas the maximally possible value would be $\log_2 3 \approx 1.55$.

Another form-related information would be defined by the question of which type a randomly chosen molecule belongs to; here we call it $H_k$. $H_k = 0$ in a population state with $\varphi = 1$; for $\varphi = 2$, $H_k$ can at most be 1, and for $\varphi = 3$, it will be at most $\log_2 3$. For a statistical development, starting with $1, 1, 1, 1, 1, 1$, $H_k$ will also increase to somewhat less than the maximum possible value.

### 8.5.3 Final remarks

Qualitatively, what have we learned from this model? In imprecisely defined states, described for example as "all free atoms" or "one single droplet," the actual information about the microstate is very high, so the potential information or entropy is very low. Indeed, this qualitative argument shows that states richer in form are richer in entropy, and thus must be more likely.

The qualitative distribution and thereby the position of equilibrium depends in the model on the total energy $Q$ of the system. For large $Q$ free atoms and small molecules are favored, for small $Q$ it are the larger molecules. If we had not introduced a binding energy at all, equilibrium would always lie on the side of free atoms. Here we have not introduced any space and momentum coordinates. The same circumstance emerges even sharper in a realistic equilibrium theory of chemical reactions. The contribution of the volume to entropy is the larger the more individually mobile molecules are present, thus is largest for free atoms. Additionally, however, there is the contribution from the volume in momentum space. At the presence of a binding energy and for fixed total energy, this is larger for molecules in which atoms release much binding energy.

Two conditions thus statistically favor a richness of form: the existence of a binding energy and sufficiently low total energy (respectively temperature). Once these are satisfied, not only will the amount of form grow far from equilibrium, as shown by Glansdorff and Prigogine, but contrary to prevailing opinion the state of equilibrium is then also rich in form. This leads to the question of how this prevailing opinion could develop and prove itself in so many empirical examples. For this we must consider more closely in what sense equilibrium is rich in form.

In the model we have only computed entropies but not transition rates. For instance, if we assume that per unit time *one* atom will always change its state of binding or excitation energy, then the atomic states change with always the same velocity into neighboring states. In equilibrium there are then always molecular forms present but ever changing. Such an equilibrium is a "teeming of changing forms." For an observer who is only interested in irreversible developments it is natural to choose a mode of description in which he no longer recognizes these individual forms. *As a consequence of this choice of description* he then calls the equilibrium chaotic. But this depends, in a sense, only on that fact that in the equilibrium situation he naturally cannot have any longer what interested him historically, namely irreversible developments. This, however, is something completely different from the assertion that in equilibrium there are no forms.

Another aspect is elucidated by assuming that in equilibrium, at the end of a development process, transition probabilities decrease to zero. Then we find the metaphor mentioned in Sect. 8.1 of equilibrium being a collection of skeletons: those forms that developed "randomly" will persist forever without any further change. For instance, in this way the development of the solar system is to be interpreted. From the original "Kant-Laplace" nebula where hydrodynamic and chemical processes occur, there ultimately emerge separate planets, acting upon one another solely via gravitaiton; the stability theorems of celestial mechanics show that this "skeleton" can exist practically forever.

The usual thermodynamic examples for the blurring of forms through growth of entropy are selected lopsidedly. They refer to cases where one quantity levels out, which according to its nature or special conditions is not capable at all of creating forms through bonding forces. Thus for example in the transport of energy in the case of heat conduction; kinetic energy ("motion" in the sense of the Spencer quotation) cannot be concentrated through bonding forces. In diffusion processes matter distributions even themselves out but only because the corresponding matter is present in the form of freely moving atoms or molecules. The thermodynamically equally relevant counterexample of the growth of crystals in a liquid is mostly forgotten. Here indeed there emerges an ordered entity whose contribution to the volume-dependent part of entropy is less than if it were dissolved; but the binding energy released overcompensated this through an enhanced contribution to the momentum part. Incidentally, the occurrence of the word "overcompensating" demonstrates that there is no sharp contrast between the third and fourth answer in Sect. 8.1. To an isolated structure one can meaningfully assign a low entropy; only the entropy of the entire system is growing during its creation. What must be emphasized by contrasting the two answers is only that processes of the creation of forms are also consequences of exactly the same structure of events as expressed in the Second Law.

In closing let me make one more remark about the concept of a document. In previous papers on the Second Law, I have repeatedly made the following consistency argument:[8] "The Second Law follows, on the one hand, from the past as being factual, the future, however, being open ('possible'). To this corresponds the idea that there are documents of the past but not of the future. This, on the other hand, also must follow from the Second Law. It follows if one takes into account that growth of entropy corresponds to a loss of information. A document is an improbable fact, hence contains much information. This implies, due to the progressive loss of information, much information about the past but little information about the future." At first sight this argument appears to be problematic if the Second Law actually asserts an increase in information. But here again it is only a sign confusion, which occurs because of the mix-up of actual and potential information. Potential information increases, actual information decreases, and for a document one deals with actual information.

---

[8] Here in Sect. 7.3.

## 8.6 Pragmatic information: Novelty and confirmation

### 8.6.1 Pragmatic information[9]

We recall the origin of the concept of information from the theory of communication. What colloquially in human communication is understood as information is not the syntactically definable multiplicity of forms of a message but what a competent listener can understand in the message. We condense this into the Thesis 1: *Information is only what can be understood* (MEI, p. 351). Incidentally, the thesis holds for the two concepts of syntactic and semantic information, only just at different "semantic levels." Syntactic information exists for the sender or recipient who can distinguish letters and is interested in them, semantic information for those interested in the linguistically transmitted contents.

The discussion by Ernst and Christine v. Weizsäcker (1972; 1974a; 1984; 1985) of the biological and social/system-theoretical definition and usage of the concept of information pertain to this distinction. First they point out that all biological systems are organized in a rather strict hierarchical way. We illustrate this here in our linguistic example of the message of a telegram. At the level of the *language* it communicates important news (high news value) and for this avails itself of the level of *letters* about which nothing else is demanded but to be recognizable. The important information for humans in this case is linguistic; for that very reason it is called "semantic" (meaningful). Letters and language are two "semantic levels," hierarchically one above the other; written language only exists if there are recognizable but otherwise uninteresting letters: letters exist only *to make* a written language possible.

The comprehension of signs is initially a conscious achievement; in this narrow sense semantics exists only for humans. Yet we also use the concept of information in biology; the entire theory about information and evolution is based on it. Here we can bring in the concept of pragmatics, important already for human affairs. *Pragmatic* information is that which *effects*. If I am not interested that the sender of the telegram will arrive tomorrow and I therefore do not react, then the telegram is without effect; its pragmatic information for me was low. In human affairs pragmatics denotes an again higher semantic level than language: the structure of human relations and actions. At this level happen the important decisions of life, and the mere medium of language is for it a serving, subordinate level, as letters are for language. To the serving functions of language belongs again their reliable comprehensibility; it must be assured so that the actual news, the appropriate action, can successfully happen. (That language itself, e.g., poetic language, can be an act of high rank is taken into account in the functional theory described here by distinguishing acting talk as *speech* from the formally analyzable *language*. For discussion of the problems of this entire classification see *Zeit und Wissen*

---

[9] This subsection follows up the essay *Materie–Energie–Information* (1969), in: *Die Einheit der Natur* III, 5; quoted as MEI.

II.6.6.2. Here, while aiming at the biological aspect, we need not yet consider these problems.)

For the functioning of organisms as well as for evolution, the actually steering information, lying above the syntactic level, must be defined pragmatically from the outset. In MEI p. 350 I have described this as the "objectification of semantics."

Last, let us justify here only Thesis 2 of MEI, p. 352: *Information is only what produces information*. This thesis is a sharpening of the above statement that pragmatic information is only that which effects. In the context of life, effecting is the creation of states and processes which, pragmatically, are only then of concern if they themselves again act upon something, thus if they in turn are pragmatic information. "Producing information" is thus the pragmatic version of "being understood." The thesis is not meant as a definition of information —then it would be circular—but as a restrictive condition: *only* if it produces information is it information in the pragmatic sense. We return to this thesis in Chap. 10.

### 8.6.2 Novelty and confirmation

E. and C.v. Weizsäcker attempt to define an approximate measure of pragmatic information from two quantities one might consider more easily measurable; one they call novelty (or surprise), the other confirmation. According to the above Thesis 2 they understand pragmatic information as producing information and therefore also call it "contagious" information (compare the title of the talk *Contagious knowledge*, Weizsäcker 1985). The interrelationship of these three quantities is interpreted as a surface in three-dimensional space, according to Fig. 8.1. A cut through this surface in one of the planes, where the sum of confirmation $B$ and novelty $E$ is constant, is shown in Fig. 8.2.

The significance of this drawing, roughly speaking, is that effective information ("information is that which produces information") is only possible if some things proceed according to law (confirmation), but something new is still going on (novelty). All novelty without law is chaos, in which nothing can be understood, while mere confirmation conveys no information (does not present any surprise).

Close to the limit of 100% confirmation every novelty can be registered. The authors regard this as the situation which presumes Shannon's theory of information. They propose to measure novelty in this limiting case directly, in terms of the information in the sense of Shannon. This corresponds to $I = E$, tangent the curve. If, however, the fractional amount of the confirmation decreases, then not all novelties can still be registered pragmatically-effectively. The curve then remains below Shannon's straight line and returns for $B = 0$ to $I = 0$. The authors suspect that historically successful systems, for instance living creatures, operate close to the maximum of the curve.

## 8.6 Pragmatic information: Novelty and confirmation

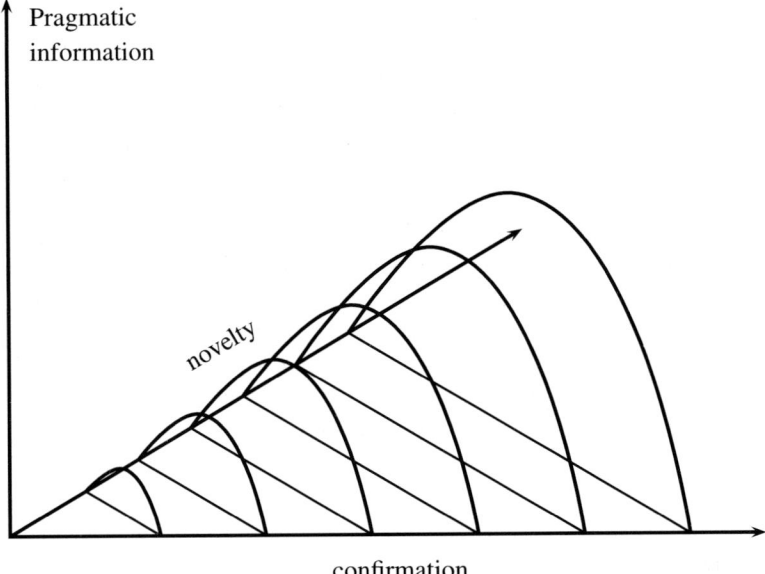

**Fig. 8.1.**

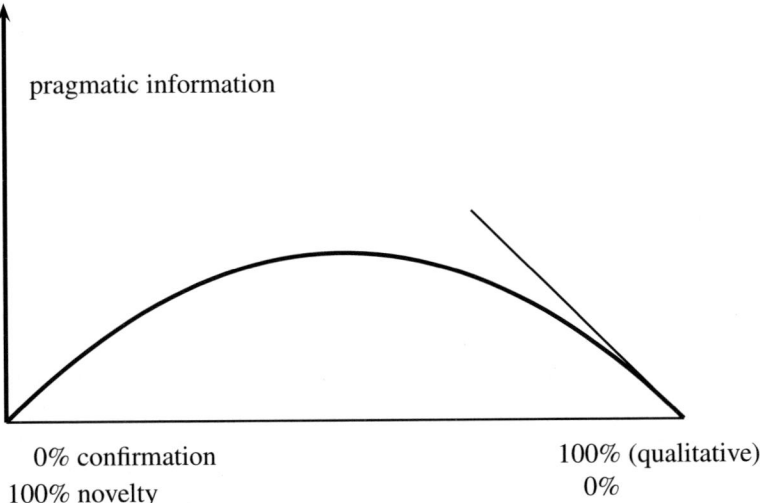

0% confirmation     100% (qualitative)
100% novelty        0%

**Fig. 8.2.**

Mere confirmation corresponds to the caricature of the specialist: he knows everything about nothing; mere novelty corresponds to the caricature of the generalist: he knows nothing about everything.

Here we can add another fundamental concept of E. and C.v. Weizsäcker: the concept of *error-friendliness*. Living beings are not faultlessly functioning machines; they cannot be and also will not. They cannot, as neither external influences nor internal functions are completely controllable. Thus, they will only survive if they are also equipped to overcome errors. System-theoretically one also calls this ability resilience, the attribute of a tumbler. They also ought not to be; an example for this is the role of mutations in selection. Mutations are, from a technical standpoint, errors of the apparatus which are often harmful, even deadly, but which also open up new possibilities of life. In a perfectly functioning apparatus there is, so to speak, no novelty, only familiarity, planned things, only confirmation. Mere novelty, on the other hand, is chaos, destroying life. Error-friendliness is close to the optimal combination of novelty and confirmation.

### 8.6.3 A model

At the beginning of these considerations (1970), to understand their structure, I once designed a very simplified model which I am printing here.

The arguments of the authors depend essentially on several "semantic levels" being considered simultaneously. They are

a) a *stream of signs*
b) a *recipient*
c) a *theoretician*

There the recipient can be thought of as being further separated into

$b_1$) a *machine*
$b_2$) an *observer*

*a)* The *stream of signs* consists of a sequence of signs, starting at a time $t_1$, which are chosen from $K$ different types of signs (e.g., letters); quantitatively, I only consider the cases $K = 2$ and $K = 3$. Suppose that at each instant $t_n$ ($n = 1, 2, 3\ldots$) of a sequence of points in time one symbol enters the receiver. Whether and when this sequence terminates is of no importance to our considerations. Every type of sign, denoted by a number $k$ ($k = 1, 2 \ldots K$), has a probability $p_k$ of occurrence in the stream of signs; $\sum_k p_k = 1$. Assume that the $p_k$ are not previously known to any of the parties involved.

$b_1$) Let the *machine* register the incoming signs. If it could distinguish beforehand the types of signs, we would then have a conventional hardware communications receiver, and would be directed to Shannon's arguments. We now introduce the point of view that confirmation is necessary for information in that the machine is only capable of detecting that a symbol arriving at time

### 8.6 Pragmatic information: Novelty and confirmation

$t_n$ is of the same type as a symbol that has arrived at a previous time $t_m$. If this is the case, it signals "No. $n$ is of the same type as No. $m$" where it, e.g., will always report the first No. $m$ when this type has arrived. Otherwise it reports nothing. Thus it has, if every type has already occurred once, a supply of $K$ different messages; before that it can happen that it "fits" an incoming signal.

$b_2$) The *observer* only knows the messages of the machine. Let $I$ be the information which he owes at each of the times $t_n$ to the reporting (or non-reporting) of the machine. Obviously, $I$ will increase with time, at least on the average, because when the machine does not report anything the observer obtains zero information. The machine must first "learn to recognize" the signals; for this it needs the "confirmation" that a signal occurs a second time. The observer will critically measure his information. i.e., from the reports of the machine he will estimate the probabilities $p_k$ and compute from this $I$ according to Shannon's formula. (This complication is not necessary for the main idea; one could also work with known a priori probabilities. But then one would not obtain the variation of the values of $E$ and $I$ required for the construction of fig. 8.1 and fig. 8.2.)

c) The theoretician knows at every time $t_n$ the stream of signs that have objectively appeared up to that time (thus to every $t_m \leq t_n$ the type of sign which has then occurred). Thus he also knows the reports of the machine. He too empirically determines the $p_k$. From this he calculates the information of the signs appearing at every instant of time and *calls* them novelty. To the theoretician novelty is thus information in the sense of Shannon. He is thought to possess arbitrary "confirmation" for the reading of the stream of signals. He furthermore *defines* a confirmation on the machine by the assertion: if the machine again receives a type of sign already received before, then for this instant of time the confirmation is 1; if it receives a sign for the fist time, then the confirmation is 0. At the beginning of the series, the "information" of the observer will be below the "novelty" of the theoretician. After the $K$ points of time have elapsed at which the "confirmation 0" has occurred (they need not follow one another uninterruptedly), information will become equal to novelty; "Shannon's limiting case" will then have been reached.

As $I, E$, and $B$ are each calculated according to a specific procedure, it is a priori not certain at all that $I$ is a unique function of $E$ and $B$. I have not checked this question theoretically but only calculated explicitly the two cases $K = 2$ and $K = 3$ and found this assumption to be factually correct.

Here it is not worth performing these primitive calculations. The approach may serve as an illustration of the discussion of subjective and objective probability and information (see the end of Subsect.8.4c above). In the sense of equating information and utility one might say that pragmatic information, in the sense of the authors, is the information for the onlooking theoretician diminished by the effort required to confirm an observation *as* an identifiable event.

More recent investigations of the authors show how large the domain of biological and societal phenomena is that can be described by these qualitative concepts.

## 8.7 Biological preliminaries to logic

### 8.7.1 Methodological

This section is not required for the direct continuation of the reconstruction of physics. Rather it introduces for the first time, as a continuation of the discussion of time and probability, a philosophical discussion to which we return at the end of the reconstruction of physics, and which will occupy the book *Zeit und Wissen*, which is currently in preparation. The methodological figure of this reflection is the *Kreisgang*. [10]: temporal logic is the basis of physics, physics is the foundation of biology, and ethology arising from biology teaches us to see structures of animal and human behavior which ultimately permit the interpretation of logic itself as a system of behavioral rules. Let it be emphasized again that in a philosophy which is not hierarchically composed, thus in a philosophy not deduced from top down, such a *Kreisgang* is not a *circulus vitiosus* of a required "proof," but merely a consistency argument ("semantic consistency"), a "circular path in a garden."

This section is only a preparation for the discussion. It refers briefly to several sections in two previous books which were already identified as "propaedeutics" for the actual philosophical question. From *Die Einheit der Natur*, Chap. III (The meaning of cybernetics), the contribution III.4: *Models of health and illness, good and evil, truth and falseness*, in the following quoted as *Models*. From *Der Garten des Menschlichen* (Weizsäcker 1977) several contributions of Chap. II (*On the biology of the subject*), here cited as GM II with the number of the contribution and section.

### 8.7.2 Life as a form of cognition

GM II.2 *The back of the mirror, reflected* reports in Sect. 2: *Evolution as a form of cognition* on the thesis of Konrad Lorenz and Karl Popper that evolution, even life itself has a structural analog to cognition. In this thesis Darwin's idea of the survival of the fittest meets the epistemology of pragmatism, which defines truth as success of action. Lorenz and Popper are, however, closer to Darwin than, e.g., to the pragmatism of William James insofar as they accept with Darwin the concept of reality in classical physics whereas for strict pragmatism as well "reality" is only an abbreviation for successful action. In GM II.2.1 (*The ontology of science*) I have tried to balance acceptance and criticism of this thesis. The title of Lorenz's book *Die Rückseite des Spiegels*

---

[10] Cf. p. XXII, fn. 14.

(Engl. edition: *Behind the Mirror*, 1966) says that the organ through which we perceive nature is itself part of nature; the mirror which reflects the universe is part of this universe and has, as an object in the universe, a non-reflecting rear surface. I easily agree with this thesis. But I gave the essay GM II.2 the title: *The back of the mirror, reflected*. The title indicates that we also see the rear of the mirror only in a mirror; in other words, the reality about which we can speak is reality *to us*. Thereby we are at the *Kreisgang* of the present book.

That evolution is a form of cognition we can formally derive from the previous sections. Evolution is growth of information; one can say the same of cognition. The question is only whether we mean in both cases the same with information and how essential growth of information is for the characterization of evolution on the one hand, and for cognition on the other. One will describe evolution as growth of *objective* information, with cognition, however, being the growth of *subjective* information. Now the concepts "objective" and "subjective" are poorly defined (see *Zeit und Wissen* I 5.5.3). The section GM II.2.3 (Information, adaptation, truth) proposes the following formulation (GM p. 201):

Information is, first and foremost, *for humans*. The concept of information originated from the theory of telegraphic communication. However, the measure of information, is in that sense already meant inter-subjectively (namely communicatively!), and it can be objectified by considering *organs* or *apparatus* as transmitter and receiver. Information is then for a *transmitter–receiver pair*. "In this formulation, what is the meaning of 'for'? We assume the receiver to be without consciousness, or at least without reflection; it can therefore not tell us or itself that there was information "for it." The scientific justification that *we* can say that there was information *for it* can be expressed by the sentence that information only exists *under a concept*. This is already true about information for man" (p. 202). Here we can refer to the previous sections. Information in general is only defined between two semantic levels. "In the same way also information for an organ. It is only defined once *we* have indicated the *function* of the organ and thereby can determine the set of yes/no decisions relevant for this function. Once *we* have understood this function *we* can determine that the organ objectively determine the concept of this function through its construction and the thereby produced effect, independent of our judgment. Organs are, if we can say so, *objective* concepts" (p. 203).

Popper (1973; see also GM II.2.2) stresses in particular, and I think with good reason, a structural analogy between the three different "levels of adaptation: genetic adaptation, adaptive learning of behavior, and scientific discovery" (GM, 197). All three cases relate to *instruction from within* and *selection*. Instruction from within is transmission of a structure (precisely that whose content we measure), be it the structure of genes, behavior practiced through trial and error, or theoretical convictions. He says "I am indeed asserting that *there is no such thing as instruction from outside the structure* or the passive reception of a stream of information which impresses itself on the sense organs.

All observations bear the imprint of theory: there is no pure, disinterested, theory-free observation." "A new, revolutionary theory functions exactly like a new powerful sense organ" (GM 199). The similarity of this epistemology to the one advocated in the present book is quite obvious, but is again a topic for *Zeit und Wissen*. Here we merely wish to point out the relationship between concept and organ.

### 8.7.3 An approach to a pragmatic concept of truth

This is a report on the essay *Models of health and illness, good and evil, truth and falseness* (*Die Einheit der Natur* III.4)[11]

Truth and falseness are the basic reflecting concepts of logic, i.e., the concepts in terms of which logic can be defined. According to Aristotle a statement is a speech that can be true or false. logic can be defined as the mathematics of truth and falseness (*Zeit und Wissen*, chapter I.6). The essay *Models* asks, in a chapter on cybernetics, about a "cybernetics of truth." "If man is part of nature then truth must exist within nature" (p. 256) It "regards man, methodologically, as a living being, the living being as a control system with feedback, and this system as the outcome of mutation and selection... The reflections are designated as models since they cannot claim to state the truth about man, about life, and about the origin of the control system. They are based on the supposition that such models can exhibit what can be objectified about man, about life, and about history. What the term 'objectify' may mean, these reflections do not ask." (p. 256) This last question is further pursued in the present book.

The essay considers three kinds of models: of health and illness, good and evil, truth and falseness.

Under the heading *Health* the concept of *norm* is discussed. "So long as we are healthy, we do not notice that we are healthy" (p. 257). This is normal, "unquestioned" life. We will find it again in *Zeit und Wissen*, chapter I.2 under the title of "simple" cognition. The concept of a norm can be interpreted cybernetically as a system of "set values" around which are playing the "actual values" of the control mechanism. The essay points out the close relationship between the concept of norm and *Eidos*, the Platonic idea. The norm "is a 'true representation'. In fact when we speak of a norm, we mean the state of affairs it expresses, not our representation of it" (p. 261). Again the analogy between a concept and the functioning of an organ.

The really interesting question, however, is: What is *illness*? It is not the usual play about a norm. If an illness can be specified by a concept then it itself is a norm. "Illness appears like a parasitic control system within the larger control system that we call the organism" (p. 263). "A highly ordered system can react to perturbations only in an orderly manner, assuming that

---

[11] Translator's note: All quotations are from the English version, *The Unity of Nature*.

it can still react at all." "Illness could thus be defined as false health. Viewed in Darwinian terms, the concept 'false' denotes diminished survivability, i.e., loss of adaptability. Speaking platonically: in a world shaped by ideas, even the Bad can come into being only in accordance with an idea."[12] (p. 264).

*Progress*, e.g., evolution, leads to a "relativization affecting the distinction between health and illness" (p. 265). What was a "false" norm in one environment or control system can become a "correct" norm in a changed environment or control system. "We can surmise that progress cannot in principle be fully expressed in norms. The theory of ideas belongs not to open, but to cyclical time" (p. 267). The short section about *good and evil* does not belong to the present book but to the chapter *Concepts* in *Wahrnehmung der Neuzeit*. Right now let us only remark that Kant's categorical imperative is interpreted as: "Will possible norms!" (p. 268) Progress is also making those norms relative. "Here we encounter the structural reason for the superiority of love over the principle of justice" (p. 269).

*Truth* is traditionally defined as "adaequatio rei et intellectus." Now I reinterpret "adaequatio" by translating it as *adaptation*. About the right behavior of animals one can say: "Rightness is the adaptation of behavior to the circumstances" (p. 271). Adaequatio is here not the similarity of photograph to object but of a key fitting a lock. "We permit ourselves the stylized use of the term 'truth' for the rightness of animal behavior" (p. 272). There we come closer to the pragmatic concept of truth as norm of successful action.

We are reminded of Nietzsche's provocative statement that "truth itself is a kind of error without which a certain living being could no longer live" (p. 272). What is this supposed to mean, if we wish to think rigorously? "If truth aligns itself with health, then falseness must align itself with illness" (p. 272). Error could then be described as "false truth." Even Plato discusses in this way the possibility of error.[13] But if Hegel is right that truth is the Whole or, pragmatically speaking, if no adaptation can be perfect in an unmeasurable multiplicity of events, then every special truth is also a falseness, but just one without which we could scarcely live.

We will address the philosophical questions that arise here only in *Zeit und Wissen*, Chap. I.6. Here we merely ask how, within the indicated context, the pragmatic concept of truth might be formally, precisely defined. There it is important to see that truth and falseness are typically not ascribed to a norm of behavior, but to a statement.

### 8.7.4 The two-valuedness of logic

A statement is defined as speech that can be true or false. Thus it is at any rate a linguistic act, i.e., a mode of human behavior. The essay GM II.6 (*Biological*

---
[12] A theological note: This idea mediates a bit between the Greek and Jewish interpretation of Bad. See *Wahrnehmung der Neuzeit*, obituary for Gershom Scholem
[13] Compare my essay *Die Aktualität der Tradition: Platons Logik*.

*preliminaries to logic*, Sect. 3: *Pragmatic interpretation of the bivalence of logic* initially goes back to a property of the simplest behavioral patterns in animals (p. 301). Such a pattern is a sequence of activities which can be triggered by an external or internal stimulus. It has an all-or-nothing-principle: either it happens, or it does not. More complicated schemes of behavior may occur in part, or stronger or weaker; the elementary schemes, however, either occur or they do not. For this one can make computer models. "... A mechanism must function which normally does not react but only to relevant changes of the circumstances. Now finite beings like organisms can only have a finite number, perhaps only a few, reliably functioning relatively complex active schemes of behavior that can be triggered" (p. 304–5). Here the positive is dominant. "If the trigger is activated, the behavioral scheme does occur. If it is not activated, 'nothing' happens and it requires further inquiry about the disappointed interest of the respective observer to find out what it is that does not happen."

The behavioral scheme is not yet a statement. "We ethologists may define a 'state of affairs for the animal' as those circumstances which have a trigger effect... Human thought is then based on a representation (idea) of possible actions and their success, be it through language, be it, e.g., through visual fantasy" (p. 301). An imagined trigger of imagined possible actions would be a "state of affairs for humans" (p. 302). Speech is initially an action which stands for another action. (See also, still independent of the linguistic form, the interpretation of cognition, imagination, and thought as "symbolic motion," "experienced motion different from the motion it represents," in GM II.3: *The unity of perception and motion*, Sect. 5: *Symbolic motion*, p. 218.) A statement is then a speech which signifies "a state of affairs for humans." "The state of affairs is defined by a mode of action appropriate for it" (p. 302). In this last sentence the pragmatic theory of truth is invested; it does not start with the naïve-realistic supposition that there were states of affairs "per se." States of affairs, like information, exist only under a concept, and the objective form of the concept is the function of an organ for animal behavior, for human behavior the complex of possible modes of action. "A state of affairs is what can be expressed in a statement" (Strawson 1959).

One cannot simply deduce from the all-or-nothing-principle of elementary behavioral modes an all-or-nothing principle for the presence of "states of affairs for man." Even in complex animal behavior—in drosophila, for example—there is an "orientation in the environment," some sort of continuum of possible complex "states of affairs for the animal" which also guides the behavior.[14] What again has an all-or-nothing-principle is just the enunciation (or thought) of a particular statement. If the statement is simply made, it again has the dominance of something positive: it is "understood as being true." From the simple pronouncement there branches out in both directions a

---

[14] Here I must modify my presentation GM, p. 302. I am indebted to Martin Heisenberg for instructive conversations about these questions.

## 8.7 Biological preliminaries to logic

spectrum of possible behavior. On the side of inexpressibility there are modes of orientation which cannot be expressed in language. On the side of expressibility there is slight or grave doubt, up to the form of a reflecting sentence. For the reflecting sentence, which thus has been in doubt, it is certain that it can be true or false and that it is now accepted or discarded.

In this spectrum there remains a difference between the mere renouncement of a statement that has come into doubt ("perhaps it is not quite that way") and the assertion of its falseness. The full symmetry of truth and falseness which had become the dogmatic starting point of logic arises only as postulate about the fully reflected statement: it *ought* to be true or false. The two-valuedness of logic is not obvious. It is a claim. The pragmatic value of this claim is obvious. Statements that can be negated permit the unlimited accumulation of retrievable knowledge, thus *power* (See GM II.5: *On power*, section II.4: *What is power?*, p. 265–9). In GM II.4 under the title *The rationality of emotions* I described orientation not based on statements in the logical sense.

We see here that the biological preliminaries can serve not only as a justification of logic but also for the relativization of the dogmatism of logic. In the logical discussion of quantum theory we will have reason to question this dogmatism.

# Part III

# On the interpretation of physics

# 9

# The problem of the interpretation of quantum theory

## 9.1 About the history of the interpretation

*Was weiß ich, wenn ich weiß?*

### 9.1.1 The task

The interpretation of physics is a philosophical task. It is a supplementary task, such as we are confronted with it. We have already indicated this in the choice of words in the titles of the first and third part of the book: "The Unity of Physics" but "*On* the interpretation of Physics." The attempt to reconstruct the unity of physics is in itself already to some extent supplementary. But it is guided by the ideal of completion, as expressed by the concept of a closed theory. Such a theory has a mathematically infinite, empirically open number of possible consequences, but its foundations are assumed to be formulated by a finite, small number of postulates. The interpretation of the theory, however, embedding it in what we call our world view, the modification of this world view by the theory—this is a task whose possible ramifications we initially do not even know. We can only consider contributions to it.

We begin therefore with a historically given fact: the debate over the interpretation of quantum theory as it actually happened in our century.

It is understandable that quantum theory provoked a debate over its interpretation. Not only is it incompatible with the world view of classical physics, but also with certain positions of classical metaphysics. The initial problem was to recognize and precisely formulate this incompatibility. Then one must decide whether to consider this incompatibility as philosophical progress or a weakness of the theory. The present book is based on the conviction that we are dealing with fundamental philosophical progress. According to this conviction, it is not quantum theory that must defend itself before the court of traditional philosophies but those philosophies themselves must stand trial—in itself a philosophical process—with quantum theory in the witness stand.

For this reason it is necessary to formulate the testimony of quantum theory philosophically as accurately as possible.

We will attempt this in four steps.

*First,* we sketch the historical course of the debate. There we can rely on the outstanding book *The Philosophy of Quantum Mechanics* by Jammer (1974), but we will also draw from the accounts of the creators of the theory and from personal experience.

*Second,* we will try to show that quantum theory is free of internal contradictions; that it is in accord with itself, semantically consistent.

*Third,* we will turn to some of the suspected paradoxes and alternative interpretations that have come up in the years after the mathematical conclusion of the theory. That the paradoxes do not signify any self-contradictions of the theory was clear early on; Einstein recognized this by 1930. However, they were seen as invitations to go beyond quantum theory.

*Fourth,* we ask how from our own interpretation such an extension of quantum theory might look. This will be the subject of Chap. 11.

### 9.1.2 Prehistory of the debate

Roughly speaking, we can divide the history of the debate over the interpretation into three periods:

1900–1924: Problems of interpretation of the unfinished quantum theory
1925–1932: Completion of quantum theory and creation of the Copenhagen interpretation
1935–present: Rearguard actions

The beginning of the second period we mark by Heisenberg's paper (1925), its end by J.v. Neumann's book (1932). The beginning of the third period we indicate by the paper of Einstein, Podolsky and Rosen (1935).

The prehistory of the debate over the interpretation took place in the first of the three periods. In Sect. 2.11 we attempted to describe the birth of quantum theory as a reaction to the impossibility of a fundamental classical theory. This is a subsequent interpretation, an attempt to see after the fact that "it had to happen this way." Contemporaries could not see it that way. The first insight into the fundamental problem of classical physics one finds with Einstein (1905), then with Bohr (1913a); it had played a role for Heisenberg (1925); however, its infeasibility as a fundamental theory had scarcely ever been explicitly expressed in all its severity. Without exception, the problems of interpretation were formulated in the language of classical physics. This origin overshadows the debate up to the present day and for a long time characterizes it as "rearguard action," admittedly a necessary one for our understanding.

With Einstein's 1905 hypothesis of light quanta the *wave–particle duality* emerged as a central problem of quantum theory. There it was still assumed to be self-evident that all physical objects exist in space. Then this duality

## 9.1 About the history of the interpretation

is a complete alternative, if we only interpret both concepts broadly enough. Particles or bodies are localized objects; fields, especially waves, are states which in principle extend over all space. According to the empirically well-founded classical theory, electromagnetism (light). and gravitation were fields, matter consisted of particles. Einstein's light quanta ascribed a dual nature to electromagnetism, de Broglie's matter waves a dual nature to matter (Broglie 1924). The resolution of this problem was one of the goals of quantum theory.

Bohr (1913) was probably the first who saw that quantum theory actually demanded a radical break with classical physics. Its relation to the empirically well established and conceptually closed classical physics became a central problem to him, which he formulated in the *correspondence principle* (see Meyer-Abich 1965). Quantum theory can only be correct if it implies classical physics as a limiting case. By that was only meant the classical field theory of electromagnetism and the classical particle theory of matter. Bohr occasionally joked: "If Einstein sends me a radio-telegram that he has now finally proven the particle nature of light, this telegram will only arrive because of light being a wave."

The reconciliation of the two models through a *statistical* theory was occasionally considered. Einstein (1917) had introduced the statistical law of emission and absorption of light quanta. Since Bohr's theory of the hydrogen atom the "quantum jumps" between stationary states had become part of the theory. The correspondence principle reinterpreted the classical intensities of radiation as "transition probabilities." Finally, the hypothesis of Bohr et al. (1924) was an elaborate attempt at a statistical theory. The authors assumed the objective existence of a radiation field which, however, was only assumed to determine the statistical frequency for the occurrence of radiation energy at matter during emission and absorption. That would have been incompatible with the individual conservation of energy of the radiation field. The refutation of this idea by experiment led to the more radical notion of quantum mechanics.

We can summarize the difficulties of this duality for the transition to quantum mechanics as follows. Light and matter exhibited both a localizability characteristic of particles, and interference phenomena characteristic of waves. With the available classical concepts for both light and matter one could think of three possible solutions:

1. There are actually only particles,
2. there actually is only the field,
3. there are both in interaction.

In the first case one must explain the meaning of the field. One could then scarcely come up with another interpretation but that it somehow described the statistics of particles. This led to the difficulty that probability densities were not supposed to show any interference. The hypothesis of Bohr, Kramers, and Slater avoided this problem by not assuming light quanta as particles but only a quantum-like exchange of energy upon emission and absorption. For them light was a pure field whose *significance*, however, lay in the probability of certain processes in matter which was described as pure particles. This was more a solution of the third type but in line with Bohr's idea of correspondence that, as in classical physics, also in quantum theory light is a field, matter corpuscular.

In the second case one must explain what is meant by particles. One could scarcely understand them as other than special configurations of waves. Einstein considered early on "needle radiation," solutions of Maxwell's equation, sharply localized within a small solid angle. Another possibility was to hope for solutions of a nonlinear wave equation similar to the example of general relativity.

A third solution, which unlike Bohr, Kramers, and Slater did not distribute both models onto light and matter but which assigned real particles *and* a real field to one and the same object (e.g., light) was then apparently not seriously considered.

Quantum mechanics offered an unforeseen way out of this paradox. But before this ("Copenhagen") solution was found, it is understandable that the three cited classical solutions were also tried out in quantum mechanics; we will briefly discuss them in subsections c, d, and e. And the intellectual difficulties were so great that to the present day, the Copenhagen solution has scarcely ever been precisely spelled out, despite the full mathematical clarity of the theory. It might very well be that without the considerations attempted in this book it cannot be clearly formulated at all.

### 9.1.3 Schrödinger

Inspired by de Broglie, Schrödinger in 1926 found his wave equation for matter. Within the next few weeks, he had shown that his theory was mathematically equivalent to Heisenberg's matrix mechanics. Understandably he assumed his waves to be the factual reality behind Heisenberg's abstract formalism. Initially he hoped to have thus conclusively solved the problem of the duality, in the sense of a solution of the second type mentioned above. In general there were supposed to be only fields, a pure physics of the continuum. He could interpret the quantum mechanical energy relations, e.g., in the Compton effect, as pure frequency relations. In a true microphysics he wanted to completely replace the "macrophysical" concepts of energy and momentum with the concepts of frequency and wave number. Then there was no longer any need for "quantum jumps" but only continuous transitions of the wave functions. Physics, so he hoped, could then be strictly deterministic again (Jammer (1974), pp. 24–33).

With these hopes he encountered the criticism of the "Copenhageners," notably Bohr and Heisenberg. From talks with Heisenberg I know something of the atmosphere and content of these discussions. Schrödinger had to explain why the electron, which inside the atom he described as a wave, outside the atom quite obviously is observed as a particle (single scintillations, triggering of counters, tracks in a Wilson chamber). He said: "From people entering a bath fully clothed and coming out in clothes it does not follow that they are also wearing clothes inside." But the argument proved too little. If in reality there were only an electron field, he had to show what the "clothes" were which allowed it to appear outside the atom in the form of particles. He regarded the electron as a wave packet and proved that in the harmonic oscillator a wave packet stays together indefinitely. But Heisenberg showed that this is only due to the uniformly spaced energy spectrum of the harmonic oscillator. Wave packets normally spread irreversibly.

All this came up at a memorable visit of Schrödinger to Copenhagen in the Fall of 1926. Schrödinger caught the flu and was devotedly nursed by Bohr and his wife, where he was staying. But upon opening the door to Schrödinger's sick room one could see Bohr sitting at the bedside, persistently saying to Schrödinger: "But Schrödinger, you *must* admit that...!" Upon departing, Schrödinger supposedly said: "If those damn quantum jumps should start all over again, then I regret having made that entire theory."[1]

Schrödinger's interpretation failed not only because of the spreading of wave packets. It was recognized that Schrödinger's wave in configuration space is something completely different from de Broglie's wave in ordinary three-dimensional space. De Broglie's wave is a classical field. Nothing stands in the way of describing, e.g., its electrostatic self-interaction by means of a nonlinear wave equation or its interaction with the Maxwell field by means of a multilinear equation in $\psi, \psi^*$ and the electromagnetic field strength $F_{ik}$. Schrödinger's description of the one-electron problem, i.e., the hydrogen atom, could be construed as a theory of a de Broglie wave. The two-electron problem (helium) which, after incorporating the electron spin and the Pauli exclusion principle, could be solved exactly with Schrödinger's method, remained inaccessible to the classical de Broglie wave.

Heisenberg moreover remained steadfast all his life that the Schrödinger equation *must* be strictly linear, in contrast to classical wave equations. Only in this way could one maintain the superposition principle which according to his conviction was fundamental for quantum theory. The Schrödinger function can be *defined* as a vector in a linear representation of the symmetry group, and the time-derivative is one of the generators of this very group, hence necessarily a linear operator in vector space. A nonlinear Schrödinger equation would require the transition to nonlinear group representations.

### 9.1.4 Born

Born chose the first of the solutions enumerated above. He held on to the particle nature of the electrons and interpreted in 1926 the intensity of the Schrödinger wave as a probability density (Jammer, pp. 38–44). He was interested in the problem of the indeterminism. He said: "The motion of the

---

[1] Heisenberg, in *Der Teil und das Ganze*, formulates this sentence slightly differently from what I remember, from his own reminiscences.

particle conforms to the laws of probability, but the probability itself is propagated in accordance with the law of causality" (Jammer, p. 40). "Orthodox" quantum theory followed him there.

Nevertheless, Born's suggestion did not yet solve the problem of the interpretation. Heisenberg once told me[2] "Born only published his interpretation at that time because he did not understand that it does not work that way." As mentioned above (*b*), the statistical interpretation was in the air. Born himself explained later (Interview 1962, *Jammer*, p. 41, footnote 33) that his interpretation was inspired by Einstein's earlier conception of Maxwell's field as a "ghost field" which guided the photons statistically along their path. A variant of this "ghost field" (with the renunciation of light quanta) was the hypothesis of Bohr, Kramers, and Slater. The failure of this hypothesis made Bohr and Heisenberg guarded against hasty statistical interpretations.

One can see the difficulty from the wording of Born's thesis. Both its first and second half are couched in language that is still too classical.

The expression "the motion of the particle" seems to reveal that Born had not yet seen what Heisenberg enunciated in 1927, namely that the concept of the *path* of a particle must be abandoned altogether. Indeed, Jammer interprets Born's view to mean that the path of the particle is everywhere defined, although statistically bent; then, e.g., Young's double slit experiment cannot be explained. Another way to characterize the difference between Born and Bohr is that Born, for the rest of his life, ascribed the deviations of quantum theory from classical physics to indeterministic deviations from the law of causality, whereas for Bohr the changed perception of reality was decisive. Einstein clearly understood the radical nature of Bohr's point of view; this was precisely what he then objected to in quantum theory. Born, in his correspondence with Einstein, tried in vain to convince his friend that the chasm was not so deep, given that in classical physics minute uncertainties in the initial conditions lead to arbitrarily large uncertainties in long-range predictions. Pauli, in a letter from Princeton, finally tried to explain to Born that his insistence on the unpredictability of events misses Einstein's point, the concept of reality in classical physics. This idea, however, seems to have remained foreign to Born.

---

[2] In *Der Teil und das Ganze*, p. 110, Heisenberg formulated this criticism in a more subdued and seemingly opposite way: "I certainly considered Born's thesis to be entirely correct but I disliked that it made it appear as if there still existed a certain freedom in the interpretation. I was convinced that Born's thesis followed directly from the established interpretation of certain quantities in quantum mechanics." This, however, implies that the statistical interpretation was for Heisenberg at that time not a novelty but already a foregone conclusion; and it was alluded to that the rigorous derivation of this interpretation from the theory would give it a slightly different form than Born had intended. A few months later Heisenberg found this form in the uncertainty relations.

At least the main point of the statistical interpretation, the reduction of the wave packet upon a measurement, is suppressed in the second part of Born's thesis. "The probability itself is propagated in accordance with the law of causality," i.e., the Schrödinger equation, only as long as no no measurements are made. This follows conclusively from the postulated repeatability of the measurement; we discuss in Sect. 9.3h the only consistent way out that has been proposed, Everett's many-worlds theory. Born however, in accordance with his view, used the reduction of the wave packet without hesitation. As to him the wave function represented probability, i.e., knowledge, he thus did not have the difficulties somebody might have had who, like Schrödinger, interpreted the wave as a "reality." It seems to have escaped Born that just this "realistic" interpretation of the wave was behind the expected causal propagation of the wave. This diminished perception for philosophical problems may also have been the reason that he never really understood why only the Copenhagen interpretation was considered the real breakthrough, not his statistical hypothesis.

### 9.1.5 De Broglie

As an attempt at a solution of the third kind we can consider de Broglie's idea of the pilot wave (l'onde pilote), or the double solution of the wave equation of 1926–27 (Jammer, pp. 44–49), which was taken up again by Vigier (1951–56). The wave equation is assumed to have two solutions: the continuous $\psi$-function and a singular function representing the particle. The $\psi$-function should then guide the particle statistically. In this way de Broglie hoped to convert the indeterminism of quantum theory into a mere expression of ignorance. The continuous wave, as it interacts with both solutions, supposedly "guides" the particle on a path which, by virtue of incomplete knowledge, is only statistically determined.

This idea did not gain acceptance, and is mentioned here only to illustrate that all three of the foregoing solutions have actually been tried. The idea probably failed as soon as the unavoidable reduction of the wave packet occurred, insofar as the wave ought to describe the probabilities for the particles. In de Broglie's theory a measurement obviously does not reduce the continuous wave function to an eigenfunction of the measured observable, corresponding to the measured eigenvalue (see also Sect. 9.2c). To get the correct further predictions from the $\psi$-function one had, like Everett (Sect. 9.3h), to renounce the reduction but show that the future path of the particle is only determined by one component of $\psi$ in the expansion of eigenfunctions of the observable, namely the one belonging to the measured eigenvalue. It is hard to see how this is supposed to follow from the required nonlinear field equation.

### 9.1.6 Heisenberg

### α) Quantum mechanics.

In contrast to the three physicists mentioned above and perhaps all of the older physicists except Bohr, Heisenberg from the outset had a skeptical philosophical disposition toward classical models of the atom. In his book *Der Teil und das Ganze* (1969) he describes his path into atomic physics. In the first chapter the eighteen year old Boy Scout reflects in a conversation with two friends on a hiking trip "in the beech-green of Lake Starnberg" on the concept of the atom. He dismisses a drawing that visualizes chemical valences by means of hooks and eyes on atoms. "For hooks and eyes, it appeared to me, are rather arbitrary structures which can be given any number of shapes according to technical expediency. Atoms, however, ought to be a consequence of the laws of nature and should be induced by these laws to combine into molecules. In my belief there could not be any choice and hence also not any such arbitrary forms as hooks and eyes" (p. 13). He had read, initially with some feeling of peculiarity, about Plato's purely mathematical models of the atom in the *Timaios*, four of the five regular solids: tetrahedron for fire, octahedron for air, icosahedron for water, cube for earth. Later in life he saw this as an early form of group representations, i.e., an explanation of the laws of nature through symmetry. But already for the eighteen year old there was "... a certain fascination with the idea that in the smallest constituents of matter one ultimately should come up against mathematical forms"[3] (p. 21). He concludes "that atoms probably are not things" (p. 25). In retrospect we can put these arguments into a simple form: if atoms are assumed to *explain* the properties of macroscopic bodies then they *may not* have those very same properties; else they repeat but do not explain.

Heisenberg later said about his teachers: "From Sommerfeld I learned the optimism, in Göttingen I learned mathematics, from Bohr physics." In his first conversation with Bohr, 1922 in Göttingen, he learned about Bohr's tentative approach to describing the counterintuitive reality of atoms. He asked at the end: "Will we then ever understand atoms at all?" Bohr hesitated a moment and then said: "Oh yes. But at the same time, and only then will we learn what the word 'understand' means." Einstein (1949, p. 44–46) describes that phase of the quantum theory with the words: "It was as if the ground had been pulled out from under one, with no firm foundation to be seen anywhere, upon which one could have built. That this insecure and contradictory foundation was sufficient to enable a man of Bohr's unique instinct and tact to discover the major laws of the spectral lines and of the electron shells of the atoms together with their significance for chemistry appeared to me like a

---

[3] According to my own interpretation Plato refers to actual mathematical forms and not material bodies having mathematical forms. It is his intention to lift the antithesis between mathematics and physics: the things of our perception *are* ideas. See *Zeit und Wissen*.

miracle—and appears to me as a miracle even today. This is the highest form of musicality in the sphere of thought" (See Sect. 2.11a).

It was given to Heisenberg and not Bohr to find the "solid ground" on which one could build. For this the Göttingen mathematical school was essential, and in particular Born's quest for a new, mathematically consistent mechanics of the atom.[4] Heisenberg's decisive paper of 1925 even indicates in the title that it constitutes a radical step: *Quantum-theoretical Re-interpretation of Kinematic and Mechanical Relations*, with the short introductory summary: "The present paper seeks to establish a basis for theoretical quantum mechanics founded exclusively upon relationships between quantities which in principle are observable." Thus already the kinematics is modified. The "values" of position and momentum are no longer numbers, but noncommuting quantities which Born (1926) soon recognized as matrices. It is in this abstract scheme that one can collect observable quantities like measurement results for transition probabilities without having to incorporate them into the unobservable classical model of the particle.

### β) Uncertainty relations.

The intent to restrict attention to measurable quantities no doubt was influenced by Mach and contemporary positivism. The positivistic mode of thought facilitated the quantum theoretical revolution, as it had already had a stimulating effect on the young Einstein. For it broke with two dogmata rooted in the classical paradigm, realism and apriorism. In this mental atmosphere it was easy to search for new laws. But it is characteristic that both Einstein as well as Heisenberg in their later years decidedly renounced positivism. They distanced themselves from the new dogmatism which wanted to put in place the experiences of the senses as the unquestioned foundation instead of the classically described material reality or the unshakable experience a priori. A talk with Einstein in the spring of 1926 was decisive for Heisenberg. It was there that he formulated his criticism of the intended restriction to observable quantities, "Only theory decides what can be observed." (*Der Teil und das Ganze*, p. 92) In other words, one can only see what one knows.

Heisenberg had fastened the hook for the comprehension of this fact precisely in the formulation that he wanted to establish a basis based exclusively upon relationships between quantities which *in principle* are observable.[5] Quantities which are unobservable in principle should be excluded, yet only theory decides what in principle is observable. In fact he had only relied on those quantities that *actually* and unquestionably had been observed; but what about those quantities that had not been actually observed, like, e.g., all the positions and momenta along the path of a particle? Where they in principle observable? B.L. v.d. Waerden, in connection with his studies on the

---

[4] See Born (1924), his book (1925).
[5] See also Sect. 4.1a.

history of quantum theory, has once asked me, quite astonished, about the meaning of a letter of Heisenberg to Pauli, late in the Fall of 1926, in which he describes the problem of atomic physics as completely unresolved. The finished mathematical theory was already at hand; then what was still missing? In the language of the present book one can answer: the physical semantics was missing.

The solution was the uncertainty relation (1927). The classical properties of a particle, position and momentum, are in principle observable but they are in principle not observable simultaneously. That was not a premise but a consequence of quantum theory. The theory decided what is observable. In the language of Hilbert space: The operators for position and momentum separately have eigenvectors but they do not have any common eigenvectors. On a classical trajectory of a particle, however, both must be defined simultaneously; therefore there never exists a classical trajectory.

Most instructive is Heisenberg's own path to this insight (p. 111–112). "We had always casually said: the path of the electron in the cloud chamber can be observed. But perhaps what one really observed was less. Perhaps one could only perceive a discrete sequence of inaccurately determined positions of the electron. Actually one can only see single water droplets in the chamber which certainly were very much larger than an electron. The correct question therefore must be: can one represent in quantum mechanics a situation where an electron approximately—that is with a certain inaccuracy—is found at a given position and at the same time has approximately—again with a certain inaccuracy—a prescribed velocity, and can one make those inaccuracies small enough not to come into difficulties with experiment?" The affirmative answer to this question is the uncertainty relation.

Heisenberg's thesis was often misunderstood as "positivistic,"[6] as if it asserted: "States with sharply defined position and momentum cannot be observed, therefore they do not exist." Only the logical inverse is correct: "According to theory these states do not exist, hence they also cannot be observed." "They do not exist according to theory": this is the above statement that they do not occur in Hilbert space. The thought experiment with the gamma ray microscope only fends off the objection: "But we can observe them, therefore they must exist." *If* both light and the particle in the microscope obey quantum theory then it is shown that such states simply cannot be observed.

The allegation that the uncertainty relations were due to the perturbation of the state by the measurement process is misleading. Using the word "state" in the sense of quantum theory, as ray in Hilbert space, then neither before nor during nor after the measurement does there exist a state with simultaneously defined position and momentum. The "perturbation" is the reduction of the wave packet, i.e., the transition to new knowledge by means of a measurement. For example, first one *knew* the momentum of the electron and

---

[6] See the analogous discussion of Einstein's relativity of simultaneity, Sect. 2.8.

therefore it did not *have* a position; afterwards one *knows* its position and therefore it *has* no momentum. Whether there still exist, beyond quantum theory, "actual" completely defined positions and momenta is the problem of "hidden variables" which we will consider later. Heisenberg's argument merely shows that quantum theory is consistent. Naturally it originates in the experience gained from the clarification of a previously incomprehensible difficulty, a clarification that was achieved through a decisive renunciation of a classical description.

**γ) Particle picture and wave picture.**

In his paper about the uncertainty relation Heisenberg added a footnote in proof. It says that according to a remark by Bohr the uncertainty of position and momentum were unavoidable because an electron could not only be described as a particle but also as a wave. In the first months of 1927 there was a technical disagreement between Bohr and Heisenberg about the conjectured correct interpretation of quantum mechanics which even led to serious personal irritations. While they were separated for a few weeks, Bohr going to Norway for a skiing trip while Heisenberg remained back in Copenhagen, each found his own solution: Heisenberg the uncertainty of position and momentum, Bohr the complementarity of wave and particle. At Bohr's return they eventually agreed to the formulation of complementarity being the cause of the uncertainty. That is the fourth possible solution to the problem of duality. Matter and light "by themselves" are *neither* particles *nor* waves. Yet if we wish to visualize them we must use both pictures. And the validity of one picture imposes limitations on the validity of the other. This is the main point of the Copenhagen interpretation.

We will discuss later what complementarity meant for Bohr's way of thinking. Now we give an account of what happened to the duality of the pictures in Heisenberg's interpretation, as he presented them in his book *The Physical Principles of Quantum Theory* (1930) and how in those days one learned it from him as a student.

Quantum theory was seen, since Bohr's correspondence principle, as the result of a transition from a given classical theory to a corresponding new theory. This transition is called quantization of the classical theory. It was accomplished, no doubt following an idea of Born, by bringing the classical theory into Hamiltonian form and then replacing the canonically conjugate variables by algebraic quantities satisfying Heisenberg's commutation relations. In the language of Hilbert space, these quantities are self-conjugate linear operators. One can then show that conversely the classical theory follows from its corresponding quantum theory in the limit of large quantum numbers. This is, up to the present day, the prevailing interpretation which one can find in all textbooks.

The question then was from which classical theory one ought to start. For matter one started with classical point mechanics and obtained through quantization Schrödinger's wave mechanics in configuration space, which one then learned to reformulate in Hilbert space. For light, Dirac (1927) started with the classical Maxwell equations; their quantization resulted in the first example of a quantum field theory. But where then was the duality of the pictures? Was the wave theory in actual fact the theory of the Schrödinger wave? Were the particle picture and the wave picture therefore in the same relationship with one another as classical and quantum theory? But the wave picture was as classical as the particle picture, and Schrödinger's wave in configuration space, as Heisenberg always emphasized, was something completely different from the de Broglie wave. Jordan and Wigner then quantized de Broglie's classical field theory of free matter and found a theory mathematically equivalent to Schrödinger's.

Heisenberg describes these mathematical facts in the appendix of his book of 1930. The same quantum theory follows from the quantization of two completely different classical theories: the mechanics of point masses and de Broglie's wave theory. This is the mathematical reason for the duality of the two classical pictures. Heisenberg pointed this proof out to me in a conversation: "This proof is important. You must understand it." I asked whether he could not tell me in simple words *why* two different classical theories lead to the same quantum theory. "I myself know nothing more. You can just prove it. Therein lies the secret."

One could make this state of affairs more approachable by reversing the line of reasoning. From quantum field theory one arrives at two different classical theories through two different limiting procedures. That this can happen is mathematically not surprising. But one strictly speaking, one has in this way already abandoned the spirit of the correspondence principle. One then treated the particular quantum theory as given and derived from it, in a second step, classical theories. Bohr, however, was never ready for such a reversal, and also Heisenberg did not like it. How should one know about the quantum theory of fields if not through experience that can be described classically? To me however, as well as many of the younger generation, a direct path to quantum theory appeared feasible and desirable. The reconstructions in Chap. 3 are the outcome of the search for such a path.

Formally one could consider the interaction-free field theory of de Broglie also as Schrödinger theory of the single particle problem of classical point mechanics. Quantizing then this field theory was already a "second quantization" (see also Sects. 3.5c and 5.4e). This terminology is now generally accepted. Heisenberg forbade me to use this expression "which is capable of making any understanding of its meaning impossible." He insisted that the Schrödinger wave in three dimensions and the de Broglie wave had completely different physical meaning: the one as quantum theoretical state vector of *one* particle and the other as classical amplitude which after quantization corresponds to *many* particles. Yet this does not explain their formal identity; therein lay the "secret." Precisely this became later the starting point of my own interpretation of quantum mechanics: the relationship between the quantum theory of a particle and classical wave theory is the quantum theoretical transcription of the relation between probability and frequency of occurrence. Repeated quantization is the same as iterating the concept of probability in the declaration that probability is the expectation value of a relative frequency. Heisenberg eventually approved of this interpretation.

According to this interpretation the duality of the pictures is, however, a secondary phenomenon. The field is the quantum theoretical probability field of the particles.[7] Furthermore, adding the ur hypothesis, then even a particle is not a last entity but a statistical distribution of urs; the particle is related to the ur in the same way as the field is to the particle.

### 9.1.7 Bohr

Bohr's thinking was never based on mathematical structures, on what physicists somewhat condescendingly call the formalism, but rather on the classical description of experience and an unrelenting examination of the meaning of concepts.

We have already discussed Bohr's earlier conceptual accomplishments: the introduction of quantum theory into the atomic model in Sect. 2.11a, the correspondence principle in this chapter. Now we must deal with complementarity.

Bohr coined the concept of complementarity in 1927, and from then on it had to him a key physical and philosophical position. As regards philosophy, it admittedly was prepared since Bohr's youth. We can recall the anecdote where he, shortly after 1927, described the philosophical consequences of the new quantum mechanics to an old friend who was not a physicist (Chievitz, I believe). The friend finally replied: "Yes, Bohr, that is all very nice. But you will have to admit that you already said exactly the same already twenty years ago." If philosophers later grumbled that Bohr had used a very special concept from physics as a model for quite different problems in, e.g., psychology or ethics, then they did not recognize the genesis of the concept. This mode of thinking had always been present for Bohr, and in 1927 his great experience was that it proved so effective even in physics.

The key to a precise understanding of the concept of complementarity, however, remains in its application in physics 1927. Particle and wave are two classical ways of describing the phenomena, both of which are forced upon us through experience, but which preclude one another when strictly applied. That is what one means by saying that they are complementary.

This is an accurate description of what in 1927 could be ascertained about the duality of particle and field. In this sense the description is correct also

---

[7] This interpretation was clearly formulated by Bopp by 1954. My investigations in Weizsäcker (1955) were partly stimulated by Bopp's work.

according to our present knowledge. We have, however, seen at the end of the previous section that from our point of view duality is not a fundamental but a derived fact. Even if one does not accept the particular interpretation we assign to the present state of knowledge, one must admit that the concept of complementarity is of no importance for theoretical physics today; it is mentioned in textbooks but more out of historical respect. Yet for Bohr the concept in itself had fundamental significance which we here want to pursue a while.

The philosophy of Bohr's later years[8] can be grouped around three basic concepts, the concepts of *phenomenon, language,* and *classical description.*

Bohr explicitly introduced his concept of *phenomenon* only in 1935 in the reply to Einstein, Podolsky and Rosen. Science ought to assert only what we, at least in principle, can know. We can only know what is linked to phenomena according to laws of nature. One should, however, not designate isolated perceptions as phenomena but in each case only the intelligible entirety of a *situation* through which perceptions obtain a direct meaning. It is no "the breaking of dawn" is a phenomenon but an open landscape at early dawn, with a human being in it for whom this break of dawn indicates the arrival of the sun.[9] It is not the deflection of a pointer on a scale that is a phenomenon, but a room with instruments which were built by the machinist of the institute, and on which the experimentalist reads off the strength of the current of a discharge. Bohr therefore acquired the habit of making drawings of thought experiments in a "pseudo-realistic" style: a wall was not represented by *one* line but by two parallel lines, hatched in between, to indicate its material thickness, etc.

One can see how this concept of phenomenon embraces as a whole all the elements that occur separately in the competing schools of positivism, realism, and apriorism. It is about sense perceptions of real objects which we interpret conceptually. Isolated sense-stimuli are unintelligible, objects about which we do not know anything are irrelevant for us, concepts only make sense when they can be applied to concrete objects. Kant says: "perceptions without concepts are blind, concepts without perceptions are empty." Bohr naturally realized how far modern physics has advanced, with instruments and hypotheses, beyond the idyll of a summer morning or a quaint laboratory. It was precisely this physics that Bohr wanted to remind of what it must always assume if it is to produce *knowledge*: real instruments, and causal chains that can be verified.

The outline I have chosen for this book is much more abstract than Bohr would have chosen. I do not know whether he would have approved of it. Nevertheless, I feel in it greatly indebted to the introduction to Bohr's concept

---

[8] See, e.g., Meyer-Abich (1965), Scheibe (1973) Chap. 1, Heisenberg (1969), Jammer (1974), my article *Niels Bohr* in Weizsäcker (1957), my (1982), and above all Bohr's own later works.

[9] This example is not by Bohr.

of phenomenon. Time should not be introduced primarily mathematically as a continuum of numbers but in the grouping of present, past, and future which forms the basis of our experience. Irreversibility is explained in the seventh chapter as the course of that event which through our experience can become a phenomenon; the difficulties many physicists have with this explanation are always due to the fact that they assume a mathematical model of the event, without careful investigation of how the concepts employed can be given a phenomenological meaning. In the third chapter we build up quantum theory from the concept of decidable propositions, the abstract formulation of what is necessary for empirical knowledge.

*Language* is necessary, as there is no science if we cannot *say* what we know. To Einstein's "God does not play dice" Bohr replied: "It is not relevant whether God plays dice or not, but whether we know what we mean when we *say* that God is throwing dice or not." For this reason Bohr liked to explain the necessity of complementary concepts as being due to *limitations of our means of expression.* "We are suspended in language" he used to say in his conversations with Aage Petersen. He was talking that way long before linguistic philosophy had become fashionable. Admittedly he never made the structure of the language itself an object of investigation. In the linguistic explanation of complementarity he only pointed out that in the description of phenomena "we always depend on expressing ourselves by means of word-paintings." If we do not have a word which unambiguously describes a phenomenon we then have to use several approximate words with mutually exclusive scopes. He usually explained this limitation not in terms of the structure of language (like, e.g., in the now familiar idea of the philosophical prejudice which lies in the use of nouns and definite articles), but by dealing with specific questions of every subject area. Thus he was speaking strictly ethically when he spoke of the complementarity of justice and love; in this domain he liked to refer to the wisdom of the psalms or ancient Chinese. An examination of the psychological process of speech led to the thesis of a complementary relationship between the analysis of a concept and its actual use.

For *quantum theoretical complementarity* Bohr found it decisive that we must describe every actual measurement by means of *classical concepts.* About this thesis Bohr remained unrelenting. Here I quote again an often told anecdote. During tea at the institute, Edward Teller and I were sitting next to Bohr. Teller tried to explain to Bohr that after a longer period of getting accustomed to quantum theory we might be able after all to replace the classical concepts by quantum theoretical ones. Bohr listened, apparently absent-mindedly,[10] and said at last: "Oh, I understand. We also might as well say that we are not sitting here and drinking tea but that all this is merely a dream." He clearly pointed thereby to the prerequisites for a phenomenon.

---

[10] Teller likes to tell the story as: "Bohr fell asleep. After I finished he woke up again and said ..."

But why cannot the concepts we use to describe a phenomenon be "quantum theoretical"?

According to Bohr's thinking in terms of the correspondence principle, one might reply that there are not any "quantum theoretical concepts" at all in this sense, that is, not any observables specific to quantum theory, but only modified laws for classical observables. Bohr's precise argument was then that on the one hand, a measurement device must be observable, i.e., we must be able to describe it in space and time. On the other hand, it must be possible to draw conclusions, from its directly observed response, about the not directly perceived behavior of the measured object in a strictly causal manner.[11] Both conditions are simultaneously satisfied if the measurement apparatus obeys classical physics. The noncommutativity of quantum theoretical observables, however, rules out both conditions to being satisfied at the same time. A device can therefore be used for a measurement only in that approximation where one can neglect the noncommutativity of the observables in the description of the measurement process.

This question belongs to those technical problems of the semantic consistency of quantum theory which have not satisfactorily been dealt with in the present theory of measurement. We discuss it further in Sect. 9.2b.

### 9.1.8 Neumann

J. v. Neumann's book *Mathematical Principles of Quantum Theory* (1932) can be described as the takeover of quantum theory by mathematics. Neumann had soon identified the totality of Schrödinger wave functions as a Hilbert space and applied to quantum mechanics the theory of Hilbert space to which he himself had made significant contributions. (I have heard the anecdote that he had not published the transformation theory of Dirac–Jordan before these two authors because he had thought that precisely this was what physicists had meant in their previous work on matrix and wave mechanics.) After the publication of Neumann's book there remained no doubt that precisely this was the mathematical structure to which the development of quantum mechanics had led, even when there still remained disagreement about details like the admissibility of the Dirac $\delta$-function.

Thereby, however, shifted the weight of the various arguments in the interpretational problem, initially perhaps almost imperceptibly. Physicists at first argued from the *motives*: will—can—may the theory assume the form into which it gradually developed? Now one could argue from the *result*: is the accepted form in the theory in accordance with certain motives or not? And if it is not, will one still hope for a change of the theory of will one adjust the motives? The mathematically versed among the leading theorists, like

---

[11] This, incidentally, is an up-to-date version of Kant's duality of the prerequisite for all experience: intuition and reason, with the law of causality belonging to the principles of pure reason.

Pauli, Born, Jordan, Heisenberg, and no doubt also Dirac, always pursuing his own independent ways, were certainly convinced from 1926 onward that the mathematical form of the theory had been found. But Neumann's book acted like the codification of a system of law: now this knowledge was available not only to the initiated but to everybody. The longer quantum theory existed the stronger this influence became. Hilbert space theory is the form in which quantum theory has proved itself now for more than seven decades; it provides the common language in which one can talk about quantum physics. Even a relatively obvious attempt at a generalization like the introduction of an indefinite metric, e.g., by Heisenberg (1959), has not been accepted as a fundamental theory; in field theory, as with Bleuler-Gupta, it is only an expansion of the mathematical formalism but not a modified fundamental theory. Therefore also all attempts at an axiomatic formulation from simple physical postulates also aimed at a reconstruction of this theory. Our third chapter also follows this tradition.

In this way, however, a leveling tendency is encouraged in the interpretation. It shows most clearly in Neumann's postulate to consider every bounded self-adjoint operator as an observable. Bohr never accepted Neumann's mathematics as the adequate form of quantum theory. For Bohr a quantity was an observable only if one could specify a possible measurement device in space and time. That would at most give the algebra of all of Neumann's observables much too broad a context within which to formulate the actual quantum theory. If the suspected triviality of the ur hypothesis turned out to be false, one would also have to interpret the theory of our Chap. 4 and 5 in the same way. "Real observables" would then be only those which can be constructed from ur alternatives.

### 9.1.9 Einstein

About Einstein's reaction to quantum mechanics and its Copenhagen interpretation there exists an overabundant literature.[12] In this section we only want to briefly characterize his place in the history of the interpretation. We have mentioned his accomplishments in the early history of quantum theory; the problem of the interpretation, from his point of view, we will discuss in Sect. 9.3d.

Einstein never approved of the form which quantum theory had assumed in the hands of Bohr and his pupils. The highly praised achievement of Bohr, from the quote in Sect. 9.1c above, "appears to me as a miracle even today." Later, in the same text, Einstein says: "It is my opinion that the contemporary quantum theory by means of certain definitely laid down basic concepts, which on the whole have been taken over from classical mechanics, constitutes an optimum formulation for the connections. I believe, however, that this theory offers no useful point of departure for future development" (p. 86).

---

[12] See, above all, Einstein (1949), Bohr [ibid.], Jammer (1974), Pais (1982).

Einstein tried, up to the Solvay Congress of 1930, to find internal contradictions in quantum mechanics. He did this mainly by inventing thought experiments in which quantities were measurable which were not assumed to be so according to quantum theory. The climax was a thought experiment involving the measurement of energy in a gravitational field that was supposed to contradict the uncertainty relation between energy and time. Bohr, who before had already demonstrated all these thought experiments to be in agreement with quantum theory, found this time in Einstein's argument a contradiction to general relativity which Einstein conceded. Bohr (1949) later gave an impressive account of this incident (see also Jammer 1974, pp. 121–136). After Einstein had lost this fight of the titans with Bohr about the interpretation of quantum mechanics, he revised his opinion that quantum mechanics, although consistent, was incomplete. It would not describe the reality of atoms but only an incomplete knowledge of them, and therefore could not make deterministic but merely statistical predictions. Einstein now pinned his high hopes on a general nonlinear field theory which might give particles and quantum phenomena as singularity free solutions. He did not succeed in solving this difficult problem. One can suspect that it would have foundered on the same problems we have mentioned above in connection with the ideas of de Broglie.

## 9.2 The semantic consistency of quantum theory

### 9.2.1 Four levels of semantic consistency

Semantic consistency of a theory will mean that its preconceptions, how we interpret the mathematical structure physically, will themselves obey the laws of the theory. This can in general only be achieved to a limited extent. For the mathematical content of a theory must be sharply delineated whereas the verbal understanding is based on undifferentiated colloquial speech. Quantum theory, however, as the most general physical theory known to us, ought to be demonstrably semantically consistent to a particularly high degree.

Quantum theory is a theory about probabilities, that is of predictions. The predictions are possible results of empirical decisions of alternatives, i.e., the results of measurements. Quantum theory must therefore make assumptions that, and how, alternatives are decided, and how measurements can be performed. Thus the theory of measurement is at the center of every check of its semantic consistency.

We go through this check in four steps.

In the *first* step (subsection *b*) we reassure ourselves of the meaning of the usual expression in which an observer, who is not described quantum theoretically, obtains *information* through measurements of an object. Here we merely consider the consistency of the description of our knowledge of the object in terms of the $\psi$-function, i.e., only the meaning of the theory, which will subsequently be applied to its preconceptions.

The *second* step (c) assumes, according to the now already traditional theory of measurement, a separation of the "subject part" into the conscious observer and a *measurement apparatus*. It interprets Bohr's thesis of a necessarily *classical description* of the measurement device as the necessity of irreversible processes during the measurement.

In the *third* step (c) we apply quantum theory to the measurement apparatus.

Finally, in the *fourth* step (d) we investigate whether quantum theory can be applied to the *observer* himself.

### 9.2.2 Gaining information by means of measurement

In the historical section of this chapter we mentioned several times the *reduction of the wave packet by means of measurement*. Now we discuss the meaning of this thesis.

In most expositions of quantum theory it is mentioned as a perplexing fact that theoretically there are two different ways the state vector ($\psi$-function) can change in time:

1. continuously, according to the Schrödinger equation
2. discontinuously upon measurement

Continuous change appears to be natural because we know it from classical physics. Why should there also be discontinuous changes, the "quantum jumps" Schrödinger hated so much? And how is the measurement, i.e., the interaction with a conscious human being, supposed to act on the state of the system? But let us leave out consciousness and restrict attention to the interaction of the object with the measurement device. How can this process—which must be completely describable by the Schrödinger equation—give rise to a discontinuous change of state?

We will remove this appearance of a paradox step by step, following the usual theory of measurement (see, e.g., Jauch 1968). In the presentation we only pay attention to the inevitability and consistency of the quantum theoretical description. We will see that there occurs no paradox at all if one applies this description consistently and without ambiguous vocabulary.

The "discontinuous" change of the known state due to a measurement is first of all a natural and unavoidable consequence of two basic rules of the theory:

$\alpha$) the probability interpretation of the wave function
$\beta$) measurements can be repeated

We have discussed above how rule $\alpha$) had been accepted historically; in our own deduction it is the starting point for the entire quantum theory. Rule $\beta$) means that one can check the results of measurements.

In the language of probabilities, it follows from rule $\beta$) that the probability is 1 to again find a measured result upon immediate repetition of the measurement. This simple formulation holds only for "measurements of the first kind" (Jauch, p. 165), that is, for measurements, like the position measurement of a particle, which do not change the value of the measured observable and therefore can be used as a preparation of the state for the subsequent second measurement. "Measurements of the second kind," for example the momentum measurement on a particle, change the value of the observable by a certain amount. For them one must distinguish between the value of the observable before and after the measurement. Repeatability then means that the value after the first measurement turns out to be the same as the value before the second measurement. To simplify matters we restrict attention here to measurements of the first kind.

Rules $\beta$) and $\alpha$) together express precisely the relationship between fact and possibility, i.e., past and future. We interpret a fact that has been found about an object as something objectively given, as an objective actuality which therefore can be found again: factuality implies necessity for the result of a (correctly performed) test of the fact. The probability interpretation, on the other hand, simply expresses the meaning of the $\psi$-function as it arose out of the mature quantum mechanics. Even Einstein did not dispute this after 1930; precisely because of it he considered quantum theory to be incomplete. But our investigation here is only about the *consistency* of quantum mechanics, not its completeness. But when, in a certain prepared state, one can find via measurement an incommensurable state with nonvanishing probability, this then implies something like discontinuous change.

This state of affairs is most easily expressed in the language of information theory. An observation is the gain of information by the observer. The discontinuity exists in the knowledge of the observer. Assume a sequence of measurements $\ldots M_0, M_1, M_2, \ldots$ on the same object at certain successive times $\ldots t_0, t_1, t_2, \ldots$ Between $t_0$ and $t_1$ he knew the result $M_0$. After $t_1$, he knows more: he also knows the result of $M_1$, and so on. The discontinuity exists, strictly speaking, only in his description of his gains of knowledge. His mental processes presumably go on continuously. But what he can *say* about them changes at intervals. The idealized point in time $t_1$ is actually only an indicator of the time interval in which he obtained information of $M_1$.

The $\psi$-function, i.e., the set of the components of the state vector, is on the other hand nothing but a list of all possible predictions he can make about the result of a future measurement, assuming the result of the last one to be known; for example about $M_2$ given the result of $M_1$. If we consider $t_2$ as a continuous variable, i.e., if we do not assume beforehand when the next measurement will be done, then $\psi_{M_1}(t_2)$ gives the probabilities for any later time $t_2$. As the object constantly changes state under the influence of forces and inertia, it is only natural that $\psi_{M_1}(t_2)$ is varies continuously with $t_2$: all these probabilities vary continuously with time. It is equally natural that the gain in knowledge through $M_2$ changes all predictions "discontinuously" in the above sense.

Thus far we have only repeated explicitly the arguments which have compelled physicists, since the introduction o the statistical interpretation, to use

## 9.2 The semantic consistency of quantum theory

the dual rule for the change of $\psi$.[13] No contradiction arises if we follow these simple arguments. But we can now see that the name "state" for $\psi$ is misleading. $\psi$ is a catalog of knowledge that follows from *one* observed fact and which determines the probabilities for an infinity of possible future events; $\psi$ is no more than that.

The past does not need a description by means of a $\psi$-function which depends continuously on time. As far as we know the past, it consists of facts which in principle one could list separately (e.g., at time $t_0$, from the results of $\ldots M_{-3}, M_{-2}, M_{-1}$). Insofar as we do not know the past, we can make hypotheses about its facts; more about this under c below. The future, however, is only known to us in the form of a probability catalog, called the $\psi$-function, a catalog whose validity lasts precisely to the next measurement, and beyond that only as a "mixture" of $\psi$-functions belonging to the possible results of the measurement.

There is only *one* necessary conclusion from quantum theory regarding the past. Every fact, now past, was once a future fact; its probability could then be determined from a $\psi$. Hence, firstly, every fact of the past must be a formally possible event, according to quantum theory, for the given object under consideration. Secondly, the relative frequency of certain past events must agree, on the average, with their probability as it can be computed from quantum theory from the previous events. In this sense the $\psi$-function expresses, between two past measurements, e.g., between $M_{-2}$ and $M_{-1}$, what predictions the observer could make after $M_{-2}$ about $M_{-1}$.

Moreover, in this sense it is equally permissible to use the $\psi$-function for "retrodictions." Let us assume the observer had made the measurement $M_{-1}$ and forgotten or never obtained the result of $M_{-2}$. Then he can use the Schrödinger equation to compute, backwards from the result of $M_{-1}$, the probability of the unknown but objective fact which is expressed in the result of $M_{-2}$. Here again the relative frequencies of such "retrodicted" results must agree with their quantum theoretical probabilities. In this way it is possible to use two completely different $\psi$-functions for the same time interval: one for the prediction of $M_{-1}$, being made by an observer who knows $M_{-2}$, the other for the retrodiction of $M_{-2}$ by an observer who knows $M_{-1}$. Operationally, the retrodiction is also a prediction. It is the prediction of what one will discover once one learns $M_{-2}$ from a document or the remembrance of a human being. This double $\psi$-function makes sense if we know that between $M_{-2}$ and $M_{-1}$ no observation was performed; otherwise none of the two $\psi$ can be defined in the manner described above.

---

[13] This is precisely the difference between quantum mechanics and the theory of Bohr, Kramers, and Slater. The latter did not assign the probability field to *one* or a fixed number of particles, did not reduce it after a measurement, and therefore could only determine statistically the number of light quanta or energy quanta.

We summarize: $\psi$ is knowledge, and knowledge depends on the information collected by the knowing subject. Knowledge is of course not dreaming, not "merely subjective." It is knowledge of objective facts of the past which will turn out to be identical for anybody who has the necessary information; and it is a probability function for the future that holds for everybody who has the *same* information, and which can be checked empirically through measurement of relative frequencies in the manner described in the third chapter. All paradoxes occur only if one interprets $\psi$ itself in some other sense as an "objective fact," a fact going beyond that at a certain time a certain observer has a certain knowledge. Facts are past events which we in principle can know today.

### 9.2.3 Measurement theory, classically

#### $\alpha$) The idea of a quantum theory of measurement

Einstein, Bohr, and Heisenberg were masters in the discussion of thought experiments. Einstein probably invented the method, although it had forerunners in classical physics. Bohr wielded it with the greatest confidence and reflected on its fundamental significance. Epistemologically, the idea of a thought experiment is to test the semantic consistency of a theory by describing, by means of the theory itself, the process of gaining information through a measurement. We reiterate that a prior understanding of how we normally describe measurements is not dogmatically assumed in a thought experiment, but is scrutinized for its compatibility with the theory. Einstein's measurement of time with moving clocks, Heisenberg's position measurement with a gamma ray microscope do *not prove* the impossibility of absolute simultaneity for distant events or the simultaneous determination of position and momentum. Rather, these thought experiments show that the theory which asserts these impossibilities *cannot be refuted* by such experiments, assuming one analyzes these experiments in accordance with the theory.

The quantum theory of measurement is, so to speak, the general theory of arbitrary quantum theoretical thought experiments. It treats the measurement apparatus and the measured object together as one combined quantum theoretical object and shows that, or in what approximation, one can obtain the same statements about the state of the object as from an isolated examination of the object. Heisenberg describes this as "shifting the cut between observer and object." There is one point of decisive importance where the just described quantum theory of measurement does not yet prove the semantic consistency of quantum theory. It is about the *objectivity of the measured result*. One assumes as self-evident that a *measured result* is an *objective fact* which can be stored in the measurement apparatus, completely independent of whether and when an observer reads it. In our initial interpretation of measurement being a gain of information this was initially only assumed about the

## 9.2 The semantic consistency of quantum theory

consciousness of the observer is that he stores the facts in his memory. For the measurement to be a phenomenon Bohr demanded that, as mentioned above, it be describable in terms of classical concepts. Precisely because of this the explicit recourse to the consciousness of the observer becomes superfluous. The measured result can then be described as objectively existing like all states in classical physics, and it makes no difference who reads it off and when. On the other hand, it is just the meaning of the uncertainty relations that this does not hold for a quantum state: the electron has a definite position only if this position has just been measured. When Heisenberg had concluded, from a first reflection on the quantum theory of measurement, that "the cut between observer and object is movable," Bohr replied in a letter that in a certain sense one can also assert the opposite, that the cut may not be moved beyond the measurement apparatus, as the measurement device must *necessarily* be described classically.

This controversy between Bohr and Heisenberg deserves closer inspection. The prevailing opinion is that the measurement apparatus itself obeys quantum mechanics but that it is only suitable for a measurement if "it does not do any harm" to describe it classically with regard to the measurement process (the "Golden Copenhagen Rule"). Bohr himself had occasionally considered that perhaps in principle quantum theory may not be applicable in the macroscopic domain. Ludwig (1954) has later sharpened this idea to the conjecture that the classical theory of macroscopic and the quantum theory of microscopic processes must be unified in a comprehensive theory, different from both. The entire plan of this book, however, has been to look for and enumerate arguments for the exclusive and strict validity of quantum theory. Therefore we will also start here with the prevailing "orthodox" view. Then we are bound to interpret Bohr's point of view according to the "Golden Rule" and also *justify* it.

We originally postulated that the measured result must be a "result," an objective *fact*. This corresponds to Bohr's postulate of a *unique* description of the phenomenon. But now we again take up the question of Sect. 9.1g: why and in what sense had the unique description to be classical? In our interpretation of quantum theory, Bohr's way of thinking in terms of correspondence is the historical form of a fundamental point of view which for our purposes we must also formulate more precisely than the word "classical" implies. To Bohr, classically given were Newtonian–Hamiltonian mechanics and Maxwell's electrodynamics. To us these are initially data from the history of science. The mechanics of Galileo and Newton replaced Aristotelian physics, and before the physics of the Greeks there was again the different rationality of the myth. Teller's question in the tea anecdote was just: What is the meaning of the word "classical" if it is to denote a feature of the physics of Newton and Maxwell that can never be replaced by a new physics?

From the perspective of Hilbert space theory, classical physics ought to be a limiting case of quantum theory. Exactly the opposite as in Bohr's argument, a classical observable ought to be a quantum theoretical observable for which one may disregard, in a certain approximation, the noncommutativity. But we do not follow Neumann's "leveling" assumption (Sect. 9.1h) that every bounded self-adjoint operator represents a permissible observable. We rather conclude from the quantum theory of measurement (see Sect. 9.2d below) that only those operators are quantum theoretical observables which may actually serve as Hamiltonian operators. The operators of interaction depend, according to classical physics, on the relative coordinates (or, for magnetism, on the relative velocities) in coordinate space, within special relativity as action at a point, hence "local." If one builds quantum theory on the classical spacetime continuum, then this is an additional empirical fact. In the ur hypothesis one chooses the more thrifty assumption that the separability of alternatives determines the common symmetry group of all objects and hence also the law of interaction. Then the structure of spacetime is a *consequence* of this symmetry and not the other way round. What then distinguishes the classical description as regards measurements?

Bohr says that a spacetime description and causality are compatible only in classical physics. He demands causality for the measurement instrument in the sense that it allows a unique conclusion about the object from the observed phenomenon, thus not to lose any information. This requirement can be satisfied if we submit the measurement apparatus and the object to the Schrödinger equation; this postulate will be the starting point in the following sketch of the quantum theory of measurement. How is it with the description in space and time? About the structure of space we have not made any assumptions in the theory of measurement but rather about the structure of time, in the sense that the measured result must be a fact. This makes it plausible to suspect that the key to the concept "classical" lies in the irreversibility of facts.

### β) The irreversibility of the measurement

The discussion between Teller and Bohr at tea in Copenhagen 1933 found its continuation a decade later on a transcontinental train in America (verbal communication by Teller). This time Teller tried to convince Bohr that an irreversible process in the measurement apparatus was decisive for a measurement. Again he was not successful. Bohr replied that the fundamental properties of a measurement cannot depend on a specific physical theory like thermodynamics.[14] Naturally we side here with Teller. If we ought to quote authorities, we recall that Einstein looked upon thermodynamics as "the only physical theory of universal content concerning which I am convinced that, within the scope of its basic concepts, it will never be overthrown" (see Sect. 2.5). From the standpoint of this book irreversibility is a consequence of the fundamental structure of time, as explained in Chap. 7. We now apply those arguments to the act of measurement.

---

[14] We can remark that Bohr himself had occasionally called measurements "irreversible."

The measurement process is irreversible in the sense of statistical physics. The measurement apparatus has many degrees of freedom. Denote by $L$ the observable of the object and assume $p(\lambda)$ to be the probability for finding the eigenvalue $\lambda$ of $L$ in the state of the object before the measurement. Assume further an observable $M$ of the measurement apparatus which corresponds, in terms of the interaction during the measurement, to the observable $L$ in such a way that $M$ assumes a specific eigenvalue $\mu_\lambda$ for $L$ having the eigenvalue $\lambda$. One must then expect that for a subsequent measurement *on* the measurement device (reading off the measured result) $p(\mu_\lambda) = p(\lambda)$. Even when the state $|\lambda\rangle$ of the object was a pure case, the state $|\mu_\lambda\rangle$ of the apparatus corresponds to an extremely large subspace of its Hilbert space. While the value $\mu_\lambda$ for the apparatus remains constant after the measurement, a complicated and unknown motion takes place in this subspace, even including an unknown interaction with the environment of the measurement apparatus. Even if we fictitiously assume the exact quantum state of the measurement apparatus to be known immediately after the measurement, this knowledge will be lost before long. There is then not a known $\psi$-function for the measurement device, only a probability $p(\mu_\lambda) = p(\lambda)$. This situation in the apparatus can be sensibly described by a density matrix $W(\mu)$. But this density matrix in turn will vary discontinuously through an observation *on* the measurement apparatus, what we call "reading off the result." There is no paradox here, as the density matrix, like all quantum descriptions, expresses knowledge. The state of the measurement apparatus is "objective" only in the sense that one cannot deduce from $W(\mu)$ a contradiction to possible predictions about observables which do not commute with $M$. This is so simply because the phases of the components of the state vector, which would be necessary for this prediction, are unknown; they are lost in the abyss of not knowing, of what one calls thermodynamic irreversibility. Therefore the assumption that a specific value $\mu_\lambda$ had been present in the measurement apparatus *before reading it off* cannot do any verifiable harm.

Here we will not study the detailed theories of the form and time dependence of $W$ during the process of measurement. Wigner (1983) emphasized that there is no possible dynamical evolution of the system which might transform a pure case into a mixture. That is evident. But we should no longer be surprised that we do not know of any dynamical development according to the Schrödinger equation which brings about the reduction of the wave packet. One does not *mean* a dynamical development in the object or the measurement instrument when one describes the loss of information through irreversible processes. Loss of information signifies loss of knowledge. The observer chooses after the irreversible process a more modest description, as he could not follow the dynamical development. It does not do any harm to *assume* that there still is "objectively" a pure state at hand but that we have lost its trace; that is precisely what we mean with a statistical mixture. Only one must not forget that the pure state itself is no more than a catalog of possible predictions, deduced from the knowledge of past facts. Because of this, neither is it a contradiction to admit that the same density matrix can be built from different systems of basis vectors; as already for the pure state, it only means that it expresses probabilities for different, mutually incommensurable observables and that it is up to the choice of the observer which of those he wants to measure in the next experiment. The "objective" existence of an unknown pure state only means that it, with all its implications for probabilities, could have been computed from the initial state of the *total system* of the object being measured and the measurement instrument *if* it had been known. The fictitious assumption that this maximum knowledge of the future had been available does not imply any contradiction to what the observer actually knows. We recall here that already in classical probability theory the more comprehensive knowledge of an observer $B$, who thereby computes different probabilities about the same event than a less knowledgeable observer $A$, does not imply a contradiction to the knowledge of $A$; for the probabilities of both observers belong to different statistical ensembles. Compare the throwing of two dice in Chap. 3.3.

It might now appear that with this mode of thinking, we have reduced irreversibility to a mere loss of knowledge, while irreversible processes in classical thermodynamics are observable facts. The answer is that it is an objective fact that a macrostate of high entropy contains more microstates than a macrostate of lower entropy. Therefore the probability of its return to the state of lower entropy is small but not zero. If we could follow the microstates on their path through time we would be able to specifically select that minority among them which will run back into a macrostate of lower entropy. But if we only know the macrostate we must be content to say that these anti-irreversible developments represent a small minority in a random distribution.

I should apologize for treating in such broad detail problems which must appear trivial once one has understood the theory. Decades of experience, however, have taught me to recognize the verbal traps that tempt us to find nonexistent difficulties.

## $\gamma$) The role of the observer in the Copenhagen interpretation

To recap, we have made it qualitatively plausible that the quantum theory of measurement proves the semantic consistency of the "orthodox" interpretation; we will present a formal model in the next section. Yet it has not diminished by one bit the necessity of an explicit reference to *knowledge*. The $\psi$-function *is* defined as knowledge. The reduction of the wave packet is not a dynamical evolution of the $\psi$-function in accordance with the Schrödinger equation. Rather, it is identical to the event in which an *observer* recognizes a fact. It does not happen so long as only the measured object and measurement apparatus interact, nor so long as the apparatus has not been read out after the measurement interaction ends; it *is* the gain of knowledge associated with reading.

The orthodox Copenhagen interpretation then says that quantum theory describes what the observer may know but it does not describe the observer itself. Bohr and Heisenberg have expressed this on several occasions. Heisenberg writes (1930, p. 44), regarding the cut being movable, that were one to move the cut to include the observer, there would be no physics left. Bohr's concept of a phenomenon ultimately means the same thing: a phenomenon is what an observer can know; his own thought processes, however, are not a phenomenon for the observer.

This cautious attitude at least avoids contradictions. However, it also imposes constraints on checking semantic consistency. If quantum theory does not describe the observer, then it cannot be applied to the assumed prior understanding of the observer either. Yet Bohr in particular always emphasized that quantum theory had removed the strict separation between the object and the observer. It had "emphatically reminded us of the old wisdom that we are both actors and audience in the great drama of life."

Classical physics too, as science, is *knowledge* about nature, as empirical science knowledge of possible observers. Yet it can make its statements in such a way that it describes actually existing facts. Here "actually existing facts" are completely independent of whether they are observed. In this sense, *epistemologically*, it strictly separates the observer from the object. The object is as it is, no matter what an observer knows about it; there is no need for the observer. But whether the observer itself, *objectively* considered, seen as an object of nature, obeys the laws of classical physics or not is irrelevant for this epistemological separation. Historically, classical physics was considered compatible with a monistic as well as a dualistic interpretation of the so-called "mind–body problem." The mind–body problem, e.g., as occasionally discussed in medicine, is actually a product of classical physics. If physics holds for all material objects and if the human body itself is a material object, then it too must obey physics; the question was then whether consciousness also yielded to physical description. Even if one analyzed, hypothetically, the knowledge of the observer according to the laws of physics there still remained the logical–epistemological independence of the objectively described facts from the knowledge of the observer; classically speaking, the facts exist no matter whether anybody knows about them.

It is this clean epistemological separation that quantum theory revokes. Bohr occasionally described this, while discussing the problem of determinism and freedom, with the words: "In classical physics the behavior of the object is strictly causally determined but the observer is described as being completely free; free to measure whatever he wants, be it the position and momentum of a particle, be it the wavelength and frequency of a wave. In quantum theory the observer is no longer completely free, e.g., he may choose whether to measure position or momentum but not both at the same time. Because of this the object becomes freer: its behavior can no longer be completely determined." This was just an aperçu. But its hard core was the epistemological inseparability of subject and object. In quantum theory the *content* of knowledge,

## 9 The problem of the interpretation of quantum theory

that which is known, namely the $\psi$-function, is only a catalog of probabilities, hence knowledge itself.

We elucidate the meaning of this comment separately for the observer and for the object. As an example we choose Young's double-slit experiment.

Through the opening $L$ of the source $Q$ there appears a wave, say a light wave that passes through both slits $A$ and $B$ in the screen $S$ and produces dark bands on the plate $P$, which we now consider specifically at the point $X$. The intensity at $X$ can be computed wave-theoretically. Let $X$ be the location of an interference minimum. If only one of the slits is open, be it $A$ or $B$, there will be a finite intensity at $X$. On the other hand, if both slits are open, the two partial waves will cancel, yielding zero. One can have light emerge from $L$ with such low intensity that there is always only *one* light quantum involved. As is well known, the interference is thereby not affected: each light quantum only interferes with itself.

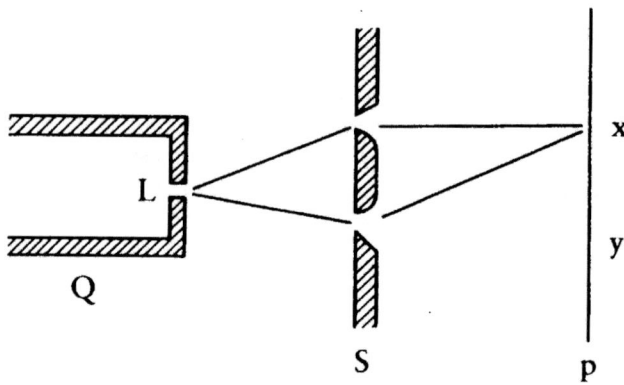

The intensity at $X$ is the probability, in practical terms, for a long series of measurements, the relative frequency of ocurrence of a light quantum at this point. $|\psi|^2$ is actually a probability density. Let there be at $X$, e.g., a silver grain of an emulsion. If only $A$ is open, it will almost certainly be blackened after a sufficiently long time $T$, likewise if only $B$ is open. If both are open it will take a very much longer time than $T$ for a blackening to occur (infinitely long if one could disregard that the intensity vanishes only along one line but that the grain is of finite size).

We first draw the conclusion for the *object*: If both slits are open, one may not say: "The light quantum passed either through the slit $A$ or the slit $B$ but we do not know through which one." Had it passed through $A$, there would be a corresponding probability distribution according to which the grain at $X$ should be blackened after the time $T$; similarly had it gone through $B$. But the grain is not blackened at time $T$. This is what Bohr called the *individuality of the process*, the indivisibility of the process: one cannot mentally dissect the event where both slits are open into the two sub-events with $A$ open and $B$ open. If both slits are open then neither the passage

## 9.2 The semantic consistency of quantum theory

through $A$ nor its passage through $B$ is an objective fact which has either happened or not happened. What is objective is the blackening that occurred for each light quantum at a point on $P$ (call it $Y$). The blackening at $Y$ is *objective precisely because* there the individual process is terminated by an *observation*. Following the discussion in $\alpha$) and $\beta$) above, we can now clearly say that the blackening at $Y$, insofar as the description could be continued by the unreduced $\psi$-function, would not be the end of the individual process. We *dispense* with this description as it is not humanly possible in practice. We can also dispense with an actual observation at $X$ or $Y$, as the blackening is an irreversible process which every observer will discover who examines the plate some time afterwards. It does *not do any harm* to consider it to be objective. Linguistically one can say that objects exist *only for subjects* to which they are "objecting." The blackening is objective precisely *because* it can be observed. In this sense in quantum theory the object can in principle not be separated from the subject.

Now the consequence for the *observer*: he is not described in the description of the experiment. Rather, he is the one who describes it. There, however, it does not depend on him as an individual person. About his perceptions, actions, and his knowledge, the only thing that is relevant for the description of the experiment, is what every other trained observer would perceive, do, or know. In the language of Kant, it depends not on the empirical but on the transcendental subject: that a subject perceives, does, or knows something, and perhaps several subjects together. But the observer thus "objectified" is not "pure mind." He must see with his eyes, work with his hands. He is observer only because he himself—the dualists cautiously say: his body—is part of the world of the phenomena. He can see his own hand, touch his own eye, is in bodily contact with the objects he describes. The question of how the observer could describe himself remains open. We return to it in Sect. 9.2e.

### 9.2.4 Measurement theory, quantum theoretically

One need not explicitly describe the measurement apparatus classically. We sketch here a quantum theory of measurement.

We begin with a remark about self-adjoint operators in quantum theory. They are used in three apparently different ways. Such an operator $H$ may represent

1. the Hamiltonian operator, i.e., the generator of the dynamical group
2. an observable
3. a density matrix

We first reduce 3 to 2: a density matrix can be written $W = \sum_i a_i P_i$, where the $P_i$ are projection operators and the $a_i$ are real numbers $\leq 1$. A projection operator is the observable which expresses the presence of a state or a subspace of the space of all state vectors. Hence the density matrix is simply a probability distribution of certain observables.

Now we try to reduce 2 to 1. How can one actually observe an observable? We assert that we can do so by making it the Hamiltonian operator of the interaction between object and measurement instrument. We demonstrate this in detail, with the intention of showing that quantum theory is consistent in the following sense. Let $p(\lambda)$ be the probability of finding upon measurement of the observable $L$, the eigenvalue $\lambda$; then there exists a property of measurement apparatus corresponding to $\lambda$. Call this property $\mu_\lambda$. The probability of finding the property $\mu_\lambda$, while observing the apparatus with a second instrument registering the property in question, should then be $p(\mu_\lambda) = p(\lambda)$.

The quantum theory of measurement must therefore describe two objects: the primary measured object $X_1$ and the measurement apparatus $X_2$. First we describe the measurement in the quantum theory of the object, thus of $X_1$ alone. Assume $X_1$ to be in the state $|x_1\rangle$ immediately before the time $t_0$. Let the observable $L$ be measured at time $t_0$ by a measurement of the first kind. The eigenvalue $\lambda$ of $L$ ought to have the probability $p(\lambda)$. We assume just the value $\lambda$ to be observed. Then $X_1$ will be, immediately after $t_0$, in a state $|\lambda\rangle$, with the condition $L|\lambda\rangle = \lambda|\lambda\rangle$.

Now we describe the same process in the quantum theory of the measurement, hence on the combined object consisting of $X_1$ and $X_2$. Let its state immediately before $t_0$ be a product state

$$|x\rangle = |x_1\rangle|x_2\rangle. \tag{9.1}$$

This assumes that there has been no interaction between $X_1$ and $X_2$ before the measurement, and that the states of both objects are known, independently of one another. $|x\rangle$ obeys the Schrödinger equation

$$i\,|\dot{x}\rangle = H|x\rangle. \tag{9.2}$$

Before $t_0$ there was no interaction, therefore

$$H = H_0 = H_1^0 + H_2^0 \quad (t \ll t_0). \tag{9.3}$$

In a short time interval around $t_0$ there is an interaction $H_i$:

$$H = H_0 + H_i \quad (t \approx t_0). \tag{9.4}$$

Later that interaction ceases again:

$$H = H_0 \quad (t \gg t_0). \tag{9.5}$$

$|x\rangle$ as a function of time before $t_0$ must be a solution of (9.2) with $H = H_0$. The additive form of $H_0$ in (9.3) ensures the continued product form of $|x\rangle$. The interaction destroys the product form. From $t \approx t_0$ on $|x\rangle$ will have the form of a sum

$$|x, t\rangle = \sum_{\mu_1 \mu_2} c_{\mu_1 \mu_2} |\mu_1\rangle|\mu_2\rangle. \tag{9.6}$$

## 9.2 The semantic consistency of quantum theory

We can assume

$$H_i \gg H_0 \quad (t \approx t_0); \tag{9.7}$$

i.e., we assume that we can neglect the free motion of both objects during the brief duration of the interaction. We will show that we can then use the eigenfunctions of $H$ as a basis in the sum (9.6). *Assuming* the product $|\mu_1\rangle|\mu_2\rangle$ to be an eigenfunction of $H$, then $c_{\mu_1\mu_2}$ will not depend on time. $|c_{\mu_1\mu_2}|^2$ will then be the probability of finding, upon a simultaneous measurement of $X_1$ and $X_2$, the objects in the states $|\mu_1\rangle$ and $|\mu_2\rangle$, respectively. Under the condition (9.7) the product $|\mu_1\rangle|\mu_2\rangle$ will be, during the interaction, an eigenfunction of $H_i$ alone. Can we find such a basis in the Hilbert space $V_1$ of $X_1$ such that the probability $|\langle x_1|\mu_1\rangle|^2$ will not be changed during the measurement interaction? This basis would indeed be a candidate for describing the entire process by means of eigenfunctions of $H$.

We now recall that this process is assumed to represent the measurement of the operator $L$. For all its eigenvalues $\lambda$ and their eigenfunctions $|\lambda\rangle$, $|\langle x_1|\lambda\rangle|^2$ is the probability to find precisely $\lambda$ before the interaction. That interaction ought to measure precisely $L$, therefore the probability to have found exactly the value $\lambda$, after the interaction but before reading the instruments, must again be $|\langle x_1|\lambda\rangle|^2$. Hence we choose the basis $|\mu_1\rangle = |\lambda\rangle$ for all $\lambda$ and $\mu_1$.

On the other hand, there must exist an observable $M$ in the measurement apparatus $X_2$ which corresponds exactly to the observable $L$ in $X_1$: such that a reading of the instrument after the interaction where an eigenvalue $\mu$ of $M$ has been found *means* that a certain eigenvalue $\lambda$ of $L$ would now be found at a second experiment on $X_1$. Let us call $\mu_\lambda$ the eigenvalue of $M$ corresponding to $\lambda$, then this condition will be satisfied if

$$c_{\lambda\mu} = \delta_{\mu\mu_\lambda}. \tag{9.8}$$

Hence $H_i$ must be an operator whose eigenvectors are precisely the products $|\lambda\rangle|\mu\rangle$ such that its eigenvectors for these products vanish for $\mu \neq \mu_k$ and assume a value different from zero for $\mu = \mu_k$. This is an abstract definition of an operator we would call a *local operator* were $L$ and $M$ the position operators of the two particles and $\mu_k = \lambda$: the two particles interact only while at the same point. Exactly such an interaction is required for a position measurement.

Relative to the basis $|\lambda\rangle|\mu\rangle$ the interaction operator must therefore have the form

$$H_i = \lambda\mu\,\delta_{\mu\mu_\lambda}. \tag{9.9}$$

As

$$L|\lambda\rangle = \lambda|\lambda\rangle, \quad M|\mu\rangle = \mu|\mu\rangle, \tag{9.10}$$

the product

$$H_i = LM\delta_{\mu\mu_\lambda} \tag{9.11}$$

will be a general expression for $H_i$. This again corresponds to a local interaction of two fields $\varphi$ and $\psi$ by means of an $H = \varphi(x)\psi(x)$.

Thus we see that an observable $L$ can be measured by a Hamiltonian operator which is a product whose object-factor is $L$ and measurement apparatus-factor has a one-to-one correspondence between its eigenvalues and the eigenvalues of $L$. In this sense an observable must be a possible Hamiltonian operator for a measurement interaction.

### 9.2.5 Quantum theory of the subject

Characteristic of Bohr's description of measurements as well as for his entire world view is his *caution*. He describes what he knows, and he refuses to make statements about what he does not know. He sharply distinguishes the *epistemological inclusion* of the observer, without which quantum theory could not be interpreted, from the *objective description* of the observer which quantum theory factually has not accomplished. But the desire for a unified world view and the progress achieved in biology and cybernetics in the six decades since the creation of quantum theory prompt us to ask whether the objective description of the observer must remain excluded *in principle*, or whether in principle it might be possible. We argue for the latter.

For Bohr the question did not arise in this way, as he already looked upon physicalism in biology with the deepest skepticism. He suspected (1932) a complementarity between a physical and a proper biological description of organisms. Also for organisms, physics should turn out to be correct for every single experiment, but specifically biological processes should require renunciation of a complete physical description when the corresponding complete physical analysis would kill the organism. Accordingly the observer would be beyond the reach of a quantum theoretical description not only because he thinks, but just because he is a living being. The desire to clarify these questions led Max Delbrück into biology. Yet thus far physicalism has always proved itself in biology. We therefore try the opposite question. We ask *how* one must carry out the factual description of an observer in quantum theory, assuming it to be possible.

Let us denote as before the measured object by $X_1$ and the measurement apparatus by $X_2$. We call $Y$ the observer who reads the measurement apparatus and $Z$ the "meta-observer" who observes the observer. How would $Z$ perceive the event where $Y$ observes the fact in $X_2$ from which he deduces his new knowledge about $X_1$? In its detail this question goes beyond our present knowledge. But the question makes sense as a "thought experiment about a theory." It scrutinizes the meaning of quantum theory by hypothetically assuming it to be applicable to the consciousness of human beings.

We initially choose, following the traditional terminology, the language of the ontological distinction between body and mind. Let $Z$ apply quantum theory to $Y$'s body. According to our working hypothesis we can describe this body like a measurement instrument that interacts with $X_2$ in the manner described above; now $X_2$ is, for $Y$'s body, the object to be measured. As long as nobody observes $Y$'s body, the interaction with $X_2$ will create in $Y$'s body a situation which $Z$ describes as a classical (irreversible) probability distribution for the results of measurement of the observable $M$ of $X_2$ by means of $Y$'s body. This probability distribution will be reduced for $Z$ whenever $Z$ observes $Y$'s body.

But how can $Z$ observe the situation in $Y$'s body? There is a very simple way: $Z$ must ask his friend $Y$ what he, $Y$, has observed. In the traditional but cumbersome and problematic manner of speaking which distinguishes body and mind (or consciousness), we must describe it by saying that $Z$'s consciousness induced $Z$'s body to emit sound waves which are received by $Y$'s body and which transmit to $Y$'s consciousness the question: "Hello, $Y$'s consciousness! What have you, by means of your body, found out about the value of the observable $M$ in $X_2$?" $Y$'s consciousness answers with the same telegraphy.

In this description the second observer, $Z$, turns out to be superfluous. The procedure works only if $Y$'s consciousness knows something about $Y$'s body. Hence we can choose from the outset $Y$'s consciousness as the observer of $Y$'s body. For $Y$'s consciousness $Y$'s body is then an ordinary measurement instrument on which facts can be observed and probabilities be used for making predictions. The reduction of the wave function of $Y$'s body occurs in $Y$'s consciousness. Nothing is changed relative to the quantum theory of measurement.

Thus let us make our last leap and assume quantum theory also to be applicable to consciousness itself. This would then be the first actual application of quantum theory to the knowing subject. The assumption that this application is possible is certainly hypothetical, but it does not contradict in any way the logical structure of abstract quantum theory. Abstract quantum theory deals with decidable propositions, independent of their special nature. The question of whether $Y$'s consciousness observed a fact or not is a meaningful alternative. The probability that an observer who has perceived a fact $x$ will then observe a fact $y$ is a meaningful probability. Thus we can introduce as observer the consciousness $Z$ (which we previously called $Z$'s consciousness) that informs itself about conscious facts in the consciousness $Y$, e.g., by having $Z$ talk to $Y$, and by making and checking predictions about future facts in $Y$'s consciousness. The consciousness $Y$ is then like any other quantum object. The probability function by means of which $Z$ describes the "state of consciousness" of $Y$ is reduced whenever $Y$ has said something to $Z$ and has been understood.

Here, however, the "realist" will protest: "But $Y$'s state of consciousness is certainly objective, as he knows himself." For a second time this leads us to eliminate the external observer $Z$ and let $Y$ observe himself. In our interpretation of quantum theory this is permissible and we have no difficulty ascribing to $Y$ a better knowledge of himself than $Z$, and consequently express this knowledge by means of a different $\psi$-function than the one used by $Z$.

Here we note that the distinction between $Y$'s body and $Y$'s consciousness has now become quite useless. According to our starting hypothesis $Y$'s body obeys quantum theory, and according to our last assumption $Y$'s consciousness obeys it. What prevents us from considering both merely as different aspects of one and the same reality? Why should not $Y$'s body be just that which $Y$ can observe of himself in space, and $Y$'s consciousness that which $Y$ is perceiving of himself through "introspection"? We return to this fundamental philosophical idea later. For the rigor of the present argument it is not necessary. In that regard, it remains a comment for now.

What can $Y$ know about himself? He may know facts: as long as his memory is reliable, he has a factual knowledge of his past experience which includes his present consciousness. He may know possibilities: he can be able to estimate probabilities about his future experience. Every time he is consciously having a new experience he will add a new fact to his collection and will reduce the probability function for his future experiences. He is unable to predict himself deterministically, neither his external experiences nor his moods, thoughts or intentions. In this way he may consistently play both roles, that of the knower and of the known.

We have arrived at the deep philosophical problem of self-awareness. We do not claim to have solved this problem. We merely assert that there is nothing in quantum theory itself that might prevent its general application, without any modification of its interpretation, to mental processes. What I know about myself is objective in the sense of facticity. What I objectively do not know about myself might in the future become accessible to my knowledge. This structure of time is no different for mental processes than for objects we call material.

## 9.3 Paradoxes and alternatives

### 9.3.1 Preliminary remarks

We have characterized the third period of the interpretation of quantum theory, from about 1935 to the present, as the period of "rearguard actions." We do not wish to imply that the Copenhagen interpretation is objectively the last word on the interpretation of quantum theory. But it should be said that in *these* last fifty years scarcely any substantial progress has been achieved beyond it. This might be due to the fact that in the debate of these decades, no forward-looking questions have been asked of the theory, only retrospective ones—i.e., that rearguard actions have been fought. The reason for this, however, is understandable.

The Copenhagen interpretation had its origins, both historically and factually, in Bohr's idea of correspondence. Classical physics is a given. It furnishes, in our terminology, the *preconceptions* for quantum theory. It is in this language that one describes measurements. The Copenhagen interpretation shows then the *semantic consistency* of quantum theory. It demonstrates precisely how one must modify the preconceptions, i.e., how one must restrict the range of applicability of classical physics so that no contradiction arises. The apparent paradoxes and alternatives which were discussed afterwards were without exception attempts to revoke, at least partially, the sacrifices one had to make in the preconceptions. These attempts were without success. This was an important "grieving process"; only then did it become quite clear how deep these sacrifices went. In this debate Einstein proved again his eminent intellectual standing. In his famous thought experiment with Podolsky and Rosen he precisely brought to light the central point of the sacrifice: the renunciation of the belief in the "objective reality" of physical objects. Therefore we devote an entire section to this grieving process, with Einstein's protest as its central part. But it remains a grieving process. None of the proposed returns to classical principles has succeeded. The psychological meaning of Freud's concept of grieving process lies in the fact of accepting what has happened, not to suppress its harshness, and thereby to clear the way to another future.

That during these decades it was not possible to take fundamental steps forward in the interpretation is understandable from the concurrent development of physics. These were the decades of the "conquest of the world through quantum theory." Quantum theory was a closed theory and there were no apparent limits to its applicability. This had two consequences. On the one hand, the productive physicists, of the generations that were younger than Einstein and Bohr as well as those still young today, were fascinated by the concrete progress, the discovery and explanation of ever new phenomena. Scarcely any of the leaders of this progress were very much interested in the problem of interpretation. This, however, was already related to the second consequence: the leading physicists understood very well that probably not much could be learned from this debate, in the form that it had taken.

Indeed, if there is no more at hand than a closed theory and a fixed, historically prescribed set of preconceptions for the theory, which through reinterpretation have been made semantically consistent with it, then one cannot get more out of this material than a mere depiction of what is already known. A broadening of the horizon is necessary if one wants to accomplish more, either an incorporation and critique of the preconceptions in a much broader and deeper philosophy or a new closed theory. The former would have demanded going beyond the scope of what had become by then rather uninteresting disputations of contemporary philosophies like realism, positivism, apriorism which were all much alike in their prejudices. A necessary but not yet sufficient requirement would have been a thorough knowledge of Western philosophy from its beginnings and thereby a critical assessment of the motives of modern philosophy. Neither could the philosophically ill-prepared physicists accomplish this nor could the philosophers who did not understand the new physics. The latter, however, a new closed theory beyond quantum theory based on correspondence, would also have been the physical prerequisite for a new direction in the interpretation of quantum theory. It would have reduced present quantum theory to the role of a set of preconceptions, and it would have offered the means to criticize those preconceptions. But new closed theories are like continents: they are discovered and cannot be produced at will.

The present book originated from a search in both of these directions. The philosophical broadening of the basis will be carried a step further in the book *Zeit und Wissen*, now in preparation. The *Aufbau der Physik* attempts to contribute physically to what can be accomplished through an *intrinsic* analysis of quantum theory and its preconceptions. There in general it is about reducing the preconceptions to their *basic elements*. As such there remain the *structure of time* and the concept of the alternative (Chap. 3). Their connection is manifested by the fact that every alternative implies a question about the *future*, and that it is decidable only by virtue of the irreversibility of the measured result, due to the past being a fact. Finally, the ur hypothesis (Chap. 4) is intended to be a technical step forward within quantum theory. It too originated from the question about the most stringent elements of quantum theory possible, but raises the hope of incorporating into basic quantum theory also the theory of relativity and of elementary particles. Related to all this is the conviction that only such a decisive *step forward* can lead to new insights for the problem of the interpretation as well.

The present section, however, is about the "grieving process" of the past debate. It mentions a few of the suggested paradoxes and alternatives that had been proposed. On the one hand, it is a reminder of why these are not paradoxes and presumably also not viable alternatives. On the other hand, it attempts to understand in each case why the question had been posed in this way; sometimes these *motives* point beyond the present form of quantum theory.

### 9.3.2 Schrödinger's cat: The meaning of the wave function

Schrödinger had to admit, after the discussions described above, that a wave theory was not suitable for describing particle phenomena. For this reason he remained ever since of the opinion that quantum theory in its present form is not an adequate theory of reality, despite all its successes. He no longer participated in its development, and turned to Einsteinian-type of problems of a unified classical field theory.

In an article from 1935 (see Jammer 1974, pp. 215–218) he treats with irony the Copenhagen point of view by means a thought experiment. Let a living cat be locked up in a box and with it a deadly poison which can be released by a single radioactive atom inside the box. After one half-life of the atom the probability is 1/2 for the cat being still alive, and 1/2 for being dead. Schrödinger describes the $\psi$-function of the system at this time with the words: "The half-alive and the half-dead cat are smeared out over the entire box."

The answer is trivial: the $\psi$-function is the list of all possible predictions. A probability 1/2 for the two alternative possibilities (here: "living or dead") means that the two incompatible situations most now be considered equally possible at the instant of time meant by the prediction. There is no trace of a paradox.

Schrödinger's reason to consider the situation as paradoxical lay in his hope to interpret the $\psi$-function as an "objective" wave field. In the implied deterministic description he saw no reason to take seriously the difference between the present and the future. I have seen from a letter he once wrote to me (after the war) how foreign the idea was to him, as well as to many other physicists, that this difference was something to be taken seriously physically, and not merely "subjectively." He had read an account I had given about my interpretation of thermodynamic irreversibility. He wrote me of having had great difficulty understanding my unusual mode of expression but that he could see now that I had simply meant the arrow of time. To me, naturally, precisely the word "arrow of time" was a mere metaphorical mode of expression for the actual, phenomenologically given state of affairs.

Schrödinger accomplished a heightening of the paradoxical impression by considering a living being as an example. The poor cat is treated here simply as a measurement instrument to illustrate the irreversibility of the measurement process by means of the striking, and to us humans, moving contrast between the states of life and death. We have argued above (Sect. 9.2e) that in quantum theory there is no inherent reason why it could not be applied to living beings. But we do not need this assumption for the discussion of Schrödinger's example. It suffices to remark that obviously there cannot be a clear description of a quantum theoretical thought experiment if on the one hand one uses living organisms as its integrating parts but on the other hand one does not take seriously the application of quantum theory, i.e., here simply the concept of probability, to the organisms.

### 9.3.3 Wigner's friend: Inclusion of consciousness

Wigner Wigner (1961) apparently, but then erroneously, was of the opinion that Bohr had maintained the applicability of quantum theory to consciousness. Many physicists got that impression when Bohr spoke of the introduction of an observer as being unavoidable. We have explained above that it was precisely not meant this way. On the other hand, we have asserted that there are no inherent reasons in quantum theory against its application to consciousness. The latter opinion should refute the thought experiment which became known as "Wigner's friend." Therefore we must find the flaw in Wigner's argumentation.

In this thought experiment the theoretician $W$ and his friend $F$ are describing the same binary experimental alternative between two states $x_1$ and $x_2$ of an object. $W$ uses a wave function $\psi = \alpha\psi_1 + \beta\psi_2$ which—according to Wigner—is valid for him until he knows what has happened. The friend $F$ is the observer. The event in the object will elicit a mental response in $F$'s consciousness. If $x_1$ has occurred then $F$ will see a flash of light, if $x_2$ has happened he does not see a flash. Hence $W$ can restrict his own prediction to a prediction about the two mutually exclusive mental events that happen to his friend: $x_1 = F$ is seeing a flash of light, $x_2 = F$ does not see a flash of light. He is now asking his friend: "Have you seen a flash of light?" Up to the instant when $W$ receives the answer from $F$ he will—so it appears initially—correctly use the state vector $\psi$ for the description of $F$'s mental state. But once he has received the answer he may ask the friend: "What did you know about the flash of light before I had asked you?" In case $x_1$ he obviously gets the answer: "I have already told you that I have seen the flash," in case $x_2$, with the same indignation: "I really have not seen it." That means that for $F$ the state was already reduced before $W$ asked him. This seems to imply a contradiction as the two states $x_1$ and $x_2$ are incompatible with $\psi$. The answer is that a conscious act, it it really has happened, also constitutes a fact, i.e., presumes an irreversible process in the consciousness. $F$ could not have answered $W$'s question at all if he had not stored the result in his memory. Exactly as in a material apparatus, this irreversible process destroys the phase relation which is described by the complex numbers $\alpha_1$ and $\beta_1$. As soon as $W$ knows that $F$ has observed the process, he no longer has any right to use his $\psi$ for a description of $F$'s consciousness in the time interval between $F$'s act of perception and $F$'s answer. He must replace $\psi$ by a mixture of $x_1$, with the probability $|\alpha|^2$, and of $x_2$, with the probability $|\beta|^2$.

Here the impression of a paradox is created by describing the consciousness naïvely and therefore inconspicuously according to the classical ontology. The correct application of quantum theory to the consciousness, as described above, rules this out for the respective future. No more is then asserted than that at the present time I do not know my own future.

### 9.3.4 Einstein-Podolsky-Rosen: Delayed choice and the concept of reality

Einstein designed this famous thought experiment *after* he had admitted quantum theory to be free of internal contradictions. Hence it was not supposed to inherently refute quantum theory but to bring to light its consequences for the concept of reality in such a way that it became clear why Einstein was not ready to accept such a theory as being final.

#### α) The thought experiment

The authors (Einstein et al. 1935; nowadays one usually calls them simply "EPR") consider two bodies $X_1$ and $X_2$ that interact with one another at a time $t_0$, and afterwards move apart from one another to a very large distance, say the distance from Earth to Sirius. While they were together in direct interaction, two mutually commuting observables of the combined object consisting of $X_1$ and $X_2$ were measured, e.g., their distance $x_1 - x_2$ in one direction or their total momentum $p_1 + p_2$ in the same direction. After their wide separation we measure at time $t_1$ on the object $X_1$, which we assume to have arrived on Earth, either the value of $x_1$ *or* the value of $p_1$. In an admissible Lorentzian reference frame, e.g., the rest frame of the center of mass of both objects, an observer measures on Sirius, at an instant that is simultaneous to $t_1$ in this reference frame, either $x_2$ or $p_2$ on the object $X_2$. No physical signal can bring the measurement result from Earth to Sirius before the measurement there had been done. But *in case* the position had been measured on $X_1$, then the observer on Earth can accurately predict what position $x_2$ the observer on Sirius will report having found himself, *if* also the position had been measured on $X_2$. *If*, however, the momentum $p_1$ had been measured on $X_1$, then conversely just the value $p_2$ of $X_2$ is predictable. But according to quantum theory $X_2$ cannot have simultaneously predictable values of $x_2$ and $p_2$.

It was completely clear to Einstein that this does not imply a logical contradiction within quantum theory. The two assumptions that at time $t_1$ just $x_1$ or just $p_1$ had been measured are incompatible; hence at most only one of the two could have been satisfied. Thus it cannot happen that $x_2$ and $p_2$ simultaneously become predictable. But the apparent paradox lies in the fact that the $\psi$-function of $X_2$ on Sirius, due to the reduction of the wave packet, is instantaneously changed through a measurement on $X_1$ on Earth. The impression of a paradox is not altered by admitting that the $\psi$-function is a catalog of knowledge which must be suddenly modified through new information. For the *content* of the knowledge changes drastically in this thought experiment, in a way about which Einstein was justified in calling irreconcilable with the traditional view of physical reality. Measuring on Earth the position $x_1$ of $X_1$, the position $x_2$ of $X_2$ is predictable. According to the traditional view of reality this means that $X_2$ must have had just this position immediately before the measurement; otherwise why would it then, with certainty, be found? If, however, one decides on Earth to measure the momentum $p_1$ instead, then $X_2$ has the predictable momentum $p_2$. If it had this value also immediately before the measurement, there arises through this *reality assumption* a contradiction to quantum mechanics in that $x_2$ and $p_2$ are simultaneously determined. If one wants to uphold quantum mechanics one must abandon this reality assumption.

For the sake of clarity let us emphasize that the assumption of the repeatability of a measurement, which might be indispensable for the postulates of quantum theory, amounts to much less than the reality assumption just introduced. Repeatability implies that one would find the same result immediately *after* an actual measurement, *if* one performed it. The reality assumption means that immediately *before* the actual measurement the result must have been there, *independently* of whether one performed it. Even the retrodiction mentioned in Sect. 9.2b, i.e., the retrogressive construction of a $\psi$-function into the past, does not change anything there. The retrogressive $\psi$-function says that one must *have found* the same result immediately *before* the measurement, if one had performed it then.

Wheeler (1978) calls the EPR experiment a thought experiment with *delayed choice*. We can paraphrase this as follows. In the spirit of the quantum theory of measurement one can regard each of the two objects $X_1$ and $X_2$ as a measurement device for a measurement on the other. Then their interaction at time $t_0$ is itself the measurement interaction. As we have described the experiment, $X_1$ would be the measurement apparatus, $X_2$ the object. The measurement on $X_1$ on Earth is then the reading of the measurement device which permits us to determine the state of the measured object $X_2$ on Sirius. The measurement on $X_2$ on Sirius is then only a control measurement whose result (if on both sides $x$ or on both sides $p$ is measured) is foreseeable for a correctly performed experiment. As between $t_0$ and $t_1$ there has not happened any irreversible process on $X_1$ (especially not a measurement), $X_1$ is in a pure quantum state before the measurement at time $t_1$. Therefore one can wait until the time $t_1$ before one *chooses* the quantity one wants to measure on $X_1$. This delayed choice of the observable to be measured on $X_1$ then also decides in which eigenstate of which observable $X_2$ will be found, if one checks the measurement on Sirius. According to the reality assumption, however, $X_2$ ought to have been in this state from $t_0$ onward. Hence the reality assumption would imply a *back influence* of the measurement on $X_1$ at time $t_1$ on the state of $X_2$ immediately after $t_0$.

All these conclusions are self-evident if one interprets the $\psi$-function as a *probability catalog*. Theoretically determined probabilities are always *conditional probabilities*. The delayed choice of the observable to be measured on $X_1$ is at the same time a choice of the conditions under which the result of a measurement on $X_2$ is to be predicted, i.e., of the statistical ensemble of which $X_2$ is to be considered to be a member at time $t_1$. Einstein's argument discloses only that the quantum theoretical structure of this probability catalog (namely the superposition principle for $\psi$) is incompatible with the reality assumption.

### β) An older thought experiment with delayed choice

Jammer (1974, pp. 178–180) pointed out that in my first physics publication (Weizsäcker 1931) there is an explicit description of an experiment with delayed choice. In a letter dated 1967 he brought this analogy to my attention, about which I do not remember whether I had noticed this in 1935 in connection with the work of Einstein, Podolsky, and Rosen. My paper, which had been suggested and supervised by Heisenberg, was a test of whether the version of quantum electrodynamics by Heisenberg and Pauli (1929, 1930) was suitable for correctly describing the Heisenberg gamma-ray microscope.

The assumed experimental arrangement was an optical lens under which there is an electron somewhere in a predetermined plane parallel to the center plane of the lens ("object plane"). In addition its momentum parallel to the plane was assumed to be known before the experiment. The $z$-coordinate of the electron is thus known as accurately as possible. A light quantum enters

## 9.3 Paradoxes and alternatives 283

from the side, is scattered by the electron, passes through the lens, and is absorbed by a photographic plate on the other side (i.e., above) the lens. There it blackens a point with the coordinates $(\xi, \eta)$. What can one deduce from this about the electron? If the plate had been put, as usually done, into the *image plane* corresponding to the object plane, then there follow from the laws of optics the coordinates $(x, y)$ in the object plane where the light quantum has been scattered by the electron. If on the other hand one had chosen the *focal plane* of the lens for the position of the plate, then there follows from $(\xi, \eta)$ the direction of motion of the electron before it had passed through the lens, hence, according to the conservation law of momentum, the momentum of the electron after the scattering. The observer can now *choose* in principle (for large dimensions and a rapidly movable plate) *after* the scattering process on the electron, in which plane he wants to slide in from one side a plate prepared beforehand. In this way, he decides only after the measurement interaction whether consequently the electron will obtain a well-defined position or a well-defined momentum.

Jammer was surprised that I had then not given more weight to this idea but that I had introduced it more or less en passant. I replied to his written question: "The problem which led to this paper was certainly closely related to that raised by Einstein, Rosen and Podolsky. Except that Heisenberg, who suggested it to me, and I as well regarded this state of affairs not as a paradox, as conceived by the three authors, but rather as a welcome example to illustrate the meaning of the wave function in quantum mechanics. For this reason the matter did not carry such weight for us as it did for Einstein and his collaborators on the grounds of Einstein's philosophical intentions. The purpose of my paper was not to bring into full relief facts which were for us self-evident, but rather to examine, by means of a quantum-field-theoretical computation, the consistency of the underlying assumptions. The work, thus, properly speaking, was rather an exercise in quantum field theory and its purpose, for the sake of which Heisenberg had proposed it to me, was rather a test of whether quantum field theory is a good quantum theory, than an additional analysis of the quantum theory itself."

To this Jammer makes in his book the fair but somewhat surprised remark: "It may well be that Heisenberg and von Weizsäcker were fully aware of the situation without regarding it as a problem. But as happens so often in the history of science, a slight critical turn may open a new vista with far-reaching consequences. As the biochemist Albert Szent-Györgi once said: 'Research is to see what everybody has seen and to think what nobody has thought'. In fact, even if it was only a slight turn in viewing a well-known state of affairs, the work of Einstein and his collaborators raised questions of far-reaching implications and thus had a decisive effect on the subsequent development of the interpretation of quantum mechanics." (p. 180)

As I read this passage in Jammer's book, after 1974, my spontaneous reaction was that I wrote in the margin next to the word "decisive" in the last sentence, "i.e., misleading." Indeed I considered and still consider this entire debate a mere rearguard action. All the same my reaction was perhaps unjust. My respect for Einstein, as I have already said in the preface of this book, has grown from decade to decade. His insistence on a position that was not granted success by history was nevertheless important, even be it merely as a "grieving process," in overcoming a certain flippancy over the question of interpretation, not by Heisenberg and Bohr, but their successors—a flippancy which arises so quickly in victorious schools. We young ones were in the thirties of unlimited arrogance as regards Einstein's reaction to quantum theory. The EPR paper did not impress us very much; the reaction was more: "Well, now also Einstein has understood what quantum theory really implies."

I must confess that I myself in those days, up to about 1954, suffered under the growing awareness that I did not understand quantum theory. Logically, so it appeared to me around 1935, four or five people had understood it, perhaps Heisenberg, Pauli, Dirac, Fermi; certainly not I. Philosophically, so it seemed to me, only Bohr understood it; nobody else understood him; and besides, even Bohr, so it seemed to me further, did not know the last word about it. Completely superficial appeared to me the subsumption of the Copenhagen interpretation under "positivism." I found it unfortunate that even Einstein, strictly speaking, counted Bohr among the positivists. How far the Copenhagen interpretation is from positivism revealed itself in the complete inability of the neo-positivistic theory of science merely to correctly represent the reasoning of Bohr and Heisenberg, not to mention of seeing them in their proper perspective or perhaps interpreting them.

The fact could not remain hidden forever that quantum theory, on the one hand, was correctly applied by the physicists themselves but never really understood, i.e., in such a way that it could be enunciated. This was the legitimate basis of the new debate over the interpretation which was initiated by Einstein and which at times, in the sixties and seventies, grew to an avalanche. But it seemed to me that the "realism," which most of the critics were hoping for, was philosophically not any better than the beleaguered positivism. *This direction of the debate appeared to me doomed from the start, and I never participated in it.* Einstein's own concept of reality had very deep philosophical roots. Before we get involved in it we should extend by one more step the logical analysis of the EPR experiment.

### γ) The semantic consistency of the probability interpretation, according to the EPR model

To simplify the expression we choose a simplified model of the EPR experiment that does not couple position and momentum measurements, but two binary alternatives. It is taken from Bohm's textbook on quantum theory (Bohm 1951, pp. 614–619; see also Jauch 1968, pp. 185–187; and my 1973a). Consider a spin-0 particle decaying into two spin-1/2 particles, without any exchange of spin and orbital angular momentum. Later these two particles are observed at two widely separated places $x_1$ and $x_2$ ("Earth" and "Sirius"). Both observers can choose whether they want to measure the spin component of their particle in the $y$- or the $z$-direction. We deal with three observers: $A$, $B_1$, $B_2$. $A$ has the initial information which we expressed in the last few sentences. $B_1$ makes its measurement at the position $x_1$, $B_2$ at $x_2$. After the measurements, they inform one another of their results and review their predictions. Initially $A$ and $B_1$ did not know which of the two possible experiments $B_2$ would decide to perform. We denote these two experiments $y_2$ and $z_2$, and their possible results $y_2^+, y_2^-$ and $z_2^+, z_2^-$. Thus, e.g., $y_2^+$ means that The spin of the particle at position 2 has been measured in the direction of the $y$-axis and found to be positive. Just as little did $A$ and $B_2$ know which of the experiments $y_1$ and $z_1$, with the possible results $y_1^+, y_1^-$ and $z_1^+, z_1^-$, would have been performed by $B_1$. $A$ now has a list of four conditional probabilities $p$, which depend on the possible choices of the two observers $B_1, B_2$.

| | p | | p | | p | | p |
|---|---|---|---|---|---|---|---|
| $y_1^+ y_2^+$ | 0 | $z_1^+ z_2^+$ | 0 | $y_1^+ z_2^+$ | $\frac{1}{4}$ | $z_1^+ y_2^+$ | $\frac{1}{4}$ |
| $y_1^+ y_2^-$ | $\frac{1}{2}$ | $z_1^+ z_2^-$ | $\frac{1}{2}$ | $y_1^+ z_2^-$ | $\frac{1}{4}$ | $z_1^+ y_2^-$ | $\frac{1}{4}$ |
| $y_1^- y_2^+$ | $\frac{1}{2}$ | $z_1^- z_2^+$ | $\frac{1}{2}$ | $y_1^- z_2^+$ | $\frac{1}{4}$ | $z_1^- y_2^+$ | $\frac{1}{4}$ |
| $y_1^- y_2^-$ | 0 | $z_1^- z_2^-$ | 0 | $y_1^- z_2^-$ | $\frac{1}{4}$ | $z_1^- y_2^-$ | $\frac{1}{4}$ |

Let us assume that $B_1$ measures $y_1$ and finds $y_1^+$. He then has two conditional probabilities for the result that $B_2$ is going to report to him.

| | p | | p |
|---|---|---|---|
| $y_2^+$ | 0 | $z_2^+$ | $\frac{1}{2}$ |
| $y_2^-$ | 0 | $z_2^-$ | $\frac{1}{2}$ |

Let us assume $B_2$ to measure $y_2$. When $A$ knows what measurements both $B_1$ and $B_2$ will perform, i.e., according to our present assumption $y_1$ and $y_2$, but before he knows the result of $B_1$, $A$ will then predict for both possible results of $B_2$, namely for $y_2^+$ and $y_2^-$, the probabilities $p = \frac{1}{2}$ according to his table. $B_1$ however, after his result $y_1^+$, knows for certain[15] that $B_2$ will report the result $y_2^-$. If quantum theory is correct, $B_2$ will indeed find $y_2^-$. He for one will now predict that $B_1$ will report the result $y_1^+$ upon measurement of $y_1$.

We can go through all the cases enumerated and will not find any contradiction. It is completely normal that different observers will, depending on their prior knowledge, ascribe different probabilities to the same experimentally possible results. Each of them will be able to test precisely his probabilities as relative frequencies, if he repeats the measurement many times on the ensemble which corresponds to his prior knowledge. This state of affairs, already known from classical probability calculus, holds as a matter of

---

[15] "Certain," as always in such thought experiments, if no errors have been made.

fact also for the $\psi$-functions. It is completely legitimate that two different observers describe the same process with two different $\psi$-functions according to their prior knowledge. In particular, $A$ may continue to predict the more distant future from his original state vector, if he does not know the results of $B_1$ and $B_2$; he must merely not forget the unknown phase change in his $\psi$-function due to the measurements of both $B$. The unreduced wave function is simply a probability catalog for an observer who does not know the results of later measurements. We return to this in section 9.3$h$ in connection with the formulation of Everett.

### $\delta$) Einstein's concept of reality

This is not the place to elaborate on Einstein's philosophy in detail. Here we simply connect his argument in the EPR paper to the metaphysical background of his concept of reality.

Jammer (p. 185) distinguishes in the EPR argument concerning the incompleteness of quantum theory two explicitly formulated criteria and two tacitly assumed premises. Quoting from Jammer, which in turn contains verbatim quotes from EPR:

"1. *The reality criterion.* 'If, without in any way disturbing a system, we can predict with certainty (i.e., with probability 1) the value of a physical quantity, then there exists an element of physical reality corresponding to this physical quantity.'
2. *The completeness criterion.* A physical theory is complete only if 'every element of physical reality has a counterpart in the physical theory.'

The tacitly assumed arguments are:

3. *The locality assumption.* If 'at the time of measurement ... two systems no longer interact, no real change can take place in the second system in consequence of anything that can be done to the first system.'
4. *The validity assumption.* The statistical predictions of quantum mechanics—at least to the extent they are relevant to the argument itself—are confirmed by experience.

We use the term 'criterion' not in the mathematically rigorous sense denoting necessary *and* sufficient conditions; the authors explicitly referred to 1 as a sufficient, but not necessary, condition of reality and 2 only as a necessary condition of completeness. The Einstein-Podolsky-Rosen argument then proves that on the basis of the reality criterion 1, assumptions 3 and 4 imply that quantum mechanics does not satisfy criterion 2, that is, the necessary condition of completeness, and hence provides only an incomplete description of physical reality."

## 9.3 Paradoxes and alternatives 287

We are interested here in the starting point, the *reality criterion*. We have already used it above under the title of the "reality assumption." It is characteristic of its logical structure that it does not define the concept of reality, but that it posits a *sufficient* condition for concrete applicability. If *at least* this condition is fulfilled, Einstein will speak of reality. Precisely this criterion is violated in quantum theory; therefore quantum theorists are not speaking *in this sense* about reality. One definition of what is real is given by Einstein (1949, p. 80): "Physics is an attempt conceptually to grasp reality as it is thought independently of its being observed. In this sense one speaks of 'physical reality'. In pre-quantum physics there was no doubt as to how this was to be understood. In Newton's theory reality was determined by a material point in space and time; in Maxwell's theory, by the field in space and time. In quantum mechanics it is not so easily seen." But this explanation also anticipates a previous understanding of what is meant. The philosophical key word is "reality" in the first sentence. Einstein himself is aware that doubts are possible as to "how this is to be understood." He therefore gives unquestionable examples: the classical physics of Newton and Maxwell. What reality is assumed to mean beyond these models is indeed the very problem.

It is characteristic of the entire philosophical tradition of "realism" that it cannot explain any further its central concept of *reality* by means of a definition, but that it assumes, unnoticed by its lesser minds but explicitly by its more critical ones, that it is self-evident in and of itself. Now behind all this is one of the basic problems of philosophy. If a hierarchical conceptual exposition of philosophy is possible at all, then there must be one or a few "basic concepts." Clearly these cannot be reduced by skillful definition (through a more general concept and specific difference) to some other concepts—then they would not yet have been the basic concepts. Greek philosophy reflected on this problem with a rigor for which modern theoreticians of science or even physicists have no eye. For them *Being* was the central concept. This is the origin of the expression "das Seiende–the Being" in the German philosophical linguistic usage which Einstein uses in the quotation above. In our century Heidegger has shown again that even for the Greeks, "Being" is not a self-evident but deeply enigmatic concept, and that lastly a hierarchical exposition of philosophy does not seem to be possible.[16] We must now ask where Einstein's concept of reality stands with respect to this tension between Greek metaphysics, classical physics, and quantum theory.

Greek philosophy started with the desire to think not only about parts of reality but its entirety. Thus its concept of Being was linked to the religious view of the whole; it was, as Heidegger says, onto-theology. "God" is the popular name for the One, what Parmenides calls the Being, Plato the Good, Aristotle the mind (nus). This background, the true eternal Being of God, then ensures the by itself imperfect, changeable Being of all individual things.

---

[16] For all this see *Zeit und Wissen*.

Formally the opposite path was chosen by classical physics. It coined the concept of Being of its objects through things of everyday life, which it stylized, for the mathematical description, as extended bodies in space and finally as systems of point masses. From things (res) comes its name for Being: reality, "Dinghaftigkeit." From Greek metaphysics it took over the belief in the unity of Being. It attempted to subjugate everything to *this* concept of being and thus created the "world view" which Einstein came to know in his youth and which in old age he described ironically by saying "In the beginning (if there was such a thing) God created Newton's laws of motion together with the necessary masses and forces. This is all; everything beyond this follows from the development of appropriate mathematical methods by means of deduction" (Einstein 1949, p. 18). Over the course of his life, through both theories of relativity, through his contributions to early quantum theory, and through the idea of a unified field theory, Einstein had destroyed almost everything that was a concrete model in this world view and improved on it. For this very reason we young quantum theorists were so perplexed not to find him at Bohr's and Heisenberg's side. We were ready to see in the work of the Copenhagen school the crowning achievement of Einstein's life's work. Actually, however, Einstein had no doubt changed the concrete models but not abandoned the concept of reality of classical physics. Under $\epsilon$) we will consider more closely the structure of this concept, in connection with the concept of space.

Einstein's decision in this conflict was ultimately metaphysically determined, and he knew this. In conversations he occasionally presented a philosophical argument using the name of God with apparent playfulness: "God does not throw dice" or "Subtle is the Lord but malicious He is not." When pressed, he answered directly: "I believe in Spinoza's God who reveals himself in the harmony of all that exists, but not in a God who concerns himself with the fate and actions of human beings" (see Hoffmann and Dukas 1972, p. 119). Spinoza's God is the God of Greek metaphysics.

It was precisely the temporal state which Einstein perceived as merely subjective.[17] I repeat here a remark that has already been quoted several times. Four weeks before his own death he wrote to the relatives of the friend of his youth Besso: "In quitting this strange world he has once again preceded me by a little. That doesn't mean anything. For those of us who believe in physics, this separation between past, present, and future is only an illusion, however tenacious" (Hoffmann and Dukas 1972, p. 302). On the one hand, this is meant in the sense of Greek metaphysics. Picht (1960) called the didactic poem of Parmenides appropriately "the epiphany of the eternal present"; for Plato in the *Timaios* and for Plotin time is "the image of eternity, moving according to number." On the other hand, the "eternal present" appears to Einstein in the image of the spacetime continuum of general relativity, thus

---

[17] See *Zeit und Wissen*, Chap. 4I.3.6, "Conversation between Einstein and Carnap about the Now."

not as a primary unity transcending mathematics but as an extended four-dimensional space.

Now precisely this combination of a metaphysics going beyond time and general relativity is directly opposite to the methodological starting point of the present book. We begin, not to criticize metaphysics, to which we return only at the end, but to build physics, with time in its modes of present, past, and future. We do not claim this structure of time to be the ultimate truth. But we assert that it forms the basis of all experience, and thus all empirical science. It is to us so "unsubjective" that, on the contrary, it enables us to formulate in the first place the difference between subject and object, and thereby a sensible concept of subjectivity. Within that context we can then also say what the interpretation of Being as "reality" means. The realist treats all that is as facts.[18] Facticity is according to our interpretation the way how the past is given to physics. Einstein's spacetime continuum, full of "real" events, also describes the present and the future as well as the past. To him possibility is merely subjective; that is why God may not "throw dice." It is the physical interpretation of time that separates us from Einstein.

### $\epsilon$) Space and object

If we ask what elements of Newtonian–Maxwellian physics are retained in Einstein's view of "physically reality," we must mention the concepts of *space* and *object*. "Object" is our technical term, Einstein also calls it "system" or "existing." Space, as just described, also embraces time. Space itself, according to general relativity, is something that is. Everything else that exists for physics is characterized by that and how it is in space. Point masses are localized, bodies are localized and extended, a field is a function of the space coordinates, i.e., has one or several variables at every point in space. How fundamental it is for these objects to exist in space can be seen from Einstein's comment about the EPR experiment on two systems $S_1$ and $S_2$. "Now, however, the real situation of $S_2$ must be independent of what happens to $S_1$. For the same real situation of $S_2$ it is possible to find, according to one's choice, different types of $\psi$-functions. (One can escape from this conclusion only by assuming that the measurement of $S_1$ (telepathically) changes the real situation of $S_2$ or by denying independent real situations as such to things which are spatially separated from one another. Both alternatives appear to me entirely unacceptable.)" (Einstein 1949, p. 85).

Bohr (1935, quoted in Bohr 1949, p. 234) replied to Einstein's criterion of reality: "From our point of view we now see that the wording of the above mentioned criterion of physical reality proposed by Einstein, Podolsky, and Rosen contains an ambiguity as regards the meaning of the expression 'without in any way disturbing a system.' Of course there is in a case like that just considered no question of a mechanical disturbance of the system under investigation during the last critical stage of the measurement procedure.

---

[18] This does not hold for Popper for whom the *reality of time* is the heart of realism.

But even at this stage there is essentially the question of an *influence on the very conditions which define the possible types of predictions regarding the future behavior of the system.*[19] Since these conditions constitute an inherent element of the description of any phenomenon to which the term 'physical reality' can be properly attached, we see that the argumentation of the authors does not justify their conclusion that quantum-mechanical description is essentially incomplete."

In the last sentence Bohr uses, as one can see, his concept of phenomenon to indicate where the expression "physical reality" may legitimately be applied. The decision to measure the position on $S_1$ defines, in Bohr's sense, another phenomenon than the decision to measure the momentum on $S_1$. The transition from one case to the other replaces a well-defined phenomenon by another and is in this sense indeed a "disturbance of the system."

It is just Bohr's 'individuality of the process' that in the final phase one must also describe the process as an integral phenomenon encompassing $S_1$ and $S_2$. Indeed, the quantum theoretical solution is precisely the second of Einstein's "unacceptable" solutions. Strictly speaking, $S_1$ and $S_2$ remain a single joint object as long as no irreversible measurement process has destroyed the phase relation between them. In the joint object the parts by themselves do not have well-defined states (see section 3.6e); in a certain sense we can say that the partial objects themselves then do not exist, unless the state of the joint object is a product of the states of its parts.

However, one cannot deny that the EPR experiment, although described without any contradiction by quantum theory, demonstrates in a perplexing and "paradoxical" way quantum theory's break with the classical concept of an object. Two grains of sand, touching each other, are distinct objects as long as we can describe their difference irreversibly; the systems $S_1$ and $S_2$, separated by the distance to Sirius, are not distinct objects so long as we can ascribe a "pure" state to both. The temporal aspect of this fact is described by Wheeler's concept of the "delayed choice." It matters neither where nor when an observer intervenes in the system: he changes the entire state.

In our opinion, Einstein put his finger on an inconsistency of the present, correspondence-like quantum theory. This theory describes space, in contrast to objects, as something existing in itself, in the sense of classical physics. For that reason it is so perplexing that objects in widely separated places ought to be *one* object. Finkelstein (1968) called present quantum theory a hybrid of a quantum theory of objects and a classical theory of space. Finkelstein tried to redress this by deriving also a quantum theory of space, in correspondence to classical geometry. Our attempt at the ur hypothesis is perhaps even more radical. It starts without any correspondence-like notions except for the concept of a decidable alternative and it develops both space and objects (particles) on the basis of the representations of the symmetry group of a system of "urs." It is then by no means self-evident that an object must be described in

---

[19] Italics by Bohr.

space at all. A Sirius-like separation follows as the classical limit of a quantum theoretical, pre-space description of heavier, approximately classical objects like Sun, Earth, and Sirius. But an object whose quantum theoretical phases are known is not in itself divided into widely separated parts but is a whole, in the EPR case with a finite probability that upon a measurement of the distance operator $x_1 - x_2$ it will show the value of a Sirius-like separation.

We thus recognize Einstein's critique of quantum theory by means of the EPR experiment to a certain extent; however, we resolve it in a way that is the opposite of what Einstein had been hoping for: the quantum theory thus far was not yet a consistent enough quantum theory.

### 9.3.5 Hidden variables

This title signifies little more than a blank in this book. I carefully considered the first of the theories of hidden variables, that of Bohm (1952). This investigation triggered, as an alternative to Bohm's suggestion, the logical interpretation of quantum theory presented in this book. Afterwards I was convinced that theories of hidden variables were in vain, even if they turned out to be formally possible. They probably would destroy the amazing symmetry of quantum theory which I had tried to reconstruct from simple postulates. I also suspected that a classical continuum theory of hidden variables as, e.g., Bohm had hoped to employ for an explanation of the $\psi$-function, would always run into difficulties with thermodynamics. Therefore I, and no doubt many members of the quantum theoretical guild, have not followed the literature about hidden variables any further and have calmly awaited for these attempts to fail. Thus below there follow only a few remarks.

As pointed out by Jammer (1974, p. 254), Einstein was sympathetic to theories of hidden variables but also had reservations for each special case. He did not hope for a mere supplement of quantum theory through additional variables but for a new step, which would be as radical as the step from the Newtonian theory of gravitation to general relativity. That indeed appears to be the only science-theoretically plausible hope. But one probably will not find such a path by looking for a mere supplement to quantum theory.

Incidentally, I was, like probably again most "members of the guild," just as little interested in a positive theory of hidden variables as in a proof that such a theory is impossible. Impossibility proofs are scarcely feasible in the case of empirical theories. How can one rule out a hitherto unrecognized modification of the empirical verifications of the present theory? The accepted opinion nowadays is that there cannot be any *internal* or *local* parameters for the object under consideration. But how can one rule out external or nonlocal parameters, i.e., ultimately an influence of the entire universe on local events?

One should, however, make a sharp distinction between *deterministic* and *indeterministic* theories of hidden variables. The proposals thus far are no doubt altogether deterministic. That is, they want to restore that the future is causally determined by the present. However, one also could imagine a hidden "facticity of the future," without future events being determined by general laws from the present state. We will deal with this possibility in Chap. 11.4.

A strong argument against the search for *deterministic* theories of hidden variables is the psychological one, that the conservative desire for them can be so easily explained historically. Classical theory is deterministic. Quantum theory can explain this determinism as a consequence of the formal determinism of the $\psi$-function according to the Schrödinger equation, if one only applies the $\psi$-function to the limiting case of large statistical ensembles. Thereby we explain the belief in determinism through its empirical success in the classical limit. But a psychologically explained belief, after one has understood the explanation, ceases after a certain time to be a convincing belief—unless the meta-belief, that it had been explained psychologically, itself could be psychologically explained.

### 9.3.6 Quantum theoretical overdetermination

We conclude the discussion of hidden variables with an argument which in principle goes beyond the correspondence-like interpretation of quantum theory. Heisenberg's uncertainty relation then demonstrates that quantum mechanics presumably is at least then free of contradictions if one renounces the assumption that the quantities of classical point mechanics, namely position and momentum, always exist. The original starting point of the theories of hidden variables was the hope that these quantities actually exist but are "hidden" for quantum theory. Then one can look upon quantum theory as an *incomplete* theory. Jammer (1974, p. 185–186) points out that one can also interpret it as being *overdetermined*. The EPR paradox occurs precisely because in quantum theory there are phase relations over Sirius-like distances which have no analog in classical point mechanics. Jammer considers sacrificing this overdetermination. But the entire empirical success of quantum theory depends on these phase relations: matter waves, stability of atoms and molecules, more specifically the explanation of superconductivity and superfluidity. One cannot sacrifice them without destroying the theory.

Conversely, we must point out that a characteristic of quantum theory, as compared to classical physics, is not less knowledge but enormously *more knowledge*. Let us try to estimate this quantitatively. Assume we could subdivide, for a certain measurement accuracy, a continuum of possible values (as visualized by a line segment on a scale) into smaller, still distinguishable intervals by means of $n$ yes/no decisions. This then yields $N = 2^n$ distinguishable measurement values. The information content of *one* measurement of this quantity is then $n$ bits. Now we subject this $N$-fold alternative to quantum theory. Each of the $N$ possible measurement results then as a probability $p_k$ ($k = 1, 2, \ldots, N$). $p_k$ must lie between zero and one, with the only restriction that $\sum_k p_k = 1$. For each $k$ this is again a continuum, and we wish to assume that, through large statistical measurements, again $n$ yes/no decisions can be made, hence again $N$ different values of $p_k$ can be measured.[20] Then there are $N^N$ distinguishable quantum states of the object. The empirical determination of *one* of them has then the information content of $n \cdot 2^n$ bits. The classically possible information is then the $2^{-n}$ fraction of the quantum theoretically possible information.

From this example we see perhaps most clearly the enormous task for a theory of hidden variables if it does not want to lose again the positive results of quantum theory.

### 9.3.7 Popper's realism

This again is only a blank, painful to me. After a first crossing of swords in 1934 (Popper 1934, Weizsäcker 1934; see Jammer 1974, p. 176–178) I considered in detail in 1971, in an article about Heisenberg, Popper's view of quantum mechanics in his *Logic of Scientific Discovery*. The problem, important to Popper, of whether the $\psi$-function, i.e., the probability, describes a single case or a statistical ensemble, is discussed in *Zeit und Wissen*, chapter II.4.4. In the seventies, after I came to know Popper personally, I intended to read and review his newer publications on the interpretation of quantum theory. I was quite convinced that I would retain my critical views of this special problem but I considered it my duty to demonstrate this explicitly. I have not found the time and strength for this. Now there is missing here the section about Popper's interpretation of quantum theory and in *Zeit und Wissen* the section about his—to me in many points plainly evident but still different from mine—analysis of the probability concept. *Zeit und Wissen* only contains a brief personal appreciation of this distinguished man. Finally, let me also mention p. 583 in *Der Garten des Menschlichen*.

### 9.3.8 Everett's many-worlds theory: Possibility and facticity

Everett (1957) proposed an interpretation of quantum theory without reduction of the wave packet. If one only verbally reformulates his theory a bit, it is no longer so revolutionary as he himself, his followers up to this day, and also Jammer had believed (see the report in Jammer 1974, pp. 507–519). Of all the alternatives enumerated here, it is the only one that does not shrink from the understanding already gained by quantum theory but goes forward and beyond. Its weakness appears to be a formulation in a language which is still too traditional and which then inadequately expresses the true radical nature of quantum theory, and just because of this is so shocking.

Everett proposes never to reduce the wave function and to interpret precisely the unreduced wave function as objective description of the real world. Where the usual theory expresses a measured result through a reduction of

---

[20] As $\psi$ is complex, it actually are $2N$ different values.

the wave packet, there occur according to Everett all possible measurement results simultaneously. Their quantum-mechanical superposition, however, will never be noticed by the respective observer, due to the irreversibility of the measurement process and the ensuing loss of *knowledge* of the phase. Everett's hypothesis then says that the observer, when he has found a certain measurement result, cannot know that he has found in the other branches of the entire wave function the other appropriate measurement results. For *him* the world collapses to all that follows from the one measurement result he has observed. *To him* it must therefore appear as if there were only this branch of the world, and consequently he will use from now on a reduced wave function. The *theoretician* however, who follows the entire process mathematically, knows that all possible measurement results were found. Hence *to the theoretician* the world has split into so many noncommunicating but coexisting "worlds" as there are possible measurement results. And so on ad infinitum.

This perplexing description has precisely the structure of the novel *The Garden of Forking Paths* by J. L. Borges (1970). A young Chinese, who is working in England during the First World War as a spy for Germany, must transmit, as fast as possible, the name of a certain village to his German spy masters; victory or defeat in Flanders depend on it. The only way he sees is to get that name into tomorrow's newspapers trough the sensational murder of the owner of a neighboring estate that by chance carries the same name. But as owner of that estate he finds an eminent sinologist, who is editing a manuscript by the grandfather of our Chinese about the said gardens of forking paths. The garden is an image of human life. Every human decision is made in every possible way at the same time. The person who has made the decision then walks simultaneously on all paths that were open to him. But on each path he only knows about the one decision that has led to just this path, and he must bear its physical and moral consequences. Now our spy asks himself whether he will murder that wise and gracious host or not. He will do both, and on both paths he will afterwards only know to have done that which led to that path.

Everett's theory makes use of all of the arguments we employed for the semantic consistency of quantum theory: the gain of knowledge by the observer, the quantum theory of the measurement apparatus, the irreversibility of the measurement, and the quantum theory of the subject. To that extent it is a complete and faultless quantum theory. By a single verbal change it can be transformed into our interpretation: instead of "many worlds" one must say "many possibilities." As discussed in section 9.3$d\gamma$) above, it is completely legitimate for different observers, who know different facts, to recognize different possibilities and probabilities. The unreduced wave function of the observer $A$ is there the knowledge he has as long as he does not know about the measurement results of $B_1$ and $B_2$. Different possibilities must preclude one another and exist at the same time: that is what one *means* by saying that several things are now possible; it thus constitutes the concept of possibility. Everett

## 9.3 Paradoxes and alternatives   295

remained conservative only in accepting from classical physics the identification of reality with facticity. Had he known about temporal logic, he could have formulated a less startling but more correct description of quantum theory.

In Chap. 11 we make essential use of a model of a quantum theory without reduction of the state. Our interpretation of the model differs significantly from Everett's; but it would not have come into being without our involvement with Everett.

# 10
# The stream of information

## 10.1 The quest for substance

> *Sinnend der Weise...*
> *Sucht den ruhenden Pol in der Erscheinungen Flucht.*
>
> Schiller, Der Spaziergang

Transience is a basic human experience.

What was the stability of Egyptian art striving for if not the preservation of life through constantly recurring death?

Ever since its Greek beginnings, science has been searching for something permanent in the fleeting appearances. The philosophical question of substance begins by not identifying this or that—solid Earth, Gods, water—as being permanent but to ask what is meant by the question of something being permanent. We seek that which is *beneath* the changing surface of appearances, always and everywhere. That which is beneath, sub-sistent, substance. That is to be the actual Being which does not arise and pass away, but *is*.

Here a basic philosophical decision was made. It is not the here and now that is transient, but an everlasting hereafter. Do not search for permanence somewhere else, and here only for transience. That which itself is in the changing things will be that which is everlasting.

Modern natural science begins with classical mechanics, which is about bodies moved by forces through space over time. In the chapter on the system of theories, we followed its development into our century. The path leading to quantum theory has dissolved everything that appeared permanent in classical mechanics—this is the reason for the "grieving process" in the debate over the interpretation of quantum theory. Only time itself appears to remain.

The structure of our book has already anticipated this situation, and has taken its beginning from the structure of time itself. We have developed the reconstruction as far as our present knowledge permits. Now we return to the question of permanency or essence.[1]

Permanency or essence—this characterizes the approach of Plato's philosophy. It is about what is, what neither becomes nor passes away, the Eidos, the form or Gestalt, just the *Wesen,* to use a term from the German language tradition. The most important examples for mathematical natural science are mathematical structures. Circles drawn in the sand appear and disappear and are not true circles; however, about the circle itself, the mathematical circle, we have insight into its eternal structure. But Eidos is also the Just, in contrast to the never ending ambiguities of our human actions. Eidos is the model of human society, of the Politeia, as the philosopher depicts it. Eidos, in the mythical language of *Timaios,* is the eternal model in whose image Heaven and Earth are created in mathematical order. The mythical language still seems to assert a separation of the here and now from the hereafter. But this only appears to be so from our ignorance which is still caught in the appearances, the shadows on the cave wall. Neoplatonists denote the unpronounceable One, the spirit eternally contemplating the One, and the soul of the world, moving itself and all things, as the Hypostases, the substances. He who has seen the Hypostases recognizes that all appearances are in truth agitated substance.

For the later natural sciences the Aristotelian concept of *ousia,* the "Beingness," became important. It is used with a double meaning. On the one hand, it can denote the form, *Eidos*, in literal Latin translation the Essentia of a thing, which our tradition calls the essence of the thing. Yet on the other hand it denotes—and that is the main meaning, the "first ousia"—the con-cretum "grown together" out of form and matter, the thing. In this sense one translates Ousia as substance. The word *Hyle*, matter in Latin, originally means wood; as technical term it means the material which assumes form, which is "informed." At the height of Aristotelian abstraction matter denotes potentiality. Potentiality exists in time; due to it there is change, kinesis, what we usually and narrowly translate with motion.[2] substance in the sense of Aristotle is thus form in matter. Concrete things of course come into being and decay as matter assumes form and loses it again. The form is eternal as ever new things assume it. The classic example is a biological species whose individuals always recreate their kind. "Species," appearance, is the Latin translation of Eidos. The material does not last forever. The material in question (e.g., this wood from which a cabinet is made) is itself a concretum of the form "wood" and the elements as matter. But the elements also have form. A "first matter" without form is a mere abstraction.

---

[1] This chapter, like Sect. 8.6a, follows with some modifications the essay *Materie–Energie–Information* (1969), in: *Die Einheit der Natur* III,5; quoted as MEI.

[2] See the essay *Möglichkeit und Bewegung. Eine Notiz zur aristotelischen Physik*(1967), in: *Die Einheit der Natur* IV,4.

Aristotelian physics, as can be seen, is comprehensive. On the one hand, it is quite close to the phenomena. It can be expressed in everyday language. On the other hand, with concepts of form and potentiality, it reaches a very high level of abstraction. The mechanistic world view of early modern physics is in both aspects more narrow. It shies away from the phenomena as well as the highest abstractions. It postulates concrete models of reality beyond the phenomena: extended bodies or point masses having only geometrical or kinematic attributes, while the sensory qualities are only created as "subjective impressions" in the consciousness of the observer.

This approach is powerful insofar as it is mathematical. By means of it one can deduce unique consequences and either confirm or refute them experimentally. Thus one obtains a rapid theoretical development of these models as we have discussed it in Chap. 2. Its twofold retreat, however, creates a twofold uncertainty. As substance it knows matter in space, later on perhaps force fields; as "entities" (which only linguistically is a more abstract version of "substances") also space and time. Sensory phenomena are shoved aside into the subjective. Descartes is consistent when he then introduces consciousness as a special substance. Thereby, however, the unresolvable mind–body problem is created. Material substance in this model is robbed of its sensory qualities. Modern natural science has neither a model for the interaction nor for the identity of both substances. The uncertainty is indeed twofold. The successful mechanical model, on the one hand, rules out the world of the phenomena as something merely subjective. On the other hand, it also avoids a more abstract and thus more comprehensive concept of substance.

Of course also in Greek philosophy the relation between body and soul was a problem. In the Aristotelian tradition one calls soul the form of a living body. In the Greek language this expression sounds natural. *Psyche* initially denoted the living breath and from there may assume the meaning of a moving and forming force as well as the meaning of living sensation and a consciousness. In the Eidos philosophy form is what is permanent and therefore recognizable. The moving soul as the form of life is, on the one hand, that which allows us to recognize every living body anew as living. Perception on the other hand belongs to life. This is a description of the phenomenon we call life, made possible by the level of abstraction of the Eidos-philosophy. A causal explanation, as sought by modern natural science, it is not. In the Aristotelian description, going down from the realm of living beings we arrive at a teleological physics (for instance, bodies have a natural place to which they strive to return). Going up, to the mind which perceives the highest ideas, we are reminded of the neo-Platonic tenets that these ideas know themselves. Plato's "unwritten doctrines" may have been a sketch of these suspected connections. For the time being we let these problems rest and read now the development of modern physics as the quest for a precise understanding of material substance.

Substance is assumed to be permanent. Thus one believed in the conservation of matter in the change of appearances. In Sect. 2.4 on chemistry we have discussed the formation and successes of this belief. In mechanics one understood mass as the quantitative expression of the amount of matter. The masses of the bodies and point masses are therefore treated as constants in the basic equations. The success justifies the approach. The importance of measuring weights in chemistry rests on the same idea.

In addition to the conservation of mass there enters in the nineteenth century the conservation of energy (Sect. 2.5). One can interpret it as the conservation of the "quantity of (actual or potential) motion" (MEI, 1.). In contrast to the conservation of mass, one was not permitted to postulate it in mechanics from the outset. It followed, rather, as an "integral of the equations of motion"; its extension over all of physics marked the belief in mechanical laws of nature, or something similar to mechanics. In retrospect we see here the development of the mathematical form of the laws of nature (Sect. 2.3). Seen from modern physics, the mathematics of the Eidos-philosophy belongs to the morphological type of laws: lawful forms standing next to one another. The type of differential equations characterizes the causal way of thinking. The conservation of energy is now a causal consequence of the assumed mechanical equations of motion. Actually explained is this conclusion only in the types of laws based on symmetry groups. According to Noether's theorem, conservation of energy is an expression of the homogeneity of time. The special theory of relativity proves the identity of mass and energy and thus of their conservation laws. One can then say that matter or energy, characterized as substance, signifies precisely the identity of the fundamental laws of nature at all times; just this is expressed by the homogeneity of time. In this special sense time proves itself as that which is actually permanent, even in classical (i.e., pre-quantum theoretical) physics.[3]

## 10.2 The stream of information in quantum theory

*Die Zeit ist selbst das Sein.*
G. Picht[4]

Classical physics has not fully answered the question of substance. It is true that energy is conserved but what distinguishes it from other integrals of the equations of motion? Only in the first law of thermodynamics (Sect. 2.5) does this distinction play a central role. There, however, it remains a mere postulate, not fundamental to the theory. Other conserved quantities are also of thermodynamic interest. The hard core of thermodynamics is the Second Law with entropy, i.e., information, as the distinguished quantity.

---

[3] Cf. the essay *Kants "Erste Analogie der Erfahrung" und die Erhaltungssätze der Physik* (1964), in: *Die Einheit der Natur* IV,2.
[4] Picht (1958); cf. *Der Garten des Menschlichen* II,7: *Mitwahrnehmung der Zeit.*

## 10.2 The stream of information in quantum theory

On the other hand, energy is a mere state variable. In point mechanics it is a property of a system of point masses, constant in time but dependent on the initial state. As "substances" one would consider in it the point masses themselves; analogously in the classical chemistry of atoms. Only the transformation of "substances" into one another, as in the classical thermodynamics of chemical compounds or in present elementary particle physics, emphasizes energy as being something unchanging. As already chemical reactions can only be explained quantum theoretically, we are referred to quantum theory by our question about substance. In quatum theory, however, neither atoms nor elementary particles are immutable. Our question thus leads further into the not yet completed physics of elementary objects.

The abstract reconstruction of quantum theory suggests treating information as fundamental, and thus the substance. We do not initially care whether the quantity of information remains constant in time, but that it forms the foundation of the conceptual structure, and in that sense is the basis for the concepts of objects and their conserved quantities. We start with alternatives. A $2^k$-fold alternative, however, is $k$ bits of information. The ur is then an "atom of information."

Yet if we wish to ascribe such a fundamental role to information, we must be certain about how it is defined. In Chap. 8 we said that information only exists under a concept, or, more accurately, between two semantic levels (Sect. 8.4). For a given alternative then the lower semantic level[5] is that in which the alternative can be put as a question; quantum theoretically speaking, the level of observables which can be realized by measurement devices (Sect. 9.2d). The upper semantic level is that of the possible answers to the question asked; quantum theoretically the level of the states, especially the eigenstates of the respective observables.

Thereby, however, the question of the amount of potential information to a given $2^k$-fold alternative is still undecided. Classically one would say its value is $k$. But the "additional quantum theoretical knowledge" states that to every alternative there may exist a continuous manifold of states in vector space. To these states there exists, however, also a continuous manifold of observables, thus of alternatives equivalent to the given one. Therefore one can lay down as a definition that to every $2^k$-fold alternative will belong exactly the information $k$; one must add, however, that the decidability of this alternative itself implies a formally infinite quantity of other alternatives and thus of information belonging to the same object. To be sure, this infinite amount is not empirically measurable, but in Sect. 9.3f we have given a finite

---

[5] According to the definition in 8.4, the lower semantic level contains more potential, the upper more actual information. To the "lower" semantic level, in this terminology, there corresponds then in each case the generic term to the concepts of the "upper" semantic level. This terminology denotes as the upper level the more comprehensive one, whereas conversely the logical "generic term" is the less comprehensive and therefore has a larger content.

estimate of it according to which, in an empirically feasible approximation, it exceeds the number $k$ (there $n$) by a factor of order $2^k$.

Besides this purely quantum theoretical argument, there is another argument in classical information theory that demonstrates that it only makes sense to speak of the information of a given system if one assumes the existence of a much greater amount of additional information. In Sect. 7.6a, following the essay in MEI, Sect. 2, we justified the thesis that information is only that which is understood. Here "understanding" is interpreted not merely "subjectively," as a conscious act, but also as "objective" action, e.g., on living beings or instruments. This action we can describe as "objectified semantics." In the essay MEI there follows a Sect. 3: *Informationsfluß und Gesetz* which, at first inconclusively, considers how much information there is in the objectified semantics of a given amount of information. How many bits are needed to understand one bit? (*Die Einheit der Natur*, p. 352ff). Here we pursue these arguments a bit further.

The example in MEI, 3, is the genetic information of an animal species. If the set of chromosomes of the species contains $n$ DNA "letters," one can then specify to a first approximation, as there are four different letters, the genetic information in this set of chromosomes to be $2n$ bits. The organism which develops from a fertilized egg "understands" this information by developing into the phenotype of the corresponding species. The phenotype would thus be an objectified semantics of the genetic information. How many bits are required for this? How many bits are needed to understand the $2n$ bits of genetic information ?

Let us call this number $N$. The essay considers two opposing answers.

First answer: $N$ is enormously large, of the order of the information content of all individual protein molecules which have been produced according to the genetic code in the respective organism. Without this information it would not be viable and capable of reproducing.

Second answer: $N = 2n$. The argument for this is based on the thesis that information is only that which produces information (Sect. 7.5a). Here we quote it literally: "For the organism develops from its genetic endowment and transmits these same $2n$ bits (apart from mutation) to its offspring. These bits are necessary and sufficient for the definition of the species; they are therefore the true amount of information in the organism. Anyone who completely understands the laws governing the functioning of an organism ought to be able to derive its form and functions simply from a knowledge of the DNA chain in the nucleus of any of its cells. He would know, therefore, that the huge amount of information arrived at by the first answer is redundant and reducible to $2n$ bits. Only the second argument subsumes the organism under the concept of a living being, which is of course appropriate to it; the first answer subsumes the organism under the concept of a physical object. The

## 10.2 The stream of information in quantum theory

excess information in the first answer is simply the information contained in the concept of a living being" (p. 353).[6]

Both answers thus explain meaningful but different concepts of information needed in the objective semantics. Both can also be related to our present question of whether information could play the role of substance in fundamental physics. The distinction of the answers thereby separates the two perceptions intermingled in the traditional view of substance: substance as being that which persists, and substance as being basic.

In the second answer information is what persists in the species. It is a measure of the number of features of the species *as* a species. It is what one must know if one already knows that there are living organisms which reproduce themselves (lower semantic level) and if one wants to know which "special" species (which Eidos) of living creatures one is dealing with (upper semantic level). In the first answer, however, information is what one needs to know if one only knows that there are atoms and molecules, and one wants to know under what conditions they will combine into a living organism at all, especially into an organism of this species. Here information is the physical basis.

One must remark that persistence exists only within the scope of the corresponding concept—here, then, within the scope of the organic species. In the actual process of life there are individual mutations, there are "selfish" genes within a species, there is evolution and extinction of species. All this can only be discussed within the scope of the first answer. But if there were no concepts at all for approximate persistence or approximate reproducibility there would be no insight; then we could not do biology and we ourselves as living beings, communicating with one another, would not be feasible.

Analogous but presumably simpler circumstances we find now in fundamental physics. We consider the approach to a concrete quantum theory of particles according to Chap. 4. To the species there then corresponds a certain kind of particle, defined by rest mass and spin. The particles of one kind "do not reproduce themselves"; they do no create their own kind. This is not necessary for them as in the approximation where one can disregard their interaction they also do not die. Their individual duration is their persistence. ("Spontaneous decay" is always interaction with a field of virtual particles). According to our hypothesis, urs comprise the basis. Their number in a given object, a region in space or a hypothetically finite universe would be the measure of the maximal information in this object, region, or universe. The urs are the lowest possible quantum-theoretical semantic level. For their number, however, there exists no conservation law (except in special models of the universe, such as anti-de Sitter space).

This mirrors the dual relationship between the concept of substance and time. On the one hand persistence is a precondition for the applicability of concepts. In that respect substance is the implemented Eidos within the realm

---
[6] p. 283 in the English translation

of the possibilities. "Time itself is the Being" means here that to be is to persist in time. In this way of thinking Kant could interpret substance as that permanent substratum "which represents time in general" (*Critique of Pure Reason* B224). Persistence, on the other hand, only exists in approximations. In the reconstruction of quantum theory via variable alternatives (Chap. 4), the openness of the future expresses itself in the emergence and disappearance of urs as atoms of information. Picht emphasizes that he would like to have the "is" in the sentence "time itself is the Being" be read as a transitive verb: the time *is* the Being insofar as it creates it. Also Heidegger spoke of this (in his *Being and Time*) when he added to the sentence "There is Being" the question of what "there" had been that made the Being, and answered: the event. The event makes the Being.

Utterances like those of Heidegger and Picht point with the means of language beyond conceptual science. But with the ur as the smallest unit of physical information, we wish to stay within conceptual science. Thereby we come to the problem of how the ur itself can be defined conceptually, i.e., with the help of something persistent. This question, applied to the ur, is the question of the information content of objectified semantics; in the simple language of physicists: how can one measure one ur? The question gains its acuity if the ur is seriously taken to be the ultimate available unit of information. Then the objectified semantic of the ur can ultimately only be based on urs. Ultimately one can measure urs only with urs.

A first answer is the reference to the essential irreversibility of every measurement (Sect. 9.2). One can measure one ur only with as many urs as are necessary such that the possible loss of information in the measurement instrument consisting of urs ensures that an objective measured result can be registered. According to our particle model one presumably can interpret the measurement of the spin direction in the Stern–Gerlach experiment as the decision of one ur alternative. Due to the indistinguishability of the urs there only remains undetermined which of the about $10^{37}$ urs in the electron, bound with suitable symmetry, is intended.

The definition of the ur depends thereby on the available and actually used instruments. The definition of the ur is related to the position, time, and state of motion of the measurement apparatus (observer). information is only what is understood. Thus the selection of the basic unit of information, the ur, depends on the means available for understanding. The relativity of the urs, however, implies that a transformation from one definition of the ur to the other is possible. observers who understand the elementary information can communicate with one another.

If this communication is also possible between different successive observations of the same observer, thus formally by means of a time-translation between different points of a world line—what then is the meaning of the creation of urs? As the simplest example we choose Castell's model of the neutrino in the lowest discrete state (Sect. 5.3b). This is a single ur at the time $y_4 = t = 0$. For $t > 0$ it is a superposition of many urs. Does it now

## 10.2 The stream of information in quantum theory

contain more information than before? The actual information is the same as before. It persists, like the genetic code of a species: the object continues to be a neutrino in the lowest discrete state. In position space, however, this state is an expanding wave packet. Thus its potential information increases; in thermodynamically terms, to the expansion in the vacuum corresponds an increase in entropy. To turn potential information into actual information one needs the interaction with another object, a measurement apparatus for position measurement. For the question "In which discrete state is the neutrino" the entropy has remained the same; for the question "Where is the neutrino?" the entropy has increased. If the ur is a unit of information, then to the relativity of the urs there must correspond a relativity of information, i.e., of entropy. These circumstances are specifically quantum mechanical, made possible through indeterminism and "additional knowledge."

By MEI 3 we had applied the notion that "information is only that which produces information" to the deterministic description that holds in the classical mechanics of simple systems such as a point mass. The information on the current position of the point mass (in phase space) "produces," by means of the motion of the point mass, the information about its position at a later time. Regarding the content, these two specifications are different but due to the determinism of classical mechanics they contain the same amount of information. In this sense the amount of information remains conserved; any specification of the location in phase space suffices to deduce, by means of the mechanical laws, any other. We can speak in this way only if the location of the point on phase space is precisely known. Specifying this position, however, due to the continuous nature of phase space, would imply an infinite amount of information. If the position at a time is only inaccurately known then already according to classical mechanics the probability packet will expand (apart from uninteresting special cases like the harmonic oscillator); its entropy increases, the actual information decreasing.

In quantum theory on the other hand, as shown by our example above, even an exact knowledge of the state permits an increase in entropy. In the absence of measurements or otherwise processes described as irreversible, the change of the Hilbert vector in time is just as deterministic as the change of the phase point in classical mechanics. But for certain alternatives the Hilbert vector always implies probabilities different from one and zero, thus not maximal information. In the abstract quantum mechanics of the infinite-dimensional Hilbert space one presumably cannot say anything general about the time dependence of this information. But concrete quantum theory couples the position space in which we perform our actual measurements to certain finite alternatives, namely those determined by a finite number of ur alternatives. Thereby a basis is defined in infinite-dimensional Hilbert space whose states fall into classes, each characterized by a finite number $n$, the number of urs, i.e., the tensor rank. Then $\log_2 n$ measures the potential information of the statement that the state belongs to the tensor of rank $n$. Starting then, as in the example above, at $t = 0$ with a state of low tensor rank (there $n = 1$), one must expect that statistically the tensor rank and thus the potential information of the state regarding measurements defined in coordinate space will increase. For somewhat more complicated situations which, due to a large number $n$ of urs, allow to define more than two semantic levels we can expect that this can be described as a growth of the multiplicity of forms as in the examples of Chap. 8. We have made use of this in discussing the expanding universe in Chap. 5. A detailed model would require a complete theory of elementary particles.

In any case we have good reason to expect that concrete quantum theory will give rise to a flow of information such that the evolution of forms follows with overwhelming probability.

## 10.3 Mind and form

*Fire!*
Pascal

Information stands for us now at the systematic place of a measure of substance. In Sect. 8.2, on the other hand, we declared information a measure of form. The development of physics seems to leads us back to understand *form* as the substance in the stream of events. Are we returning to the Eidos-philosophy?

Our answer contains a Yes and a No. Yes: the Eidos philosophy signifies a level of abstraction compared to which the mechanical models of physics have been lagging behind since the seventeenth century. Abstract quantum theory compels us to return to this level of abstraction. No: Separating that which is the basis from what persists signifies a recognition of time, of the stream, precisely what the Eidos philosophy wanted to avoid. We can only speak of form if we also accept the evolution of forms.

We interpret the Yes and the No more closely.

Yes: In our time one has every so often established information, and thereby form, as a third reality accompanying the "two realities" of matter and mind (Sect. 7.2). According to our analysis this is an inconsequence, a half measure. Experiment and theory, as we know them today, no longer provide any reason to postulate matter and mind (*res extensa* and *res cogitans*) as independent "realities," i.e., as substances in the classical meaning of the word. Form is not a additional third, but their common basis.

That form is the basis of matter we have elaborated in detail: alternatives as the starting point for the reconstruction of an object.

Mind is not the theme of this book about physics. But we emphasized in Sect. 9.2e that in abstract quantum theory there is no inherent reason why it could not be applied to the self-knowledge of the mind. There, for the precision of the argument, we abstained from making the connection between the body and the mind of an observer a central theme. We have just *not* asserted that *because* quantum theory is applicable to his body it also must be applicable to his mind. Just the converse, we have said, as far as decidable alternatives exist in the self-awareness of the mind, they must be subject to abstract quantum theory as the theory of *all* possible alternatives.

Proceeding now with the transition to concrete quantum theory, we will also build up the alternatives in the self-knowledge of the mind from ur alternatives. From this we have then to *conclude* that one must also be able to describe the human mind as a body in three-dimensional space. The conclusion lies at hand that this must be the human body.

Of course this only poses the problem.

First of all the description of the mind through decidable alternatives is a scientific stylization. I know myself as a being that is wanting and drifting, waking and sleeping, that is suffering and enjoying, loving and hating, attached to a checkered, fostering and threatening environment, with understanding and ignorant partners, as a member of a society and still able to withdraw from society, environment, desires and decisions, into an inner self of unfathomable depths. What else do I learn about myself when I attempt to build up the structure of these experiences from decidable alternatives?

To this we must add, however, that we also know our environment and our body, thus the so-called material world, in terms of a multitude of qualities which we normally do not reduce to decidable alternatives. Also in Greek philosophy Eidos signified by no means only mathematical form. We mentioned other meanings of Eidos at the beginning (Sect. 10.1): the just, the perfectly beautiful in Plato, the soul itself in Aristotle. The reduction of knowledge to mathematical structures is in that regard not a return to Greek philosophy but a radicalization of modern natural science. Of course it follows there a tendency already present in Greek logic, mathematics, astronomy, theory of music which became effective in the distinguished role of mathematics with the Pythagoreans and Plato. We can say: What can be subjected to a logical-empirical decision at all can be described as mathematical structure. That we question consciousness from the standpoint of decidable alternatives means no more than that we expose it to the same kind of questioning as nature in natural science. How much we can learn from this remains to be seen.

Another objection comes from the fundamental role of consciousness for science. Loosely following Kant, it might be formulated as follows. Science is cognition, hence—insofar as we can use the concept of consciousness—content of consciousness. Matter is an object of cognition. Forms are, in scientific usage, concepts. Thus it appears to be sensible to explain matter in physics by means of forms, i.e., concepts. Physics is just that which we can know about matter. Consciousness, however, is a prerequisite for cognition; thus it would be circular to explain it in terms of the means of cognition, the forms. Precisely because of this, according to Kant, the knowing subject is not to be described as a substance, for substance itself is a category, i.e., a concept.

This objection is based on the argument of being circular, thus on the hierarchical approach of traditional philosophy (cf. *Zeit und Wissen* I.5.2.3). But we are moving in a *Kreisgang*.[7] Physics describes what happens according to laws of nature, which in classical physics is called "matter." With the same right, abstract quantum theory may attempt to describe lawful regularities of the mind. The physical preconditions, in addition to those in the consciousness of our knowledge, are utilized from the outset, but also described later on in the *Kreisgang*. The claim to provide thereby a *complete* description of reality might be infeasible; but the claim to provide a *consistent* description in the given approximation (separability of alternatives) is legitimate.

Consciousness as we know it emerges in the course of evolution from the sea of life. In Sect. 8.7 we sketched how the structures of our logic are based on preliminary stages in animal behavior. The way how animals themselves experience their behavior is the harder for us to imagine the farther their behavior is from ours. The transition is continuous into what is for us completely incomprehensible. The complete absence of "cogitatio," i.e., thought, sensation, experience in the Cartesian extended substance is merely a postulate of the philosopher who had decided to accept only that which he clearly and distinctly—and here that meant de facto mathematically—could think about.

We thus have no reason to consider "life as a form of cognition" (Sect. 8.7b) a mere analogy. The salient characteristic of human knowledge is the ability to introspect, mainly through language, to "look at oneself in the mirror." The reflection makes possible the Kantian distinction of the transcendental, i.e., cognitive, subject from the empirical, i.e., recognized subject. Inasmuch as the act of cognition is described psychologically or ethologically, it is "seen in the mirror" and thus belongs to the empirical subject. *That* it is cognition means that it is the image of the subject itself seen in the mirror. This achievement of the cognitive subject is now just what it has in common with animals. It is decisive to see that animals take part of the transcendental subject, only without reflection. Animals recognize the Eidos in the individual case (*Der Garten des Menschlichen* II, 6.4, p. 312); humans who, reflecting, distinguish the Eidos from the individual case, recognize the individual case *as* individual case and the Eidos *as* Eidos.

The topic of the classical Eidos philosophy, however, is not the lower levels but the highest ones. We can interpret the behavior of animals from human consciousness, but from where do we interpret human consciousness? Philosophy demands here that we rise from our everyday uncomprehending usage of concepts to their true perception. In Neoplatonism the highest pronounceable level is the mind (the divine *Nous*). It is the realm of ideas that know themselves, highest energy of motion which returns to itself, and thereby at the same time highest rest.

Are we returning to this form of Eidos philosophy?

---

[7] Cf. p. XXII, fn. 14.

No: we must put time at the top of the reconstruction of our science. We know an evolution of forms. For us, form is not the basis, but time.

We respond at once to an objection. We must recognize the evolution of forms realized in concrete things. But the pure forms themselves, what in this book is often called "formally possible," exist independent of their actualization in things. We cannot take part in the full philosophy of Aristotle according to which every Eidos, particularly in the world of the living beings, is eternally present in ever new individuals. But by utilizing mathematics we share in the Platonism of mathematics. The structures themselves are eternal. Following Cantor: For the sequence of numbers which actually can be reached by counting to be potentially infinite, the set of all possible numbers actually must be infinite. The possibilities are what is timeless.

We will only be able to seriously address this question in *Zeit und Wissen*, Chap. I.5.5: *Was ist Mathematik?* Here it suffices to say that in the present philosophy of mathematics, it is precisely this thesis is denied by intuitionists or constructionists. Also mathematical concepts need to be (mentally) constructible. The realm of mathematical structures does not appear to be a closed infinity, but open. Be that as it may, at least in physics the mental survey of an infinity of formal possibilities (e.g., the full tensor space of urs) is only an abstract tool, a survey over possibilities of thought. What we actually need are the respective present and constantly changing actual possibilities founded in the respective facts (Sect. 4.2a). This incidentally should be in full accord with the Aristotelian theory of mathematics.

Our entire reconstruction illustrated the problem how science is possible in open time. Everything is in flux, nothing in time remains forever. Generally applicable concepts, however, exist only for what is permanent or recurring. It is not enough to deny flux—that simply would not be true. Just as little does it suffice to declare concepts and thereby science as fiction—that would leave its success not understood. Science must demand of itself to make comprehensible the very approximation in which it can be successful. We have discussed this under the title of the separability of alternatives. In the history of thought that which one can conceptually comprehend and that which is lost that way has been alluded to in a multitude of similes. A simile is an appropriate linguistic means for what goes beyond concepts.

"You cannot step into the same river twice, for fresh waters are ever flowing in upon you." (Heraclitus). The rapids are a form that continually restores itself, but always comprises different drops of water. Another simile is the rainbow. The fixed arc in the sky is always due to the refraction of sunlight by ever-changing falling droplets. Seeing the rainbow in a waterfall in the mountains or in a fountain, one can easily observe that the rainbow moves with the viewer, which becomes very conspicuous if one does rapid knee-bends. The rainbow is a "subjective phenomenon"; every observer sees a different arc, at a slightly different location.

In the explanation of these examples, however, we assume that something persists. Water continues to flow, but does not cease. The rapids, the waterfall, the fountain have a constant form as the water is guided through a fixed environment. Yet chemistry established that stone and metal can be transformed, explaining this in terms of the unchanging atoms. Our reconstruction reduces the changeability of atoms to urs which themselves are created and destroyed and are differently defined for every observer, as the rainbow in the waterfall. What is constantly restored is the decidability of the alternatives. Decidability depends of the facticity of the decision, i.e., on the irreversibility in the world of phenomena. The uncounted urs or atoms which make irreversibility possible are, as it were, the falling drops in the rain or waterfall, underlying the observable phenomena which we then express in the language of quantum theory.

The great simile is fire. It is the simile of destruction: it is nourished by the consumption of what is solid. It is the simile of life: also for science nowadays fire and life depend on a transformation of matter into a temporary form; both propagate where they find nourishment. It is the simile for visibility: all light emanates from celestial or terrestrial fires; in antiquity one believed that the fire of the sun and the stars does not consume itself; for us, however, the analogy to nuclear or chemical reactions is near. Light is the simile of truth. So is fire the simile of the spirit. The strength of this simile encompasses almost effortlessly consciousness as well as material events. Also truth is a process that needs to be nourished. The divine spirit is just as much the eternal light in which all forms become visible as well as the claiming and consuming fire. Pascal turns away from the God of philosophers who guarantees the persistence of Being. He speaks of the God of Abraham, Isaac, and Jacob, about the God of a unique historical bond where nothing can remain as it was; who also transforms and consumes his, Pascal's, own life.

# 11

# Beyond quantum theory

*Nebo.*
5. Mose 34,1

## 11.1 Crossing the frontier

We wish to know what lies beyond the limits of our knowledge. This desire is characteristic of humans. It is particularly characteristic of humans of the Western culture which was molded by the inquisitive people of Greece.

For the reconstruction of physics, as attempted in this book, this desire poses a particular dilemma.

On the one hand, we have reconstructed our science from a logic of time, from an understanding of the open future. Structurally, we have compared the formation of human knowledge with evolution. One basic principle of the corresponding philosophy is that we philosophize *today*. We cannot understand the phenomenon of the present without the facticity of the past and the anticipation of a possible future. When thinking about our present knowledge we are already thinking of future knowledge which will surpass the present knowledge.

On the other hand, we have formulated quantum theory as a comprehensive theory of empirically decidable human knowledge. It is a closed theory in the sense of Heisenberg. It cannot, so it would seem, be further improved via small modifications. But can one imagine the progress of science passing through an infinite succession of closed theories? If not, then one of them must be the last, the final one. Why not quantum theory? But can there be definitive knowledge? If so, can it have the form of the theory?

Our question can be subdivided into three questions. We can have meant three different things with the question:

a. physics beyond quantum theory
b. human knowledge beyond physics
c. Being beyond human knowledge

If we wish to address these questions with precision, we must make them hard for us. Only the true conservative can be a true revolutionary (Heisenberg). It is convenient and not very fruitful to say: "Just around the corner something unknown awaits us." We wish to ask whether our present knowledge has not already conceptually anticipated the unknown.

### 11.1.1 Physics beyond quantum theory

We have characterized abstract quantum theory as a general theory of probabilistic predictions about empirically decidable alternatives. What physical knowledge could there be which would not fall under this characteristic? At any rate, the great empirical success of the theory, combined with its mathematical simplicity, permits the conjecture that it is a correct, perhaps the final realization of the so characterized program.

"On this side" of an abstract theory is the world of the phenomena to which it is applied. In the succession of closed theories these phenomena are initially interpreted in terms of the preconceptions that furnish the semantics to the mathematically formulated theory (cf. Sect. 2.7). The ideal of semantic consistency then requires the so interpreted theory to explain the preconceptions. In the language of the chapter on the stream of information (especially Sect. 10.3), this means that the theory itself determines the forms which present themselves to us in the phenomena. An example of such a general theory is mathematics (cf. *Zeit und Wissen* 5). Besides the concept of number and set, it does not assume anything but contains an unlimited number of deductions. And as all physics is mathematically constructed, the mathematical structures derived in this fashion, and only they, are the possible structures of physical states or processes. Our reconstruction of physics is set up such that it will determine precisely that which mathematics leaves undetermined about actual nature: which of the mathematically possible structures are "formally" possible in nature, and under what circumstances they are realized.

For this abstract quantum theory is only a framework. Concrete quantum theory, on the other hand, is designed with the intention of determining all formally possible physical objects and processes. In its implementation the task would be as endless as the completion of mathematics. But the theory of elementary particles and their interactions provides a basis for all further physics. Deriving the theory of elementary particles is therefore a natural goal of concrete quantum theory. The higher forms can only be deduced in a limited way with our scientific means. On the other hand, there are general concepts about their origin within the theory of evolution. From our approach to time, the theory of evolution also emerges as a natural consequence. If concrete quantum theory turned out to be feasible and then also in agreement with experience, it would from our present knowledge be hard to see what kind of physical experience should remain closed to it. It is in correspondence with the methodological meaning of the idea of a closed theory to postulate in this case at least heuristically its unlimited validity. Only in this way does one arrive at the necessary "conservative" rigor for future empirical questions which at most could lead to the point that the theory might empirically be falsified and replaced by a better one.

The actual revolutionary elements in Kuhnian revolutions turn out to be, at least in hindsight, in general the conceptual problems, self-contradictions or insurmountable ambiguities of the older theory. Precisely to suffer from them in a fruitful way is what is meant to be conservative in the sense of Heisenberg. As such an indigestible lump there remains at most indeterminism from the debate over the interpretation of quantum theory. On the one hand, it is a way to express the quantum theoretical "additional knowledge," the immeasurable superiority of the amount of quantum theoretical information over its classical limit; to renounce this again is impossible. On the other hand there remains the question of whether the negative characterization of this additional knowledge as indeterminism does not merely signify the egg shell of the classical origin of quantum theory. Could not events turn out to be determined if one freed them from their inadequate description by means of classical physics?

In Sect. 9.2c, we justified the classical description on the basis of the irreversibility necessary to create a measured result as a fact. Irreversibility, however, is interpreted in physics as loss of information. Semantic consistency demands to ask how quantum theory would look like if, at least in thought, we would describe the information lost in the irreversible process as being objectively preserved.

Related to this is that quantum theory starts with the approximation of separated objects and alternatives, which in itself can be shown to be erroneous. Here as well, quantum theory presumes in its basic concepts a loss of information which it ought not permit in its consistent development.

Finally the same mistake is made if one describes time as a real parameter, as done in quantum theory. Time can be measured with clocks. The point in time is a fiction. Again, it can only be determined by means of an irreversible process, and there only with finite accuracy.

With these unresolved questions quantum theory surely does not point back into classical physics but beyond itself. We call this the *self-criticism of quantum theory*. We pursue it below.

### 11.1.2 Human knowledge beyond physics

Again we initially assume a methodologically conservative attitude. What knowledge could there still be beyond physics if through quantum theory it has become the general prognostic theory of empirically decidable alternatives?

That the science of all inorganic things is unified on the basis of physics we surely may take for granted. The historical sciences of inorganic things like cosmology, evolution of stars and geology fit effortlessly, especially into our temporal approach.

In the life sciences, the physical outlook has gained more and more ground in recent decades. We have assumed it in Chap. 8, and the abstract interpretation of quantum theory may serve as its justification.

That quantum theory is also valid for our knowledge of consciousness, insofar as it can be reduced to decidable alternatives, we attempted to justify in Sect. 9.2e.

It is important to see that this view is fully in accord with an acknowledgment of the hermeneutic character of the humanities (cf. Gadamer 1960). It would be a misunderstanding of the methods of natural science to base social sciences and the humanities on scientific laws. The higher the richness of structure generated by evolution, the less we can expect to describe the underlying interesting traits of these through general laws. General laws must be simple: the richness of an individual or specific structure, however, lies in its complexity, its high value of information. Therefore, a high degree of structure of the observer is also necessary to understand a high degree of structure of the observed. Only persons can understand persons. This does not contradict the scientific approach, but rather follows from it if one implements it consistently.

Yet again we find in just this argument another reason for a *self-criticism of physics*. We have reconstructed physics in terms of the concept of decidable alternatives. This is an extraordinary simplification of the actual activity of human perception. Our perceptions of humans, also of ourselves, and our nonhuman environment are almost always affective and alluding to our wishes. They are scarcely ever completely expressed in words. They rather are immediately transformed into action or they serve an inarticulate orientation (cf. Sect. 8.7 and *Zeit und Wissen* 2). The linguistic figure of an assertive sentence is a highly specialized product of culture (Sect. 8.7d and *Zeit und Wissen* 6). It is a logical thesis that propositions, i.e., assertive sentences, can be interpreted as answers to alternatives. The two-valuedness of logic is, ethologically seen, a highly useful construct. But what is thereby lost in information cannot be brought to light again, especially not by logical construction.

Ethology thus advocates that in human achievements like art, myth, culture, social manners we find a form of knowledge whose very nature might be lost at an attempt at a logical (and consequently physical) analysis. When Bohr found complementarity in physics itself, the main reason for this was that just the loss of possible information, invariably associated with every measurement, reminded him of the loss of understanding brought about by every logical decision.

Thus, while studying the self-criticism of quantum theory, we can perhaps also consider at the same time the self-criticism of the logical character of physics.

### 11.1.3 Being beyond human knowledge

The question as usually posed is about something beyond what we can know. But at the same time it means something which is vitally important to us. The tradition of human cultures searches for such a Being mostly in religion. In our Christian tradition, our relation to it is called faith. This is then known as the question of the relationship between faith and knowledge.

This question arose here for us not from religious tradition but in direct logic from the reconstruction of physics. When we ask what we can know beyond quantum theory or physics, there arises the compelling question of whether there is something that we cannot know at all. And just not something irrelevant but something of fundamental importance.

We will approach this question in three steps: now, in direct connection with the debate on the interpretation of quantum theory; in the final chapter, with a view to the tradition of metaphysics; and in the closing chapter of *Zeit und Wissen*, which thematically will devote itself to philosophical theology.

We recall Einstein's remark about death:"For us believing physicists the separation between past, present, and future has only the significance of an admittedly stubborn illusion" (Sect. 9.3d).

For Einstein the word evidently contained a deep consolation. The suffering from time is a primeval human experience. We have dedicated the previous chapter to the struggle about what is permanent in events. The future is the unknown, hoped for or threatening. The past is remembrance, the irrecoverable; we own it only as something we have also lost. Einstein's word consoles us: all this is only a stubborn illusion. It appears to give us the solace of an eternal present.

Einstein speaks about us as "believing physicists." Deliberately he is downplaying the traditional controversy between religion and science. But he is not looking for the sometimes desired cheap appeasement. He does not mean that we are physicists and moreover believing. He obviously means that we are believing because we are physicists, and what we believe is just the deep, actual truth of physics itself. This truth reveals the unity of reality, in view of which the separation of the three aspects of time is merely a stubborn illusion.

In the section on Einstein's thought experiment, we discussed this point of view and related it to the historical content of Western metaphysics. Spinoza's God is the God of philosophers who first revealed himself to Parmenides.[1] Greek philosophy is indeed a non-superficial reconciliation of religion and science, because in its self-understanding it *is* both as unity. It is the epiphany, the revelation of the One, which first had revealed itself, imperfectly, in the Gods of traditional religion and which entails the unity of thought, science.

Here, in the first of our three steps concerning the question of metaphysics, we do not start with its philosophical history but with present-day physics: from the knowledge and thereby perhaps the possible delusion of our century.

---

[1] Cf. G. Picht: *Die Epiphanie der ewigen Gegenwart*, 1960.

That which Einstein hopes to overcome is just the starting point of our entire reconstruction of physics: time in its three modalities. If our reconstruction is successful, then Einstein poses the question of that which transcends all present scientific knowledge: Being beyond our knowledge.

We now ask, in the sense of pure physics: theoretically, can our hope of overcoming the trinity of temporal modalities be reconciled with our present knowledge (specifically, as expressed in the form of the reconstruction in this book)?

Thereby we do not mean to leap into the complete unknown, which one cannot think about, and thus not rule it out by argument. We instead pose three more modest questions, as conservatively as possible. Is it possible to use the type of reality of one of the three temporal modalities as a basis and reduce the other two to it? Hence we ask the three questions: is it possible, while retaining all insights of quantum theory, to describe all processes in the language of

either A) classical facticity,
or B) quantum theoretical modality (i.e., the $\psi$-function),
or C) direct present?

We assert that all three solutions are conceivable in principle if we make certain sacrifices in the traditional interpretations of facticity, possibility and present. We devote the three following sections to these three questions.

## 11.2 Facticity of the future

We have represented the past as factual. If all three modalities of time are to be described in the language of facticity, we are then close to the world view of classical physics and its concept of reality (cf. Sect. 9.3d). But we wish to retain all insights of quantum theory, including its indeterminism. The question is whether behind the probability predictions of quantum theory there nevertheless could hide a reality in which the future must also be represented as factual.

*At first glance*, there does not appear to follow any logical contradiction from such an assumption. There we adhere to the full quantum theory *as* theory of knowledge, thus also to our notion of "additional knowledge" of quantum theory compared to classical physics. We thereby completely dispense with the hypothesis of a *causal* determination of the future through hidden variables. This is the necessary sacrifice compared to the traditional, classical interpretation. In this assumption we describe the future as *factual* but *not* as *necessary*.

As an illustration we again consider our description of the past. We have seen in the chapter on irreversibility, and in the description of the measurement process (Sect. 9.2b) that we can describe the past as the embodiment of objectively existing facts, distinguishable from one another in principle, perhaps even discrete. Also in the retrodiction from the present state of an object according to the Schrödinger equation, the past facts about this object do not follow with necessity. The retrodictive probability for a certain past fact, like the result of the measurement $M_{-1}$ on the basis of the known result of the later measurement $M_0$, is in general not 1. The facts of the past do not necessarily follow from the present state of the object. But we can deduce the past facts from states that have irreversibly occurred in other objects, i.e., from documents. We say: the past, seen from today, is not causally necessary but it is factual.

We recall the example of an ancient solar eclipse. Our astronomy can calculate that 585 BC a solar eclipse must have occurred and been visible in Asia Minor. This is causally necessary, under the usual methodological assumption that we have correctly observed and calculated, and under the astronomical assumption that the orbit of the Moon had not been disturbed in the two and a half millennia since that year through facts unknown to us (like the impact of planetoid on its far side). Historical documents report that in the decisive battle between Persians and Lydians in that year a solar eclipse had occurred, which incidentally had been predicted by Thales of Miletus. The historical documents attest to a fact and thereby also that the astronomical assumption about an unperturbed lunar orbit during these millennia was correct. The solar eclipse was thus necessary (from our present knowledge of the lunar orbit) *and* factual. That on the other hand an electron, which enters my counter now, had shown at a previous scattering experiment the momentum $p$ is factual, if that is the case, but *not* necessary. Whereas, if we predicted a solar eclipse for the year 3500, it is under the above mentioned premises necessary but, at least for us, not factual. Indeed, in the next two and a half millennia there could occur a perturbation of the Moon's orbit and let this solar eclipse not happen.

Logically, nothing prevents one from applying the same arguments to future facts as well. These would then be objective, but unknown to us. *We* only know their prospective probabilities according to the Schrödinger equation. That we do not know them beforehand would only be due to the fact that according to thermodynamics there are no documents about them.

The last sentence may perhaps evoke an objection. In Chap. 6 we have justified the fact that there are no documents for future events on the grounds that the past is factual, the future possible. If we now also declare the future as factual, where remains then the difference between future and past? We answer this objection in two steps.

The first answer is as hypothetical as the entire discussion in which we find ourselves. It leads the argument into its *second mental approximation*. It says: "There are documents of the future, namely the phenomena of prophecies." Prophetic revelations, such as they are known among humans, represent the future, often under a veil of similes, as accomplished facts. We only give two examples: Nostradamus, a highly erudite French physician, wrote about 1550 in his book of prophecies the half-sentence:[2] "The year 1792, which will be held for the beginning of a new era..." 1792 was the storming of the Tuileries, the year 1 of the Republic during the French Revolution. The Mühlhiasl, a psychopathic and uneducated miller's apprentice from the Bavarian Forest, said in the eighteenth century, among other visionary utterances: "When the iron dog will be barking in our valley for the first time, the Great War begins." In 1914, the railway was opened in this valley.

In a physics text one cannot use an argument like this without stirring up emotions. Most scientists (historians and sociologists perhaps more passionately than physicists) will say: "How can one base a scientific hypothesis on pure superstition!" The enthusiasts of the occult will triumphantly claim this utterance for themselves. In my opinion, which I should perhaps note as my subjective perception—thus a real perception but with the special gage of my psychological disposition—according to my opinion both reactions, as they present themselves, are unjustified, even indefensible but both have a deep-seated legitimate reason of which they are mostly unaware. In my opinion the reaction of the scientist is factually unfounded but morally well justified, and conversely the reaction of the occultist morally deeply problematical but with some reason in the facts. These are four theses, to be briefly discussed.

The reaction of the scientist, in its unreflected directness, no doubt is factually unfounded. Natural science boasts of proceeding empirically. But scarcely any scientist who repudiates prophecy as superstition has taken the pains to scrutinize it empirically. He perhaps knows a few prophecies which were not fulfilled. Examples of prophecies fulfilled are so rare that he has no trouble classifying them as chance agreements. In the daily life of a scientist it is a completely legitimate question as to what fraction of his time he may sacrifice to the investigation of incredible hypotheses. But here the self-evident premise is offered that the hypothesis that there exists true prophecy is just not trustworthy. Why is it not trustworthy? Because the physicist in *his* world view cannot imagine a mechanism by which prophecy could become possible. Here precisely the "formal–factual" nature of the prophecies is what is inexplicable. One can causally-retrodictively calculate that in 585 BC there was a solar eclipse visible in Asia Minor. But nobody can compute causally from the present state of the universe that then there was a war between Croesus and Cyrus, and a philosopher named Thales predicted solar eclipses; all this one can only know as a fact through documents handed down. How could have Nostradamus predicted, to the year, the French Revolution or the Mühlhiasl the railway, unknown and incomprehensible to him? If prophecy of this kind is possible, then one must be able to perceive the future, at least in isolated fragments. How is that assumed to happen physically? Now we contemplate precisely the question of whether there is something beyond the limits of the best physics today. *For this* one cannot deny prophecy unexamined.

---

[2] I am quoting here from memory.

Yet I must admit that I have also made no attempt to test prophecies empirically, and that I have a deep aversion to doing so. Here perhaps a deeper *moral legitimacy* of the scientific objection reveals itself. Shall we know the future factually? Could we bear it? I illuminate this with an anecdote which I have already told several times in connection with politics. About 1960 an old school friend asked me: "Do you think that the atomic war is coming about which you talk so much?" Without a thought, without hesitation I answered: "I do not know." Still without hesitation: "That I mustn't know." Then I thought about it. Indeed: If I knew that war is not coming then I would not strive to prevent it; I would then have better things to do. If I knew for certain that it is coming, then neither would I make *this* effort any longer, but a different one, like damage control. But I *must* strive to prevent it. Analogously, if it is really coming, I must not know beforehand the exact nature and magnitude of the damage. In other words: Human action is only possible under the assumption that it accomplishes something. However, it is in the nature of our conscious mind, at least in our culture, that we need for it the open horizon of a future which still has possibilities. To be sure, if we wish to act responsibly, we *should* consider this possibility and its limits as accurately as possible. Every sound assessment of the future is morally justified, indeed demanded. A factual knowledge of the future, however, would paralyze our morally controlled will, such as we are conditioned. The physicist safeguards this morality, perhaps shortsightedly, by denouncing the belief in prophecies as superstition.

How justified this moral reaction is one can see if one has the opportunity to observe the *moral illegitimacy* of an infatuation with the occult. I cannot remember ever having seen that the attempt to make one's life dependent on prophecies had not brought harm to the people who made it. This is also very well known from mythical and artistic thought which often believes in prophecies. Oracles are deceptive. Croesus asked, before 585, the oracle at Delphi whether he should cross the river bordering the Persian empire of Cyrus. He got the answer: "If you cross the Halys you will destroy a great empire." He crossed the Halys and thereby destroyed his own empire. It rather seems that the priests at Delphi were well informed or capable of sound political judgment; thus arguably their prophecy was in this case causal and not factual. But are the present interpreters of Nostradamus better off?

As I have not studied these things in detail I cannot demonstrate to the scientist the *factual basis* prophecy has in my opinion. Prophecy occurs spontaneously as perception, usually with a highly charged affect. In itself all elementary perception is at the same time affective, alluring, warning, soothing. Extrasensory perception does not seem to be ruled out. Often enough it is a premonition. It often perceives what is vitally important to the perceiving person or to one in his charge, be it a friend, relative, patient. What is morally illegitimate is the fearful desire of power of those who succumb to this domain, the unholy attempt to discover God's design and thus become the master of one's own destiny.

Precisely this danger is nourished if the future is represented as a hidden collection of facts. We thereby enter the *third mental approximation*. We return to physics and ask *how* the future must be described *if* we ascribe to it "something like facticity." We return again to the objection from the end of the first approximation: If also the future is factual, where is then the difference between past and future ?

Even if there is prophecy, the phenomenological difference remains: prophecies show "scraps" of the future, which can really be interpreted only after the actual events; these scraps, as long as they are still in the future, do not join together to a coherent story that can be told, like the fragments of knowledge about the past: our information about the future remains much more meager than about the past. Furthermore the latitude of its being non-interpretable is so wide that it often remains a question of the good will of the interpreter whether a prophecy has been fulfilled for him or not. Finally, the purpose of a prophecy, hidden from the prophet, can be to make its own occurrence superfluous, by anticipating through the word the moral shock the catastrophe would cause; this is the subject of a wise novel which found its way into the biblical canon as the *Book of Jonah*. Suppose *we* attempt, we, who do not have this painful gift of prophetic insight or only rudimentarily, we who can trust this gift perhaps only then when we can make the events causally plausible— thus suppose we attempt to formulate how a visionary, highly gifted and also intellectually well trained, must describe his own knowledge about the future, we perhaps must say that he sees the facts of the future as living images; not like documented facts of the past but as *possibilities*, but with a far greater probability than could be obtained from a causal prediction. Incidentally, visionaries sometimes report having seen the past in such detail about which they had no historical evidence.

This phenomenology of visions would suggest adding to the two scientifically accessible modalities of temporal logic, to facticity and possibility, a third modality, which one perhaps might call *time-bridging perceptibility*. The question is then whether this assumption, even if it does not follow from quantum theory, is logically compatible with it.

In this terminology we could retain that there are *no documents of the future*. Prophecy is, even when it exists, something fundamentally different from documentation[3]. We could then maintain the entire theory of irreversibility of Chap. 7. However, this is then, as we already asserted about quantum theory, essentially to be interpreted as a *theory of human knowledge*. If we are willing to consider also prophecy as possible human knowledge, then we must qualify this statement even further. The theory of irreversibility and quantum theory is then a unified *theory of empirical-rational knowledge in the sense of present*

---

[3] When we say that the future is factual we say about documents only that for such a future fact, *when* it has happened, there will also be documents.

*Western scientific civilization.* Here, empiricism means to know documents, rationality means being able to think conceptually about possibilities.

A member of our scientific civilization, once he has taken seriously the possibility of time-bridging perception, will soon be tempted to devise a scientific model for it. It would thus be an inclusion of this kind of perception and—perhaps only to a certain degree—of what it perceives into the theory of human knowledge. For this a highly self-critical notion of knowledge would be necessary. We have learned about facticity and possibility as conditions of human knowledge. The desired theory must probably transcend these two basic concepts, similarly to quantum theory transcending the basic concepts of classical physics.

## 11.3 Possibility of the past

We return to the self-criticism of quantum theory (section 11.1a). Quantum theory, as we have interpreted it, assumes a twofold ignorance:

1. the irreversibility of the measurement
2. the separability of the alternatives

The irreversibility of the measurement is an integral part of the Copenhagen interpretation (Sect. 9.2c); in our reconstruction it is presumed as facticity of the past. That the separability of alternatives implies ignorance is only pointed out in our reconstruction (Sect. 3.6e). In the formal description of objects, the irreversibility of the measurement manifests itself as the reduction of the wave packet (Sect. 9.2b). The separability of alternatives is inherently related to the formation of tensor products of the state spaces. In the history of the interpretation of quantum theory the reduction of the wave packet was a difficulty for many authors; scarcely anybody has in principle taken offense at the separability of alternatives. We will deal with both problems separately: irreversibility in this section, separability in the next. The present section discusses irreversibility within the quantum theoretical formalism, utilizing the inherent relation to separability. The next section poses the question of whether and how separability beyond quantum theory is to be corrected.

As regards irreversibility, we must smooth out one rough spot in our own presentation.

We have introduced (Sect. 2.12) the individuality of processes in the sense of Bohr as the sole starting point of quantum theory independent of the modalities of time. It says, according to the Copenhagen interpretation that a process which is interrupted by a measurement is no longer the same process. This is a pertinent statement—using once again Bohr's language—only due to the finite value of the quantum of action; the interaction with the measurement apparatus cannot be made arbitrarily small but, as a law of nature, has a lower bound. In our reconstruction the same follows from the finiteness of the alternatives. An individual process is described in quantum mechanics through the development of the state vector according to the Schrödinger equation. Thus described, it lasts from one measurement to the next. This representation we have chosen in Sect. 9.2b-c. In this representation the $\psi$-function is an expression of human knowledge and nothing else. We have called it the "Copenhagen golden rule" that with this mode of presentation one does not run into contradictions.

Now, however, we join sides with those physicists who take offense at the reduction of the wave packet. We assert that our representation of irreversibility in Sect. 9.2c$\beta$ has already tacitly proved them right. There we said: the "irreversible" process is an objective fact in the sense that the microstate migrates in the majority of the cases into a macrostate which contains more microstates than the previous macrostate, hence has higher entropy. For an observer who only knows the macrostate, this is a decrease in information available to him, i.e., a loss of knowledge. The objective course, however, remained the same for an observer who knew the microstate. But that means, translated into the quantum theoretical description, that the correct description of the measurement process, which we posses in the Copenhagen interpretation, cannot be harmed by assuming that the $\psi$-function is never reduced. Thus it should be possible to describe the measurement in the language of the unreduced $\psi$-function. This proposed mental exercise, to the best of my knowledge, has thus far not been resolved in the debate over the interpretation.

The description by means of the unreduced $\psi$-function assumes initially that the measurement apparatus and also the observer, including his consciousness, are described quantum mechanically. That this must be possible in principle—of course not in practice—we have postulated in Sect. 9.2d-e. The problem does not lie there but in the relation of the $\psi$-function to the only thing we know, to the events themselves. The $\psi$-function, in the usual interpretation, specifies the probability of events. For this reason we have have called it an expression of a futuric modality, a possibility.

First of all, we remark that the quantum theoretical description of the measurement process does not change anything about it. The measurement interaction breaks up the separability of the object from the measurement apparatus. The EPR thought experiment shows that the object which originated from two objects that were once in interaction retains its individuality (in the sense of Bohr) even if for a long time there is no "physical" interaction taking place between the two parts. The uncertainty of the later measured result on one of the parts is thereby not removed; only a necessary correlation between the results of certain measurements on both parts is created.

Everett has drawn from this the conclusion that *all* possible events occur and comprise alternative worlds. We have answered (Sect. 9.3h) that his theory goes over into ours by changing a single word: one must speak of *possible* instead of actual worlds. There we have not made use of the argument that his theory, at least how it is presented most of the time, is not fully consistent precisely due to the only approximate character of irreversibility. His "forking paths" remain to that approximation in contact with one another in which the probability of the reversal of an irreversible process is different from zero.

Our real argument against Everett's dictum is a different one. He treats quantum theory like a truth revealed and postulates, as he cannot find it realized in the phenomena the way he reads it, i.e., without reduction of the state, the existence of a multiplicity of events ("worlds") which in principle can never become phenomena to us. Now quantum theory originated from the attempt to describe and predict phenomena, as consistent as possible. Arguably there must then be one interpretation of quantum theory which at least in principle can be demonstrated in the phenomena. Otherwise one would prefer to consider quantum theory, historically developed less than a century ago, not to be correct physics. In any case, in our reconstruction which as a matter of principle is based on empirically decidable alternatives, this conclusion would be compelling.

The question is thus about the meaning of the unreduced $\psi$-function in our reconstruction. For this I made a proposal at a meeting in Trieste in 1972 (see Weizsäcker 1973a) which I want to outline here. We can refer to it here the Trieste theory.

It will be seen that at its core the Trieste theory is a theory of the self-knowledge of consciousness. Indeed, the problem we discuss here originated with the question of how the possibilities we formulate in the $\psi$-function are related to facts we know, or to events we experience. Thus it can be permitted to anticipate here a discussion elaborated in *Zeit und Wissen* on facticity and reflection. To begin with, it shows how one can also represent past events, i.e., facts, as possibilities, namely as possibilities of future knowledge about them. In *Zeit und Wissen* II 6.7.5, *Faktizität und Reflexion*, we refer to Husserl's description of memory, specifically retention, the continuing knowledge of what just happened. There we refer to the concept of an objective document as condition for the knowledge of past facts (Sect. 7.3); a recollection is also a "document in consciousness." We discuss the difficulty that in consciousness apparently there must be immeasurably many recollections, recollections of recollections, etc. Following Husserl we solve the difficulty with the remark that all these different levels of recollection are not different facts but different possibilities of reflection based on one single document. "The same actual circumstance is potentially a document for a great many past events. The facticity of the past is a form of potentiality. One cannot understand formally possible perfectic statements if one does not see them as potential futuric statements about past events." A simple example, which we discussed in Chap. 7, is the application of the concept of probability to past events. "It is probable that it rained tonight" means: "It is probable that upon inspection one will find that in fact it rained tonight."

These arguments, however, do not spare us a sacrifice of familiar notions, no less radical than in the previous section, only in the opposite direction: the sacrifice of the basic concept of facticity. We present this emerging "theory of events" as a thesis, but in the guise of a fairy tale.

An event in the strict sense is a presentic event, something that is happening here and now. Quantum theory as a conceptual general theory can only describe formally possible events. In principle they are always represented by state vectors, even if we only know statistical mixtures of such vectors. The probability function $p(a, b)$ defines the probability that $b$ happens, under the twofold condition that $a$ has just occurred and that $b$ has been made possible through interaction with the environment. An event that has happened is a fact. A fact is the futuric possibility of a class of events documenting this fact.

The just mentioned second condition for the meaning of $p(a, b)$ shows that events can only happen if there is interaction. A completely isolated object is not part of the universe; it is a non-event. What we idealize as an isolated object is operationally defined in terms of a past interaction and the possibility of future measurements on it. In this way it is a fact, thus a state vector.

An event that occurs in an interaction with a small number of objects can disappear again, i.e., not become a fact. As long as the state vector varies reversibly, the event itself is reversible, can be canceled. We can speak of it theoretically but we cannot know it. One can describe it by combining all participating objects into one isolated composite object; but that means, after what we have already said, that it is interpreted as a "non-event."

If on the other hand an event distributes an amount of energy over many objects, the probability that it "disappears again" will become very small. Then we call it a fact.

Even when a fact originates in this way, the development of the state vector does not lead to one well-defined fact but to a probability distribution over several mutually exclusive facts. In abstract form this is well known.

We visualize this by means of a simple example.[4] Suppose a light object, say a matchbox, is situated at some elevated location. A Stern–Gerlach experiment directs a silver atom in such a way in the direction of the matchbox that it falls down to the right if the atom had spin $+1/2$ (say in the $z$-direction), to the left for $-1/2$. Let the probability for both cases be $1/2$. But when it is found that a single atom of the intact matchbox has fallen to the right then the probability that any of its other atoms goes to the left is practically zero. We say now that the event that the silver atom assumes either spin $+1/2$ or spin $-1/2$ actually occurs. This is the linguistic agreement of our "theory of events." The event becomes irreversible through the fall of the matchbox. Through the interaction with the matchbox the knowledge of the phase of the silver atom (which continues in flight) is lost. If after the interaction with the matchbox the two beams are geometrically combined again (through an appropriate choice of the inhomogeneous magnetic field), then they are superposed incoherently. But if no matchbox were there to interact with the two parts of the wave function of the silver atom, the two beams can be recombined

---

[4] G. Süßmann used this example in our discussions in Hamburg about 1960.

coherently. The coherent beam, after the recombination, may correspond to a spin orientation orthogonal to the $z$-direction determined by the experiment, say the $y$-direction. Then the event of the decision of the $z$-component of spin has "disappeared" again. In this interaction free case one can say that the wave function of the single object "silver atom in a fixed environment" has developed unperturbed according to the Schrödinger equation.

Thus we can say that events occur if things interact, with probabilities according to the Hilbert space theory. In principle all events are reversible. The simplest mode of reversal (cancellation) of an event is that it does not even happen for want of an interacting partner. Quantum theoreticians have an expression adapted to this interpretation. They say a state is virtually all the states from which it can be superposed. These states represent the events that could occur on the object. The expression "virtual existence" or "virtual event" expresses the situation clearly. An event in which many objects participate then acquires a very small possibility of reversal.

This can then also be applied to human consciousness. Thereby we return from our story-telling to that which we presume that we know, to our knowledge of ourselves. We continue in the style of the fairy tale for three more sentences. If a person is a participant in an event he can say that it happened. But this conscious act is also not absolutely irreversible. We cannot rule out the possibility that it is, although with very low probability, reversed, so that afterwards it can be said that it "did not happen." Only in a Cartesian ontology of consciousness must this appear to be impossible. But it can be discussed within the way of thinking characterized by the phrase "consciousness is an unconscious act."[5]

This phrase of William James, which was passed on to me by Niels Bohr, expresses the impossibility of self-knowledge assumed in the modern philosophy of consciousness, i.e., the falseness in principle of the sentence "savoir c'est savoir qu'on sait," "knowing is to know what one knows," quoted by Sartre (*L'être et le néant*, 1943). Sartre's phrase, as Sartre knows, is not logically evident. This can be seen by splitting the game between two parties: That I know what you know is not the same as that you know that I know it.[6]

The postulate is not evident that to me this cannot happen with myself. Knowledge is initially knowledge about something. If I ask myself whether I know it, then actually this is usually an expression of doubt. Perhaps I can remove the doubt, perhaps not, perhaps I am mistaken to think that I have removed it. Viktor v. Weizsäcker once told me: "If you ask me: 'What are you

---

[5] Cf. *Wahrnehmung der Neuzeit* (1983), in *Bohr und Heisenberg. Eine Erinnerung aus dem Jahr 1932* and the first section of the essay *Begriffe*; additionally *Zeit und Wissen*, Chap. 2.

[6] There was a man new to the zoo
who was put in charge of the gnu.
The gnu knew he was new,
and he knew the gnu knew,
and the gnu knew he knew the gnu knew.

thinking right now?' and I reply, then I am already lying." We will follow up this question in *Zeit und Wissen*, Chap. 2. There we will arrive at the result that a relative (partial) doubt is always possible but that a living person de facto is not living in absolute (total) doubt. The concept of knowledge used in physics we call there the "belief of the physicists."

The interpretation of the "fairy tale" with this view of knowledge in the background must initially sound like a trivial generalization of the Copenhagen golden rule. Events we can claim to know are facts, thus past events. Making deductions about them with the unreduced wave function, the result will not be different from what we found for classical statistical thermodynamics in Chap. 7. If we let the development of the wave function begin *before* the event $E$ in question and if $E$ is one of several possible decisions of a measurement process, then it follows from the "Trieste theory" that $E$ or another event $E'$ competing with it must have occurred but, if $E$ had happened, the probability of its disappearing again was very small. The present reading of the measurement instrument produces a document in the sense of Sect. 7.3 from which we can further deduce with large probability that $E$ happened. What we have added to the Copenhagen rule is only to assert this with appropriate probabilities and not absolute certainty.

But the Trieste theory show this trivial side only when we restrict attention to processes which with sufficient probability can be described classically. Its actual implication is that quantum theoretical phase relations, even if we do not know them, remain in effect even through irreversible processes—naturally only insofar as this follows from the quantum theoretical description of the measurement process. One can assert that an individual process retains its individuality also through all measurement processes—naturally only as process of the entire object, including the measurement instrument and if need be the observer. It is essential that this conclusion not change if the observer, on the basis of his observations, makes a transition to another wave function than the one previously used. For if both wave functions were correctly deduced from measurements, then the new measured result must have had a probability different from zero on the basis of the old wave function. But then the new wave function must have occurred with finite amplitude in a linear combination representing the old wave function. The phase relations between the parts of the whole object expressed in the old wave function must then also be contained in the new wave function. The thought experiment of Einstein, Podolsky, and Rosen is an example of precisely this: the measurement on object $B_2$ decides what measurement on $B_1$ will produce a predictable result. In a way the Trieste theory is an attempt to *say* how far one must loosen the concept of a fact to vindicate Einstein's analysis and Bohr's interpretation of the experiment.

## 11.4 Comprehensive present

The two previous sections are in a way mirror images of each other. The thesis of the facticity of the future represents—hypothetically—the present in the usual form of the past. The thesis of the possibility of the past, or, as we also may say, of the modality of events, represents—conceptually—also the past in the quantum theoretical perception of the future. They both presume the present as being, in effect, beyond discussion, and rather diminish its role: on the real "time axis" the respective present is a point and in particular when past and future are described as being in principle of the same kind, the location of this point seems to be irrelevant.

For a careful phenomenology of time each of the two interpretations ought to criticize the naïve assumptions of the other. The modality of events deprives conceptually-analytically the concept of a fact of its "realistic" self-evidence. Then one also may not describe the conceivable knowledge of an acausally determined future simply as a totality of facts. In the second section, in the third mental approximation, we have already taken a step in this direction. Then the present is also not simply factually there. But the "facticity" of the future makes it hypothetically clear that perhaps in some way the future is already here. Then the present might include more than one point in time.

In the reconstruction of quantum theory we several times came across the question of whether from a quantum theoretical point of view the real parameter of time is not an inconsistency. If time is measurable, there ought to be an operator corresponding to it. Now we ask the opposite question of how actually the notion of a point-like present can be justified within our interpretation of quantum theory; the limitations of this justification will then present themselves automatically.

We can consider the logic of presentic propositions,[7] in particular the dialog schemes for the implicative law of identity, $p \to p$. In a formulation of $p$ that does not label any point in time, e.g., "the moon is shining," $p \to p$ can only be defended by relating $p$ to the entire time span of the dialog and presuming the "permanency of nature," which one can formulate more strongly as the continuity of events. The present, for which a presentic proposition holds, is then necessarily an although short time interval, perhaps arbitrarily short. If on the other hand we substitute for $p$ an objectively specified point in time, e.g., "the moon was shining at 10 o'clock in the evening of June 23, 1963," then the proposition, if verified, is already a perfectic proposition. The sharp point in time of an event is itself a fact. Now we have convinced ourselves that facts presume irreversibility. Thus the notion of the present being point-like can only be interpreted in terms of irreversibility. This, however, indicates two limitations of the concept of a point-like present. On the one hand, the limit

---

[7] Editors' note: For a general overview on Weizsäcker's conception of temporal logic the reader may consult *Aufbau der Physik* 2 (not included here), *Zeit und Wissen* I 6.4–5, and, in brief, Weizsäcker 1973a.

from a short time interval to a point cannot be performed. An irreversible process always contains only a finite number of participating objects, and thus permits only the determination of an admittedly short time interval of finite length (quantum-theoretically this follows from the uncertainty relation between time and energy, which does not presume the existence of a time operator but only the relation between the width of a function and the wave number of its Fourier components). On the other hand, we have characterized absolute irreversibility as the expression of a subjective ignorance. Especially from the standpoint of the modality of events, it is completely unjustified to describe time as a straight line consisting of points.

In relativistic quantum theory space and time coordinates are not observables but group parameters of the symmetry group and only through this, in our reconstruction, coordinates of a homogeneous space. Still, positions and times are considered to be measurable. The observable of position belongs first to a particle: it denotes the property of an object. Classically, space and time coordinates denote a property of an event. Thus for interaction there should exist an observable for time. But our interaction is not local and in general does not preserve the identity of a particle. An event will thus approach the point-like localizability in spacetime better at higher energies, i.e., when more virtual objects come into play. Indeed, it is precisely then that a description of an event being irreversible will be more accurately possible.

A reconstruction of quantum theory which assumes a real time coordinate can therefore not be rigorously semantically consistent. The error might be factually identical with the error of assuming separated objects, i.e., separable alternatives. Now the separability of alternatives appears to us a requirement for a conceptual description of reality. If our ideas are correct, then it appears that we have marked with them a limit which our conceptual thinking cannot transcend. The limit, however, we have described such that we have spoken, in a language accessible even to a present-day scientist, about what may lie beyond. This is the dialectic of the concept "limits of knowledge" already discussed by Hegel;[8] one can mark it only by mentally transcending it.

To begin with, we pose a question still within the physical knowledge available today. To what extent is the separability of alternatives semantically consistent? Already when we introduced it, we pointed out that it is only an approximation (Sect. 3.6e). In the Sect. 11.3 above, we just said that only interaction makes events possible. Deciding an alternative always breaks the separation of this alternative from the rest of reality. The first question no doubt must be why the approximation of separability is nevertheless so good.

Within physics we can answer: because space is practically empty. Take the estimate of Sect. 5.5d: In a universe of volume $V = N\lambda^3$ there are about $N^{2/3}$ particles, each of which fills the volume $v = \lambda^3$. Hence the fraction $N^{-1/3} \approx 10^{-40}$ of the universe is filled with matter, the rest is empty. For this

---

[8] Cf. G.W.F Hegel, *Wissenschaft der Logik*, Chap. 2, B h.

reason there are in space asymptotically free particles, and that just means asymptotically separable alternatives.

In the approximation of separable alternatives we could reconstruct quantum theory with a real time coordinate, thus under the fiction of a point-like present. Yet events occur only due to interaction; knowledge of events is always a new event, interaction once again. True events exist only in a world open to unlimited reflection, thus interaction. This openness includes not only space but also time. One might risk the following formulation: True events exist only in a comprehensive present.

What does "comprehensive present" mean phenomenologically? It certainly does not mean the decomposition of time into discrete points or time intervals; nor presumably is it an operator of time with such discrete eigenvalues. As an example of the phenomenon of a comprehensive present one can consider a melody. It is not the individual notes that comprise the melody, but their complete sequence when present in the mind. The comprehensive present thus comprehends a "complete" event which on a clock fills a stretch of time. *In* the comprehensive present there is sequence, there is presence of what has already happened and anticipation of what is to come, the disappointment of which is a break, a destruction of the "individual process." The physicist tends to describe this phenomenon as "merely subjective." Indeed it is a mental act, perhaps a regulatory process in the brain. Our proposal now would be to describe reality itself as not being decomposed into point-like events but rather, in the spirit of the individuality of processes, objectively after the model of the phenomenologically comprehensive present. That our mental apparatus succeeds in hearing a melody would then be not merely subjective but its ability to perform this would be founded in the objective structure of mental processes. The individual process on *one* object is interrupted at any given time by the event of its interaction with another object; but this interaction itself is again an individual process of the object as a whole; if one could include the entire world in the description, its history would turn out to be one single individual process in an all-encompassing present.

Following this presentation we can say to what extent the concept of an individual process is independent of the modalities of time. The individual process on an object contains the sequence of events possible on it, always to be established through interaction. Between the events there is *for this object* not a sequence of time but quasi comprehensive present. Modalities of time are established through interaction. Facticity of the past means that past events are stored on this object in their effect upon other objects, thus in the sense of possibility of the past through its action on events in the world possible today: through documents. Openness of the future means that one cannot discern from *this* object what events will happen through interaction with other objects. Facticity of the future is not logically ruled out if the individual process of all events in the universe could be included in the prediction.

This view of all events in the universe is implied in the formalism of quantum theory only as possibility. Extending the state space to more and more objects does not abolish the quantum theoretical uncertainty of the prediction. If our reconstruction does correctly describe the reason for the success of quantum theory, then likewise nothing else is to be expected. We started with the separability of alternatives. Combining several objects into a single aggregate object does not invalidate the principle of separability, it merely enlarges the alternative, which is still considered to be separable. Within this reconstruction the individuality of processes is described by the postulate of kinematics: The unperturbed process leaves invariant the probability relations which define the state. We have seen why separability turns out to be a very good approximation *if* one starts with postulates which assume this separability in principle; i.e., we see the good approximation in which this assumption is self-consistent. Reality, however, is not strictly separable. Local events may only approximately be described as real. The true course of the universe is likely neither local in space nor local in time.

This idea we can express verbally-conceptually only through negation. The physics available to us does not describe it.

## 11.5 Beyond physics

Can we think beyond our physics to understand the universe in a comprehensive present? For this we can think of several further paths which must be classified through an intersection of two alternatives:

A: finite and infinite knowledge
B: conceptual and non-conceptual understanding

The distinction A originates in the tradition of metaphysics. Man is a finite being. Its knowledge and power have limits. There is much we do not know, perhaps will never be able to know. Compared to this we can think of infinite[9] knowledge, the knowledge of an infinite, divine subject, encompassing all that can be known. Thereby it is apparently assumed that the divine intellect accomplishes everything the human intellect does but incomparably better. Starting with this alternative, one can evidently choose two paths: one can either assume the existence of such a divine intellect or one can leave it aside, the latter by considering it as undecidable or denying it outright.

If one leaves aside the idea of infinite knowledge, one can persist in a pragmatic attitude: some things we know, some things we do not know, with the border between the two changing over the course of time, and there is no more we can say about it. In normal science, under a fixed paradigm, this is a

---

[9] The concept "infinite" here is not identical to the mathematical concept, like that of an infinite set. It rather will surpass human limitations, thus also the mathematical knowledge of humans.

common and fruitful attitude. But, to be sure, it has never produced a scientific revolution, a closed theory. It precludes the conservative from feeling the pain due to the inconsistencies of the existing. For this reason scientists with concerns about the fundamentals have often asked themselves the question of how what is unknown to us might appear to an all-knowing mind. As a model for this, however, they only have at hand the instruments of their own finite knowledge, especially the one available at their time. From this arises the methodological fiction of a knowledge of everything, which above all is quantitatively different from human knowledge. Such a fiction is, e.g., the idea of a $\psi$-function of the universe. Our analysis of the modality of events should instead not represent the quantitative proliferation but the conceptual clarification of quantum theoretical knowledge, to bring forward precisely the unresolved basic questions. How would an all-knowing mind think about modalities? Would there be an open future for it? Or would there be, as had been thought in classical metaphysics, an all-comprising presence, an omnipresence?

Before we can ask ourselves this question we must consider case $B$. Here we cannot develop a full theory of conceptual thinking (for this see *Zeit und Wissen*). Under the title of biological preliminaries to logic we have classified conceptual thinking as a special ability in the larger, comprehensive cognitive form of life. Perception is predicative, perception and judgment are in general affective. Affects are mostly related to possible actions. The assertive sentence presumed in logic, which can be true or false, is a product of culture; in the form described in logic it itself is a product of logic. Even scientific intuition reacts to structures which the scientist most of the time cannot express through concepts. Thus one can very well imagine that an understanding which surpasses our knowledge is of essentially non-conceptual form. In the second section we have hypothetically described in this way time-bridging perception. In a heuristic picture we probably should ascribe to superhuman knowledge not a form restricted to concepts. Already for human reason ("Vernunft," which in the terminology of classical German philosophy is distinct from the conceptually operating intellect or understanding, "Verstand") the idea of a holistic perception suggests itself.

Quantum theory, as we have reconstructed and interpreted it, is fully compatible with the idea of reality being a non-spatial individual process which we must describe as mental, using words familiar to us. It is an old tradition that our own personal consciousness is only one manifestation of a comprehensive mind. We return to this in the final chapter.

By venturing such a point of view, we have not refuted or "overcome" physics. Rather, we have compelled ourselves *as* physicists to reverse again the order of the arguments. We began with the unquestionable success of physics and asked what may lie beyond physics. Now, beginning with the idea of a comprehensive mental reality, we must ask ourselves why physics is so successful. Thereby we enter the *Kreisgang*[10] only at another point than

---

[10] Cf. p. XXII, fn. 14.

before. We know ourselves empirically as beings with only finite knowledge. We can know that this is only the surface of a deeper, "infinite" reality. But with our appearance as finite beings the rules are imposed by means of which we can mirror reality in finite knowledge, with finite alternatives. Those, so it would seem thus far, are the laws of quantum theory. No genuine achievement of understanding is sacrificed by the path which leads us beyond physics. Only the criterion of the authenticity of rational arguments must be handled with care.

# 12

# In the language of philosophers

## 12.1 Exposition

This chapter will, as a supplement, outline a sketch of the philosophy already contained in the reconstruction[1] of physics.

Philosophy as an afterthought is already a philosophical program. In its classical tradition, philosophy considered itself basic science: in its contents as ontology, *also, for example, as ethics*; methodologically as logic and epistemology. The other sciences, as a matter of fact, have long since ceased to recognize this claim. Through this disbelief, however, philosophy is devalued below its true rank. In the idea of obtaining a hierarchical structure of knowledge lay only a historically scarcely avoidable self-misunderstanding of classical philosophy, a specific philosophical mistake. Greek philosophy developed simultaneously and in interaction with the deductive form of mathematics. This led to the expectation that philosophy is also a deductive science and precisely as such the basis of the other sciences. Actually, however, the specific philosophical process is the Socratic query: Do you really know what you are doing? It was then the dream of philosophers to penetrate with this retrospective query to the unshakable foundation of a deductive reconstruction. But in the historical development thus far this dream remains unfulfilled.

The intention to sketch that philosophy which already is contained in the reconstruction of physics is precisely the Socratic query about our own work. We have, e.g., in Sect. 8.7, depicted the path of philosophy as a *"Kreisgang"*.[2] There it was about a "great" *Kreisgang*: from temporal logic to physics, from physics to the evolution of life, from the modes of cognition of life to the preliminaries of logic. But at every step of our reconstruction there are also

---

[1] Editors' note: Weizsäcker uses the word "Aufbau" in this chapter—alluding to the book title and exploiting its connotations of both "reconstruction" and "structure."

[2] Cf. p. XXII, fn. 14.

"small" *Kreisgänge*, individual queries: Physics is based on experience, experience happens in time; how do we speak about time in physics? Quantum theory makes probability predictions; how do we define probability? To make the structure and content of these Socratic elements stand out in our reconstruction is now our goal.

The layout of this chapter is close to the Aristotelian subdivision of theoretical philosophy into logic, physics, and metaphysics.

Instead of logic we say *philosophy of science*. It is about the *methods* of science.

*Physics* to Aristotle is the science of motion, more precisely of that which carries the cause of its motion in itself. In the reconstruction of physics we begin with time, i.e., with the medium of motion. Physics encompasses, for us as for Aristotle, the foundations of all empirical sciences.

*Metaphysics*, according to old tradition, is the name of books which in the Aristotelian canon came *after* the books about physics. In that respect the name serendipitously points to *philosophy as being retrospective*. Here we use the name in this very general sense without committing ourselves to certain theories of Western metaphysics.

## 12.2 Philosophy of science

The philosophy of science of our century can be considered a deliberately retrospective philosophy. The claim of trustworthy knowledge, which once gave wings to philosophy, has been redeemed only by the positive sciences. The philosophy of science asks what enabled the sciences to do so.

To it the sciences are thus given as historical facts. The sciences seem to fall into two large groups: the pure sciences of structure, i.e., mathematics and logic, and the empirical exact sciences, from physics and astronomy through the biological fields to sociology and psychology. It is characteristic that by the time these questions are raised, the interpretive humanities find no real place (cf. *Beyond quantum theory*, Sect. 11.1b).

What is now the foundation of the knowledge of the sciences? The exact sciences rest on *reality*; they know about it from *experience*. The structural sciences do not appear to be in need of reality and experience; they discern structures *a priori*. Exact sciences without logic and, where they become precise, without mathematics do not seem to exist: the general laws of nature are essentially mathematical. According to the weight given to these three foundations of knowledge, one can very roughly classify the schools of philosophy of science into *empiricism, realism,* and *apriorism*. Our reconstruction contains elements of all three modes of thought and does not commit itself to any one of them. We follow here its systematic arrangements and discuss, only for illustration, its relationship to the prevalent traditions.

The starting point we have in common with all three schools as they are presented today is that physics is based upon experience. Also Kant's apriorism also starts that way, but soon continues: "But, though all our knowledge begins with experience, it by no means follows that all arises out of experience" *Critique of Pure Reason*, B1). Empiricism, on the other hand, seeks to base the very certainty of science on experience. We follow Plato, Hume, Kant and Popper by realizing that the general laws cannot be logically deduced from thus far always incomplete experience. In particular we follow Hume by emphasizing that general laws of nature are always referring to the future which empirically is unknown to us. In view of the question of how our belief in laws of nature can then be justified we choose a path programmatically not described thus far. It avoids the opposing traditional viewpoints by learning from both of them: we can describe it as the *pragmatic* and *absolute* position.

The pragmatic position is content with the oft-repeated success of empirical science. It considers it superfluous and presumptuous to also want to explain this success. This is the attitude of most natural scientists. The absolute position searches for an unshakable foundation, thus for knowledge that with certainty will not be in need of any corrections. In our reconstruction we do not act as absolute: we consider science a process in an open history. We call the fundamental knowledge we describe the "belief of the physicists" (section 11.3 and also *Zeit und Wissen* I 2.4). But we are convinced that the pragmatic renunciation of an explanation of the success misses precisely one of the most interesting and scientifically most fruitful questions.

To explain this we first formulate a critique of the absolute approach that was common to the three attitudes toward the philosophy of science, at least in their beginnings. None were yet with sufficient consequence a "retrospective" philosophy. On the one hand, they assumed the historical fact of the sciences. On the other hand, they conceived of being able to justify this success, which would no longer depend on progressing factual results of science. Apriorism announced this as a program; the phrase *a priori* refers in Kant, though not to the historical, but to the order of justification. Realism itself was not quite clear about its own absolute position. It considered certain prejudices of classical physics as scientific results it wanted to abide; hence its distress with quantum theory which offers different results. We are interested here above all in the absolute position of empiricism. It searched for experience. But it meant to know, more or less a priori, what experience was; e.g., "inductive" logical conclusions from sensory perceptions. A consistent empiricism, however, should first learn from experience what experience is. Here most relevant is the discovery of innate forms of experience through ethology (Lorenz 1942). For scientific experience the most fruitful area of exploration is the history of science.

Thomas Kuhn took the important step by distinguishing normal science from revolutions. Heisenberg, however, had already referred to the decisive structure earlier, more specifically and more precisely, with his concept of a closed theory. The relationship between factually progressing science and the philosophy of science—or, to put it more fundamentally, to philosophy—may then be formulated as follows. In the phases of normal science philosophy is dispensable for the factual progress, even detrimental. Normal science succeeds *because* of its pragmatic attitude; it progresses rapidly as it does not linger to understand the reason for its own success. The great revolutions, however, originate from just this question about the reason. In these revolutions philosophy is indispensable, and science and philosophy are inseparable. A philosophy of science which describes only normal science or at most bygone, no longer dangerous revolutions, is itself normal science; its fate is to be overtaken in the next scientific revolution.

The seminal question missed by the pragmatic position is therefore how scientific revolutions are possible. For this a comparison with biological evolution is instructive. Konrad Lorenz[3] says that in evolution there occur "fulgurations," flash-like mergings of other structures into a more complex new structure which thereby also accomplishes a new simplicity of performance at a higher level of integration. From the previous structures the new level is unpredictable. That also happens in science, both in great revolutions and in many individual discoveries. Philosophically most important in science are therefore precisely not the always and everywhere applicable *methods* of science but its unique *factual* problems and results.

Our reconstruction begins with the question: how are closed theories possible? Here too we soon go into specifics. We suspect that a closed theory will be the more amenable to a fundamental explanation the later it occurs in the succession of theories, for the more generally valid it will be. In particular this leads us to quantum theory. We suspect that nowadays the best criterion of a philosophy of science is whether it can make quantum theory understandable.

This question has the same epistemological structure as Kant's question: how is experience possible at all? As regards content, with our understanding of theory, both questions are almost synonymous. Experience in the sense of Kant is experience that can be expressed conceptually. We have learned that scientific concepts obtain a precise meaning only in the context of a closed theory. "Only theory decides what can be measured" Einstein told Heisenberg, and in a theory it is the concepts which decide the meaning of measurements. Still we must consider the genesis of theories. They always come with preconceptions, in terms of which they must initially be formulated. Here Kuhn's weaker concept of paradigm has its place. Only insofar as the theory itself is considered "semantically consistent" does it define the meaning of the concepts in experience. In this sense then Kant's question of what makes experience possible, so far we can attempt to answer it at all, is just our question about the reason for the feasibility of the most recent theory accessible to us.

Both questions having the same meaning offers hope of finding the reason for the success of physics. For a reader versed in mathematics one can formulate the basic assumptions of quantum theory on one printed page; its range

---

[3] 1973, p. 48; also *Garten des Menschlichen*, II, 2.1, p. 189.

of applicability, however, according to our present knowledge, is unlimited. This would explain itself in the sense of Kant: *As* the theory in its abstract generality merely expresses conditions *of* the possibility of experience, it must therefore be valid everywhere *in* experience.

This conjecture now calls for a test. If—or to whatever extent—this conjecture is correct, then or thus far it should be possible to conceptually reconstruct the theory from a direct formulation of the prerequisites for experience in general. Naturally one will scarcely formulate these preconditions correctly if one is using an understanding of experience as it was already available before the advent of quantum theory. One must probably use the semantic consistency of quantum theory (insofar as we have attained it) to formulate the preconceptions necessary for its reconstruction. This reconstruction will also have the character of an inherent consistency check. But one should be able to formulate it in a language which is also intelligible to a contemporary non-expert.

The key concept of quantum theory, as usually formulated, ought to be the concept of probability. In physics probability has a prognostic meaning. Physics is based on experience; but experience means having learned from the past for the future. Thus on top of the conditions which make experience possible is the structure of time itself, in its modalities of present, past, and future.

With it, we are at the beginning of the reconstruction of physics.

## 12.3 Physics

We reconstruct physics as the general theory of empirically decidable alternatives. The predicate "general" has for the concept of a general theory a much farther reaching meaning then for the logical concept of a general proposition. A general proposition is characterized by its logical form. In the simplest case a predicate is announced about all objects subsumed under a certain concept: "All humans are mortal," "all $S$ are $P$." The generality of the proposition reaches only as far as there are objects which fall under the chosen subject $S$ (e.g., "human"). A general theory however, in the strict sense of a closed theory, defines its own concepts. We considered several mathematical forms of such theories in Sect. 2.3. For example, a differential equation determines the set of all its possible solutions. The program of concrete quantum theory aims to classify all in principle possible empirically decidable alternatives.

Let us illustrate by means of a few examples and counterexamples what the phrase "in principle" is supposed to mean. The mere concept of a chemical compound does not determine what chemical compounds exist. The mere concept of animal species does not determine what animals there can be. In contrast, the logical operations and the concepts of set and number presumably determine all mathematical structures investigated thus far. The set of all such definable mathematical structures which are possible "in principle" turns out to be infinite. Nowadays one does not doubt that all of chemistry obeys quantum theory, that all possible chemical compounds are solutions of the Schrödinger equation. In this way quantum theory determines all compounds "in principle." Still one learns about these compounds much more easily through experience than by means of arbitrarily large computer programs. If physicalism is correct, then a family of howler monkeys in the jungle is also "in principle" a solution of the Schrödinger equation; nobody will attempt to deduce this by calculation from the equation.

It is perhaps instructive to compare the generality of this outline of physics with the generality of the Aristotelian design (cf. section 10.1). Physics according to Aristotle is the basic study of everything that carries a beginning of its motion within itself. "Beginning" or, as we said above, "origin" is a literal translation of "arche"; this term is translated into Latin as "causa," English "cause." "Motion" in Greek is "kinesis," which not only stands for change of place (phora) but change of any kind, including growth and decay. "A" beginning: Aristotle distinguishes four "archai": substance, form, origin of motion, purpose of motion (telos). All four must act together for a thing to exist. From the origin of motion, thus a thing or event through which the motion gets started, i.e., the "causa efficiens," developed the modern concept of causality.

The structure of Aristotelian physics is most easily visualized by examples from biology. Take the oak. Its substance is wood, one form is the oak tree. Its motion is growth. The origin of this motion is the acorn, its aim is to be an oak tree. The form "oak tree" is eternal, present in always different oak trees. So is the aim *in* the tree: en-tel-echeia, entelechy, "possessing the aim in itself." Analogously the physics of inanimate objects. A falling rock in the mountains: ts substance is stone, its form rock, its origin of motion the rain which has broken it off, aim of the motion is the natural place of everything heavy: down, as close to the center of the Earth as possible. Or the planet Ares (Mars). Its substance an ethereal material similar to fire, its form a bright star, the origin of motion its being everlasting since the beginning of time, its goal the eternal proximity to the perfection of the divine unmoving mover. The correlative of physics is technology: the study of those moving things which have the origin of their motion outside of themselves, namely in human thought and action. A shoe: its substance is leather, its form shoe, its motion being produced, its origin the shoemaker, its goal being worn on a foot. The human soul knows about causes and is able to utilize them. It itself is the form of humans.

This is an outline of the unity of nature. The universe is finite. At its center is the motionless Earth, surrounded by air and the concentric heavenly spheres. Beyond the sphere of the fixed stars there is no space. For the abstract concept of space does not exist in this philosophy; space entails a relationship among bodies. All motions on Earth are ultimately maintained by heavenly motions, above all through the daily and yearly motion of the Sun. Circular motion is the most perfect motion as it, returning upon itself, can be eternal.

So too is heavenly motion created by the eternal desire for the unmoving mover. All motion is eternal; the universe is forever equivalent to itself.

This outline stays close to the phenomena. It is in accord with the appearances then known. But it is not a naïve acceptance of the appearances. Theoretical alternatives were known long before. Atomists reduced motion to collisions of atoms in empty and infinite space. Myth just did not know any eternal universe but a beginning of Heaven and Earth, and a possible end of the reign of Zeus. Pythagoreans had the hypothesis of the Earth moving around a central fire; Aristarchus proposed only 100 years after Aristotle the heliocentric system. All decisions in the Aristotelian system are justified by the requirement that they actually be conceivable. If the universe had a beginning, where would the motion of its creation come from? A creator explains nothing; where would the creator come from? One cannot think about emptiness; something being empty is unthinkable; space "is" only as relationship among existing things. Infinite extent cannot be described and is superfluous for the explanation of what is describable. Without the law of inertia one could not understand physically how the Earth should be performing rapid rotations which do not noticeably affect any objects on it.

This is an almost perfectly rounded system of ideas. Over two millennia different experiences necessitated partly to modify it through intrinsically foreign ideas, partly to replace it by less consistent but more successful theories. Christianity included the world of Greek philosophy and astronomy, laid out in this world for eternity, between what it would have considered the inappropriate limits of the Creation and Last Judgment. But it expressed the experience of the historical nature of humans. Modern natural science sacrificed the totality of the description of reality in favor of a respective mathematization. It forced dualities: space and matter (Sect. 2.2), thought and extension (Sect. 10.1), mechanics and life (already present, unrecognized, in mechanics and chemistry: see Sect. 2.4 and 2.11). It ripped open, telescopically and microscopically, the distinct limits of the world. Many of its successes were paid through inconsistencies which only resurfaced as problems in later revolutions. Thus, e.g., only the theory of relativity again attains the Aristotelian standard of the problem of space, as understood by Leibniz (Sect. 2.8–2.10).

Western philosophy has always demanded of itself a general theoretical consistency. It is not identical with the too specific model of deductive science. Our reconstruction would not have occurred without the hope that modern physics, at the end of its development, could again achieve the consistency which the system of classical philosophy, though too narrow in its contents, had possessed. How close have we come to fulfilling this hope?

The basic difference from the classical system lies in the central role of time. With this difference in mind we can compare the elements of Aristotelian physics with what has become of them today.

First the four "causes."

The relation between substance and form we have already discussed in Chap. 10 on the stream of information. We found in today's physics form in the role of substance. But forms are not eternally present. They arise in evolution (section 10.3). Only in retrospect can one say that already before the formation of the planets, rock crystals were "formally possible," or trees and monkeys at the beginning of the evolution of life. This formal possibility is an abstraction, a creation of *our* concepts.

Substance, matter, is for Aristotle one aspect of potentiality. Aristotle defines movement in terms of the pair of concepts of actuality and potentiality (energeia = being-at-work; dynamis = capacity, ability). Both are further subdivided, such that there is a "dynamis tu poiein" and a "dynamis tu paschein": being able to do and being able to adopt. Motion is defined as the actuality of potential being.[4] In our terminology potentiality is the mark of the future, actuality the mark of the present. Facticity is past actuality, which can be preserved in documents. Stylizing, one can then say: "Motion is the presence of the future."

Modern science has expressed its anti-Aristotelian pathos above all by pitting the "causa efficiens" against the "cause finalis." In terms of popular philosophy, this turns against Aristotelianism, reinterpreted by the belief in the Creation. The anthropomorphic Lord of creation may set goals for himself like any human. Against this one plays the origin of all things according to scientific laws. To accomplish his goals, the God of modern design must then create at the very outset causes and laws according to which the goals are then ultimately reached, being causally necessary. The entire controversy has nothing to do with Aristotle. The eternal God of Aristotle forever knows the forms which in the course of the universe eternally reproduce themselves and appear in it as causes as well as goals. This is, in the diction of metaphysics, a clean description of the self-reproduction of life. The egg is the means of the chicken to recreate chickens. Modern geneticists have their fun, as one might well imagine: the chicken is the means of the egg to recreate eggs; the organism is the mechanism whereby the "selfish gene" can create genes identical to itself.

The break with Aristotle occurs only with evolution, with the origin of forms. In return our world view acquires a unity not possessed by the Aristotelian: now all organisms can descend from a common origin. Their relationship to one another is not only the ecological biocoenosis, as understood by Aristotle long ago, but their physical relationship. And they originate from inorganic matter; also with a stone and a star we are materially related. But to actually think this way and not merely to assert this, we must understand the content of the mathematical form of the laws of nature (Sect. 2.3). Morphology corresponds to a philosophy of eternal forms; the differential equation in time corresponds to the modern causal approach; the extremum principle shows the general equivalence of a causal and finalistic description; symmetry

---

[4] Cf. "*Möglichkeit und Bewegung*" in *Die Einheit der Natur* IV,4., p. 428ff.

groups point to the common origin of the laws in the separability of alternatives. Mechanical causality turns out to be a derived concept. Therefore the causal paradox of the law of Inertia with which Aristotle was already struggling without success in his theory of projectile motion (cf. Wolff 1971). inertia and force both stem from the openness of the future (Sect. 4.3c). Let us point out again (cf. the beginning of Sect. 10.3) that a complete philosophy of mathematical laws of nature requires a philosophy of mathematics in which mathematics itself appears as an "art," a perception of Gestalt through creation of Gestalt (*Zeit und Wissen* 5).

For humans, however, one thing is lost in this view of the grand unity of nature: the coziness offered by the eternal and finite world of Aristotle as well as the garden of Eden in the Bible. "Am I not the fugitive, the homeless one...?" (Goethe, Faust 3348). Our astronomical home, the Solar system, is a form that had a beginning and presumably will also end. It is one among billions of systems in a universe which we are yet unable to enliven with our notions of sense and meaning. Our neighboring planets appear to our satellite observations to be deserts of ice or blistering heat. And the theory of evolution teaches us how many thousands of creatures and species must perish to enable us, offspring of victors in the struggle for survival, to step into a short existence. Among the traditional perceptions of life, the one of the Buddha before his enlightenment appears to natural science to be the most realistic: life is thirst and suffering. According to Darwin it is successful because it is thirst and suffering.

What have we just done in our portrayal of the larger unity of nature? We have burst the frame of a picture. We have expanded the limits of the Aristotelian-scholastic world view in time and space in ways incomprehensible to our perceptive faculties. Astronomy and atomic physics have empowered us to do so, indeed compelled us. But we have not pursued another crossing of frontiers: the relativization of the concepts of object, extent in space and time, which began in quantum theory and which has not been interpreted yet. We should have learned that the reduction of space, time, and matter to empirically decidable alternatives is an aspect of the foreground of physics, brought into our view by classical physics. If we have read quantum theory correctly, it teaches that this aspect of the foreground is only possible *because* it rests on the foundation of a much larger richness of structures of a different kind; where we have become partially aware of these structures, we spoke of a quantum theoretical "additional knowledge."

What lies beyond physics is traditionally called metaphysics. Our arguments about what lies "beyond quantum theory" have left us with the impression that quantum theory itself shows structures which we have not yet been able to interpret in our classical manner of speaking. The limit between "physics" and "metaphysics" itself turns out to be a still unsolved problem. It seems that physics can only be because it has an open door to metaphysics.

## 12.4 Metaphysics

Metaphysics begins with the Greeks as the philosophical study of the one God. The anthropomorphic images of religion are discarded. Philosophers are speaking in the neuter. Parmenides speaks of "the Being," Plato of "the One" and "the Good." Similarly abstract the Asian speculations: "the Non-duality" of the Advaita-Vedanta, "the void" of Buddhism.

Metaphysics initially presents itself as an ascent: in Parmenides through the poetic vision of an ascent to the gate of truth, in Plato through the philosophical simile of the turning around of one's view and the ascent from the cave, in Aristotle in terms of the composition of three books: physics, metaphysics, and the writing on the soul which factually belong together and all end with a chapter about the divine spirit.

This motion is of a dual nature. The ascent is hard, long, tedious. Then, only possible in one leap, the transition to the vision of the One which is the Whole. "Look!"—thus the goddess of truth begins her speech in Parmenides, at the end of the journey. "Suddenly," in Plato, the pupil sees, after many years together with the teacher, that which they always had been talking about. The descent, the deduction, is an important postulate which, however, for all the early great thinkers is nowhere passed down as written doctrine.

The indescribable which is seen appears to be the same in all cultures. The toil of the ascent is of argumentative nature; it is thereby bound to culture just like all forms of rationality.

We have planned a three-pronged approach for the ascent. In the chapter *Beyond quantum theory*, we asked how one might formulate, with the means of quantum theory, the conjecture of a Being lying beyond our familiar notion of the modalities of time over a linear time parameter. In the present chapter we compare these arguments with the traditional arguments of metaphysics.[5] Only in the final chapter of *Zeit und Wissen* will philosophical theology itself become an issue.

Once more we begin with Aristotle. Why does not physics, the study of movement, comprise theoretical philosophy in its entirety? Because there are things not in motion: mathematical concepts, pure forms, God. Mathematical concepts of course are mere abstractions; they are the forms which remain of things in motion if one disregards their motion. Also pure forms have their actual existence only in concrete things. But God is an unmoving knowing Being. In Aristotle's three basic books dealing with theoretical philosophy God is disclosed in a threefold way: in *physics* as origin of all movement, in *metaphysics* as Being in full reality, without unfulfilled possibilities, in *De anima* as knowing spirit, in which all forms are present.[6] This is a philosophy of eternal movement, as its origin does not change; of eternal presence, as

---

[5] Cf. *Parmenides und die Quantentheorie*, in: *Die Einheit der Natur* IV,6, and "Wer ist das Subjekt in der Physik?" in: *Der Garten des Menschlichen* II,1.

[6] Cf. Rudolph (1983).

## 12.4 Metaphysics 343

there is no future (possibility) yet to come for God, of eternal knowledge, as the forms are eternal.

How, in comparison, must *we* argue? The origin of movement we have called time in its modalities. That step beyond the usual formulation of quantum theory which can still be formulated quantum theoretically we have referred to as the comprehensive present. For human knowledge the phenomenal present is not a point but unlimited and comprehensive. The idea of an eternally present infinite knowledge appears as a conceivable generalization of this. It is then natural to say: Finite knowledge exists only for finite objects; our science is such finite knowledge.

The actual mental task, however, is not achieved by such conjectures. In the relationship between the eternal and time we are compelled to deviate from Aristotle and the entire classical metaphysics. Metaphysics in itself contains at this point an unsolved problem that becomes evident when we attempt to incorporate the modern view of time into metaphysics. For Parmenides, who makes the leap into the eternal present, the world of becoming and perishing is "doxa," what Picht quite appropriately translates with "Erscheinung" (appearance). But how does such an appearance come about? Where in Being is that which creates time and appearance? Plato explains time as an image, "aionically" progressing according to a number, of the "Aion" persisting in the One.[7] "Aion" in Greek denotes a reasonable time span, like a season or a lifetime. Traditionally it is translated as *eternity*; then time is the eternal image of eternity. In the Aion itself there must be the movement. For us there is in the present the future as possibility, the past as possibility of documents. Thinking of an all-embracing present we also should include this structure in our thoughts. This we have not done in the above conjecture. We must proceed more slowly.

The leap into abstract theology indeed skips one sphere of decisive importance. It skips the question of the quantum theoretical "additional knowledge." This knowledge refers to mathematically specifiable structures which are not separate objects in space. The universe of objects in space is only one aspect or—in a perhaps more appropriate simile—only a surface of reality. Against this reality quantum theory accomplishes a dual or cyclical achievement: On the one hand, it provides at least a mathematical model of this reality, starting with predictions about decidable alternatives in the universe of objects in space. On the other hand, it shows, starting with this model, that the so described reality must present itself, in the approximation of separable alternatives, just as the universe of objects in space, with the known laws of physics.

Furthermore we have seen that nothing prevents applying quantum theory, in addition, to decidable alternatives for psychic or mental processes. Thus from the present knowledge of physics there stands nothing in the way of a

---
[7] Editors' note: Cf. Plato's definition of time as the moving image of eternity in Timaeus 37d.

philosophy which would dare to interpret the reality which is referred to by the additional knowledge of quantum theory as an essentially psychic or mental reality. The question is only whether we know what we mean by expressions like "psychic" or "mental." They too are anthropomorphic.

We know ourselves as communicating, feeling, striving, thinking beings. We are bound to ascribe also to animals communication, affect, perception, drive. But the farther a being is removed from us the less we can empathize with it. The experience of human loneliness in a universe without any comprehensible reason is thrust upon us. The study of the biological background of consciousness, the "reverse side of the mirror," teaches us about the most complicated organic actions which enable the phenomena of consciousness. Individual psyche we only know as function of the spatial universe of objects. Hence we must be lonely in the universe of objects, where it has not developed these organs.

Quantum theory now teaches us that the bodies themselves are based on structures that are not further describable spatially. Then this must also hold for psychic phenomena. Thus one is led to suspect that essential psychic or mental phenomena must escape a perception that restricts itself, out of naïveté or scientific desire for power, to the spatial universe of bodies and the psychic processes manifest in it. Here communicative, artistic, and meditative perception will pursue other avenues.

Quantum theory, accordingly, appears to be initially open to a "near metaphysics." It will be ready to admit psychic experiences which lie beyond the classically describable sensible experience of objects in space. Such experiences have always been familiar to humankind before the era of natural science. But it means something else to acknowledge them after science has erected a coherent world view; in particular to acknowledge them not as a denial or demolition of this world view but as a condition of its feasibility.

The hunch that quantum theory could open such access to the psychic background of the universe has frequently surfaced since the theory was completed about eighty years ago. Important physicists like Pauli or Jordan have expressed such views, not necessarily thinking about the range of applicability of quantum theory itself, but more of a formal analogy to it in our relationship with a suspected, more extensive realm. Bohr's idea of complementarity acted there as stimulant; Bohr himself, however, remained very reserved with respect to such a perhaps too "direct" application of his idea. Denoting parapsychic phenomena as "Psi" originates presumably from a direct reference to the $\psi$-function. In the tradition of these ideas there are the recent widely read books by Capra (1975) and other authors who attempt to build a bridge to the experiences of Asian meditation.

This is probably a pack on the right track. That thereby also some uncritical infatuation with the occult slips in cannot come as a surprise. This movement depends thus far on its good nose. It has not reached the consistency of science, probably not even considered the rigor of philosophy as a possibility. From quantum theory one must ask whether consistent theoretical models of such experiences have already been seriously considered. For it is not easy to see what one there must search for. There are many indications that, e.g., the broadly based parapsychological investigations pursue a less promising path. They want to objectify the sought-after phenomena empirically and scientifically. But this method is most likely bound to create phenomena in the sense of classical physics; it is easily possible that it just causes the prerequisites for the actual psychic phenomena in the substratum of the universe to disappear. These phenomena, whenever they occur in the midst of life, are charged with affect or a hidden symbolism reminiscent of dreams; they are in their meaning related to the particular circumstances of life. It is an old experience that superstitious persons are tricked by their omens, as if these omens wanted to withdraw from human availability. Scientifically pursued parapsychology therefore easily assumes the character of methodologically produced superstitions. One can also destroy a love affair by approaching it with the distrustful inquisitiveness of the scientist.

Besides, collecting empirical material is only a preliminary stage of science. Presumably hard preliminary theoretical studies must be done before one knows what to look for empirically. A first inquiry into this field we have attempted in the chapter *Beyond Quantum Theory*. If quantum theory is assumed to open a new path for us here, we must first ask how it can describe psychic phenomena. These phenomena are meaningful in life; but how do we describe the structure of such a meaning? An example of the thereby occurring problems is offered by the attempts of computer science to construct "artificial intelligence." The computer is related to human thought presumably like the automobile to the human body: it accomplishes, as instrument, just that which a human does not. A computer is constructed according to logic. logic, however, was invented not for describing human thought but to correct it.

The question of the structure of thought or perception will not be answered without self-perception. Self-perception is a matter of active life, a matter of spontaneous or artistic self-representation, a matter of meditation and reflection. Reflection is the attempt to say what I have thought, it is the Socratic query. He who in our cultural tradition gets involved with it will be led by it on the path which philosophy has been led. One can attempt to break out from this path into foreign, perhaps Asian traditions. But just these traditions are challenged nowadays on a path of contest with our science, thus our tradition. Our individual abilities are limited, everybody is free to join wherever his abilities lead him. But he who is knowingly going the path of philosophy will be led with it into metaphysics.

Now we no longer speak of the "near metaphysics" of psychic perceptions, but rather the ascent to metaphysics itself; but this ascent now also leads through the mountains "beyond quantum theory." It was the great idea of metaphysics to pose the question of what knowledge is, not from a description or disputation of our always incomplete factual knowledge. Metaphysics rather asks about the criterion that we already always use when we question alleged knowledge whether it can prove itself as knowledge. Socrates, who knows that he does not know, evidently knows therewith what he means when he asks about knowledge. The God of metaphysics represents just this knowledge without which there would be no factual knowledge.

At the end of the reconstruction of physics we ought to say what notion of knowledge guided us when we described physics as knowledge in time. We have not only assumed knowledge. We also have described, following a *Kreisgang*,[8] its contents, although rudimentarily. And in this *Kreisgang* we have pursued semantic consistency; this is a formula for our theory to be knowledge. And now, in our last step beyond quantum theory, we have not ruled out the objective possibility of supra-individual knowledge. We have even postulated it, although with a question mark, as a precondition for the possibility of finite knowledge.

The reflection of what we have thereby achieved is only possible in a renewed *Kreisgang* which we will give the title "Zeit und Wissen."

---

[8] Cf. p. XXII, fn. 14.

# References

Editors' note:
See also the updated lists of main book publications of C. F. von Weizsäcker and of books on Weizsäcker's philosophy of physics at pages XXXII and XXXIII.

Abbreviation:
*QTS: Quantum Theory and the Structures of Time and Space*, volumes I (1975), II (1977), III (1979), IV (1981), V (1983), edited by L. Castell, M. Drieschner and C. F. v. Weizsäcker; volume VI (1986), edited by L. Castell and C. F. v. Weizsäcker. Munich, Hanser


Barut, A. O. (1984). Unification based on electrodynamics. In *Symposium on Unification*, Caput.
Boerner, H. (1955). *Darstellungen von Gruppen*. Springer, Berlin.
Bohm, D. (1951). *Quantum theory*. Prentice Hall, New York.
Bohm, D. (1952). A suggested interpretation of the quantum theory in terms of "hidden" variables. *Physical Review*, 85:166–179, 180.
Bohr, N. (1913a). On the constitution of atoms and molecules. *Phil. Mag.*, 26:1–25, 476–502, 857–875.
Bohr, N. (1913b). Über das Wasserstoffspektrum. *Fysisk Tidsskrift*, 12:97.
Bohr, N. (1935). Can quantum-mechanical description of reality be considered complete? *Phys. Rev.*, 48:696–702.
Bohr, N. (1949). Discussion with Einstein on epistemological problems in atomic physics. In *Schilpp (1949)*.
Bohr, N., Kramers, H., and Slater, J. (1924). Über die Quantentheorie der Strahlung. *Z. Phys.*, 24:69–87.
Bopp, F. (1954). Korpuskularstatistische Begründung der Quantenmechanik. *Z. Naturforschung*, 9a:579–600.
Bopp, F. (1983). Quantenphysikalischer Ursprung der Eichidee. *Annalen d. Physik*, 40:317–333.

Borges, J. L. (1970). *Sämtliche Erzählungen*. Hanser, Munich.
Born, M. (1924). Über Quantenmechanik. *Z. Phys.*, 26:379–395.
Born, M. (1925). *Vorlesungen über Atommechanik*. Springer, Berlin.
Born, M. (1926). Quantenmechanik der Stoßvorgänge. *Z. Phys.*, 38:803–827.
Born, M. and Jordan, P. (1930). *Elementare Quantenmechanik*. Springer, Berlin.
Broglie, L. d. (1924). *Thèses*. Masson et Cie, Paris.
Capra, F. (1975). *The Tao of Physics*. Berkely.
Castell, L. (1975). Quantum theory of simple alternatives. In *QTS II, 147–162*.
Courant, R. and Hilbert, D. (1937). *Methoden der mathematischen Physik*, volume II. Springer, Berlin.
Dingler, H. (1943). *Aufbau der exakten Fundamentalwissenschaft*. Edited by P. Lorenzen, Eidos, Munich, 1964.
Dirac, P. A. M. (1927). The quantum theory of the emission and absorption of radiation. *Proc. Roy. Soc.*, A(114):243–265.
Dirac, P. A. M. (1933). The Lagrangian in quantum mechanics. *Phys. Z. Sowjetunion*, 3:64–72.
Dirac, P. A. M. (1937). The cosmological constants. *Nature*, 139:323.
Drieschner, M. (1967). *Quantum mechanics as a general theory of objective prediction*. PhD thesis, Univ. Hamburg.
Drieschner, M. (1979). *Voraussage-Wahrscheinlichkeit-Objekt*. Springer, Berlin.
Dürr, H.-P. (1977). Heisenberg's unified theory of elementary particles and the structure of space and time. In *QTS II, 33–45*.
Eckert, M., Pricha, W., Schubert, H., and Torkar, G. (1984). Geheimrat Sommerfeld – Theoretischer Physiker: Eine Dokumentation aus seinem Nachlaß. Abhandlungen und Berichte des Deutschen Museums München, Sonderheft 1.
Ehrenfest, P. and Ehrenfest, T. (1906). Über eine Aufgabe aus der Wahrscheinlichkeitsrechnung, die mit der kinetischen Deutung der Entropievermehrung zusammenhängt. *Math.-Naturwiss. Blätter*, 11:12.
Eigen, M. (1971). Selforganization of matter and the evolution of biological macromolecules. *Natwiss.*, 58:465–523.
Einstein, A. (1905). Über einen die Erzeugung und Verwandlung des Lichts betreffenden heuristischen Gesichtspunkt. *Ann. d. Physik*, 17:132–148.
Einstein, A. (1917). Quantentheorie der Strahlung. *Phys. Zeitschrift*, 18:121–128.
Einstein, A. (1949). Autobiographisches. In *Schilpp (1949)*, pages 2–95.
Einstein, A., Podolsky, B., and Rosen, N. (1935). Can quantum-mechanical description of reality be considered complete? *Phys.Rev.*, 47:777–780.
Everett, H. (1957). Relative state formulation of quantum mechanics. *Review of Modern Physics*, 29:454–462.
Feynman, R. P. (1948). Space-time approach to non-relativistic quantum mechanics. *Rev. Mod. Phys.*, 20:367–387.
Finetti, B. d. (1972). *Probability, induction and statistics*. J. Wiley, New York.
Finkelstein, D. (1968). Space-time code. *Phys. Rev.*, 185:1261.
Franz, H. (1949). Master's thesis, MPI Physik, Göttingen.
Gadamer, H. G. (1960). *Wahrheit und Methode*. Mohr (Siebeck), Tübingen.
Gibbs, H. W. (1902). *Elementary principles in statistical mechanics*. Yale University Press, New Haven.
Glansdorff, P. and Prigogine, I. (1971). *Thermodynamic theory of structure, stability and fluctuations*. J. Wiley, New York.

Gödel, K. (1949). An example of a new type of cosmological solutions of Einstein's field equation of gravitation. *Rev. Mod. Phys.*, 21:447–450.

Green, H. S. (1953). A generalized method of field quantization. *Phys. Rev.*, 90:270–273.

Greenberg, O. W. and Messiah, A. M. L. (1965). High-oder limit of para-Bose and para-Fermi fields. *J. Math. Phys.*, 6:500–504.

Gupta, S. N. (1950). Theory of longitudinal photons in quantum electrodynamics. *Proc. Phys. Soc.*, A(63):681–691.

Gupta, S. N. (1954). Gravitation and electromagnetism. *Phys. Rev.*, 96:1683–85.

Haken, H. (1978). *Synergetics*. Springer, Berlin.

Hawking, S. W. and Ellis, G. F. R. (1973). *The large scale structure of space-time*. Cambridge University Press, Cambridge.

Heidenreich, W. (1981). *Die dynamischen Gruppen SO(3,2) und SO(4,2) als Raum-Zeit-Gruppen von Elementarteilchen*. PhD thesis, TU Munich, Munich.

Heisenberg, W. (1925). Über die quantentheoretische Umdeutung kinematischer und mechanischer Beziehungen. *Z. Physik*, 33:879–893.

Heisenberg, W. (1927). Über den anschaulichen Inhalt der quantentheoretischen Kinematik und Mechanik. *Z. Phys.*, 43:172–198.

Heisenberg, W. (1930). *Die physikalischen Prinzipien der Quantentheorie*. Hirzel, Leipzig.

Heisenberg, W. (1948). Der Begriff abgeschlossene Theorie in der modernen Naturwissenschaft. *Dialectica*, 2:331–336.

Heisenberg, W. (1967). *Einführung in die einheitliche Feldtheorie der Elementarteilchen*. Hirzel, Leipzig.

Heisenberg, W. (1969). *Der Teil und das Ganze*. Piper, Munich.

Heisenberg, W., Dürr, H.-P., Mitter, H., Schlieder, S., and Yamasaki, K. (1959). *Z. Naturforschung*, 14a:441.

Heisenberg, W. and Pauli, W. (1929). Zur Quantendynamik der Wellenfelder. *Z. Phys.*, 56:1; (1930) 59:168.

Helmholtz, H. v. (1868). Über die Tatsachen, die der Geometrie zugrunde liegen. *Nachr. Königl. Ges. Wiss. Göttingen*, 9:193–221.

Hilbert, D. (1915). Die Grundlagen der Physik. *Nachr. Königl. Ges. Wiss. Göttingen*, 395; (1917) 201.

Hoffmann, B. and Dukas, H. (1972). *Albert Einstein*. Viking Press, New York.

Jacob, P. (1979). *Konform invariante Theorie exklusiver Elementarteilchen-Streuungen bei großen Winkeln*. PhD thesis, MPI Starnberg.

Jammer, M. (1974). *The philosophy of quantum mechanics*. J. Wiley, New York.

Jauch, J. M. (1968). *Foundations of quantum mechanics*. Addison-Wesley, Reading, Mass.

Kant, I. (1781). *Kritik der reinen Vernunft*. Riga.

Kant, I. (1786). *Metaphysische Anfangsgründe der Naturwissenschaft*. Riga.

Kuhn, T. S. (1962). *The Structure of Scientific Revolutions*. University of Chicago Press.

Künemund, T. (1982). *Dynamische Symmetrien in der Elementarteilchenphysik*. Master's thesis, TU Munich.

Künemund, T. (1985). *Die Darstellungen der symplektischen und konformen Superalgebren*. PhD thesis, TU Munich.

Lorentz, K. (1973). *Die Rückseite des Spiegels*. Piper, Munich.

Lorenz, K. (1942). Die angeborenen Formen möglicher Erfahrung. *Z. Tierpsychol.*, 5:235.

Ludwig, G. (1954). *Die Grundlagen der Quantenmechanik*. Springer, Berlin.

Mehra, J. (1973a). Einstein, Hilbert, and the theory of gravitation. In *Mehra (1973b)*.

Mehra, J., editor (1973b). *The physicists's conception of nature*, Dordrecht. Reidel.

Meyer-Abich, K. M. (1965). *Korrespondenz, Individualität und Komplementarität*. Steiner, Wiesbaden.

Mittelstaedt, P. (1979). Der Dualismus von Feld und Materie in der Allgemeinen Relativitätstheorie. In Nelkowski, H., editor, *Einstein Symposion Berlin*, Berlin. Springer.

Neumann, J. v. (1932). *Mathematische Grundlagen der Quantenmechanik*. Springer, Berlin.

Neumann, J. v. and Morgenstern, O. (1944). *Theory of games and economic behavior*. Princeton University Press, Princeton.

Pais, A. (1982). *Subtle is the Lord*. Oxford University Press, Oxford.

Picht, G. (1958). Die Erfahrung der Geschichte. In *Picht (1969)*.

Picht, G. (1960). Die Epiphanie der ewigen Gegenwart. In *Picht (1969)*.

Picht, G. (1969). *Wahrheit, Vernunft, Verantwortung*. Klett, Stuttgart.

Popper, K. R. (1934). Zur Kritik der Ungenauigkeitsrelationen. *Naturwiss.*, 22:807–808.

Popper, K. R. (1975). The rationality of scientific revolutions. In Harré, R., editor, *Problems of scientific revolution*, Oxford. Clarendon.

Rudolph, E. (1983). Zur Theologie des Aristoteles.

Sartre, J. P. (1943). *L'être et le néant*. Gallimard, Paris.

Savage, L. J. (1954). *The foundations of statistics*. J. Wiley, New York.

Scheibe, E. (1964). *Die kontingenten Aussagen in der Physik*. Athenäum, Frankfurt.

Scheibe, E. (1973). *The logical analysis of quantum mechanics*. Pergamon Press, Oxford.

Schilpp, P., editor (1949). *Albert Einstein: Philosopher-scientist*, Evanston, Ill. The Library of living philosophers, volume VII.

Schmutzer, E. (1983). Prospects for relativistic physics. In *Proc. of GR 9*, Berlin. Dt. Verlag d. Wissenschaften.

Schrödinger, E. (1935). Die gegenwärtige Situation in der Quantenmechanik. *Naturwiss.*, 49:823–828.

Segal, I. E. (1976). Theoretical foundations of the chronometric cosmology. *Proc. Nat. Acad. Sci USA*, 73:669–673.

Sexl, R. U. and Urbantke, H. K. (1983). *Gravitation und Kosmologie*. Bibliographisches Institut, Mannheim.

Shannon, C. F. and Weaver, W. (1949). *The mathematical theory of communication*. Urbana, Ill.

Snyder, H. S. (1947). Quantized space-time. *Phys. Rev.*, 71:38–41.

Strawson, P. F. (1959). *Individuals*. Methuen, London.

Stückelberg, E. C. G. (1960). Quantum theory in real Hilbert space. *Helv. Phys. Acta*, 33:727–752.

Thirring, W. E. (1961). An alternative approach to the theory of gravitation. *Ann. Phys. (N.Y.)*, 16:96–117.

Vigier, H. P. (1954). Structure des micro-objets dans l'interpretation causale de la théorie des quantes. Paris.

Weizsäcker, C. F. v. (1931). Ortsbestimmung eines Elektrons durch ein Mikroskop. *Z. Physik*, 70:114–130.
Weizsäcker, C. F. v. (1934). Nachwort zu einer Arbeit von K. Popper (1934). *Naturwiss.*, 22:808.
Weizsäcker, C. F. v. (1939). Der zweite Hauptsatz und der Unterschied von Vergangenheit und Zukunft. *Ann. Physik*, 36:275. Reprinted in Weizsäcker, 1971a.
Weizsäcker, C. F. v. (1943). *Zum Weltbild der Physik*. Hirzel, Leipzig.
Weizsäcker, C. F. v. (1948). *Die Geschichte der Natur*. Hirzel, Stuttgart.
Weizsäcker, C. F. v. (1949). Eine Bemerkung über die Grundlagen der Mechanik. *Ann. Physik*, 6:67–68.
Weizsäcker, C. F. v. (1955). Komplementarität und Logik I. *Naturwiss.*, 42:521–529 and 545–555. Reprinted in Weizsäcker (1957).
Weizsäcker, C. F. v. (1957). *Zum Weltbild der Physik*. Hirzel, Stuttgart, 7th edition.
Weizsäcker, C. F. v. (1958). Komplementarität und Logik II. *Z. Naturforschung*, 13a:245–253.
Weizsäcker, C. F. v. (1971a). *Die Einheit der Natur*. Hanser, Munich.
Weizsäcker, C. F. v. (1971b). Notizen über die philosophische Bedeutung der Heisenbergschen physik. In Dürr, H.-P., editor, *Quanten und Felder*, Braunschweig. Vieweg.
Weizsäcker, C. F. v. (1971c). The Unity of Physics. In Bastin, T., editor, *Quantum Theory and Beyond*, Cambridge. Cambridge University Press.
Weizsäcker, C. F. v. (1972). Evolution und Entropiewachstum. *Nova Acta Leopoldina*, 206:515–530. Reprinted in Weizsäcker (1974b).
Weizsäcker, C. F. v. (1973a). Classical and quantum descriptions. In *Mehra (1973b)*.
Weizsäcker, C. F. v. (1973b). Comment on Dirac's paper. In *Mehra (1973b)*.
Weizsäcker, C. F. v. (1973c). Probability and quantum mechanics. *Brit. J. Phil. Sci.*, 24:321–337.
Weizsäcker, C. F. v. (1974a). Der Zusammenhang der Quantentheorie elementarer Felder mit der Kosmogonie. *Nova Acta Leopoldina*, 212:61–80.
Weizsäcker, C. F. v. (1974b). Geometrie und Physik. In Enz, C. P. and Mehra, J., editors, *Physical reality and mathematical description*, pages 48–90, Dordrecht. Reidel.
Weizsäcker, C. F. v. (1975). The philosophy of alternatives. In *QTS I, 213–230*.
Weizsäcker, C. F. v. (1977). *Der Garten des Menschlichen*. Hanser, Munich.
Weizsäcker, C. F. v. (1979). Einstein. In *Weizsäcker (1983)*.
Weizsäcker, C. F. v. (1982). Bohr und Heisenberg. In *Weizsäcker (1983)*.
Weizsäcker, C. F. v. (1983). *Wahrnehmung der Neuzeit*. Hanser, Munich.
Weizsäcker, C. F. v. (1985). Werner Heisenberg. In Gall, L., editor, *Die großen Deutschen unserer Epoche*, Berlin. Propyläen.
Weizsäcker, C. F. v., Scheibe, E., and Süßmann, G. (1958). Mehrfache Quantelung, Komplementarität und Logik III. *Z. Naturforschung*, 13a:705–721.
Weizsäcker, C. v. and Weizsäcker, E. v. (1984). Fehlerfreundlichkeit. In Kornwachs, K., editor, *Offenheit, Zeitlichkeit, Komplexität*, Frankfurt, New York. Campus.
Weizsäcker, E. v. (1974a). Erstmaligkeit und Bestätigung als Komponenten der pragmatischen Information. In *Weizsäcker (1974b)*.
Weizsäcker, E. v., editor (1974b). *Offene Systeme I*. Klett-Cotta, Stuttgart.
Weizsäcker, E. v. (1985). Contagious knowledge. In Ganelius, T., editor, *Progress in Science and its Social Conditions—Proceedings of the 58th Nobel symposium, Stockholm 1983*, New York. Pergamon Press.

Weizsäcker, E. v. and Weizsäcker, C. v. (1972). Wiederaufnahme der begrifflichen Frage: Was ist Information? *Nova acta Leopoldina*, 206:535–555.

Weyl, H. (1918). *Raum-Zeit-Materie*. Springer, Berlin.

Wheeler, J. A. (1978). The "past" and the "delayed choice" double-slit experiment. In Marlow, A. R., editor, *Mathematical Foundations of Quantum Theory*, pages 9–48. Academic Press, New York.

Wigner, E. (1939). On unitary representations of the inhomeogeneous Lorentz group. *Ann. Math*, 40(1):39–94.

Wigner, E. (1961). Remarks on the mind-body question. In Good, I. J., editor, *The scientist speculates*, London. Heinemann.

Wigner, E. (1983). Realität und Quantenmechanik. In *QTS V, 7–18*.

Wolff, M. (1971). *Fallgesetz und Massenbegriff. Zwei wissenschaftliche Untersuchungen zur Kosmologie des Johannes Philoponus*. de Gruyter, Berlin.

Zucker, F. J. (1974). Information, Entropie, Komplementarität und Zeit. In *Weizsäcker (1974b)*.

# Index

Action at a distance, 5, 24, 35
Alternative, 7, 72, 85, 105, 107, 260, 337, 341
anti-de Sitter group, 97, 171
Apriorism, xxii, 82, 251, 256, 277, 334, 335
Aristarchus, 37, 38, 339
Aristotle, xv, xxx, 24, 26, 29, 32, 236, 287, 298, 307, 309, 334, 338–343
Ashtekar, A., xxvi
Atomism, xxi, 25, 34, 89, 103, 131, 133, 152

Böhme, G., 184
Barut, A. O., 136
Becker, J., xvii
Being, 25, 287–289, 297, 298, 304, 310, 311, 315, 316, 342, 343
Bekenstein, J., xxix, 167, 169
Bellarmin, R., 38
Berdjis, F., xvii
Besso, M. A., 288
Biology, 211, 214, 215, 219, 229, 234, 274, 303, 338
Black hole, xxix, 167–169, 177
Bleuler, K., 259
Body, 5, 20–26
Bohm, D., 284, 291
Bohr, N., xiii, xiv, xvi, xix, xxxi, 8, 9, 15, 33, 43, 52, 53, 57, 83, 87, 137, 142, 145, 244–248, 250, 251, 253–261, 263–266, 269, 270, 274, 276, 277, 279, 284, 288–290, 314, 321, 322, 325, 326, 344

Boltzmann, L., 4, 113, 187, 190, 201–204
Bolyai, J., 36, 141
Bopp, F., 79, 136, 255
Borges, J., 294
Born, M., xiv, 247–249, 251, 253, 259
Boscovich, R. J., 21
Bose statistics, 69, 91, 92, 112–114, 117
Brahe, T., 39
Broglie, L. V. de, xiv, 53, 70, 245–247, 249, 254, 260
Buddha, 341
Buffon, G. L. L., 29

Cantor, G., 309
Capra, F., 344
Carnap, R., 66, 288
Carnot, S., 218
Castell, L., xvi, xxiv, xxv, 100, 112, 118–120, 123, 126, 127, 304, 347
Causality, 6, 27, 30, 86, 133, 181, 182, 206, 248, 249, 258, 266, 338, 341
Chemistry, 5, 15, 31–34, 52, 54, 55, 89, 106, 131, 137, 250, 299–301, 310, 338, 339
Chievitz, O., 255
Clarke, S., 40, 41
Classical physics, 5–9, 14–16, 20, 23–25, 27, 29, 40–44, 51, 55, 57, 81, 82, 84, 87, 105, 106, 141, 152, 182, 234, 243–246, 248, 258, 261, 264–266, 269, 276, 277, 287, 288, 290, 292, 295, 300, 308, 313, 316, 321, 335, 341, 345
Clausius, R., 218

Clifford screw, 95, 99, 103
Closed theory, 1, 2, 13, 14, 36, 41, 53, 55, 72, 243, 245, 277, 278, 309, 311, 312, 331, 335–337
Cognition, xxii, 27, 72, 234–236, 238, 307, 308, 333
Columbus, C., 94
Complementarity, xx, 78, 253, 255–257, 274, 314, 344
Compton, A. H., 138, 145, 246
Condensation model, 223
Conditional, 19, 20, 39, 40, 66, 73–75, 87, 183, 184, 207, 285
Configuration space, 69, 70, 73, 142, 247, 254
Confirmation, 229, 230, 232, 233
Connes, A., xxvi
Conservation law, 15, 34, 154, 158, 283, 303
Copenhagen interpretation, 244, 246, 249, 253, 259, 268, 269, 276, 284, 321, 322
Copernicus, N., 37, 38, 50
Correspondence principle, 55, 70, 71, 83, 245, 253–255, 258, 276
Correspondence principle, 84
Cosmology, 37, 49–51, 57, 95, 108, 109, 122, 143, 146, 147, 149–151, 154, 155, 159, 169, 177, 201, 313
Courant, R., 28
Covariance, 45, 47
Croesus, 318, 319
Cusanus, 38, 47

Dürr, H. P., xvi, 134
Dalton, J., 32
Dark matter, 167
Darwin, C., 212, 215, 219, 234, 341
Delayed choice, 280, 282, 290
Delbrück, M., 274
Democritus, 25
Density matrix, 267, 268, 271
Descartes, R., xv, 29, 44, 299
Determinism, 87, 90, 183, 269, 292, 305
Dingler, H., 41
Dirac, P., xiii, 6, 30, 147, 156, 259, 284
Distinguishability, xxiii
Drühl, K., xvii, 111

Drieschner, M., xvi, xxii, 75, 77–79, 123, 347
Dukas, H., 288
Dynamics, xxi, xxviii, 5, 6, 31, 35, 39 73, 76, 78, 83–85, 90, 94, 99

Ebert, R., xvi
Eddington number, xxviii, xxix
Ehlers, J., 143
Ehrenfest, P., 187, 190, 191
Ehrenfest, T., 187, 190, 191
Eidos, 5, 236, 298–300, 303, 306–309
Eigen, M., 221
Einstein space, 95, 108, 118, 120–122, 126, 127, 138, 140, 142, 144, 145, 169
Einstein, A., xiii, xiv, 6, 8, 9, 30, 33–35, 38–53, 55, 72, 84, 102, 120, 132, 133, 137, 140–146, 156, 170, 206, 207, 244–246, 248, 250–252, 256, 257, 259, 260, 262, 264, 266, 277, 280–284, 286–291, 315, 316, 326, 336
Electrodynamics, 25, 35, 45, 51–53, 129, 132, 135, 137, 144, 265
Elementary particle, xxx, 8, 52, 54, 55, 89, 90, 108, 111, 131, 138, 145, 152, 156, 167, 169–171, 178, 217, 278, 301, 312
Ellis, G. F. R., 209
Empiricism, 82, 149, 150, 334, 335
Energy, 34, 300
Energy condition, 159
Entropy, xxix, 4, 34, 167, 169, 187, 190, 192–194, 200–206, 211, 212, 214–217, 219–223, 225, 227, 228, 268, 300, 305, 322
EPR thought experiment, 280, 282, 284, 286, 289–292, 322
Euclid, 36
Event, 2–5, 9, 60, 63, 64, 66, 74–76, 87, 138, 181, 193, 198, 199, 206, 213, 216–218, 268, 270, 274, 279, 304, 324, 326–329, 338
Event horizon, xxix
Everett, H., 249, 286, 293–295, 322, 323
Evolution, xvii, 4, 86, 190, 211, 212, 214, 215, 218–221, 229, 230, 234,

235, 237, 267, 268, 303, 306, 308, 309, 311–314, 333, 336, 340, 341
Expectation value, 4, 60, 62, 64–66, 68, 70, 71, 103, 216, 217
Experience, 1–3, 54, 60, 61, 71, 72, 195, 251, 255, 289, 334, 335, 337
Extremum principle, 3, 6, 28, 185

Facticity, 4, 9, 57, 88, 193, 194, 200, 209, 276, 289, 292, 293, 295, 310, 311, 316, 320, 321, 323, 324, 327, 329, 340
Falsification, 43, 46
Faraday, M., 35
Fermat, P. de, 29
Fermi, E., xiii, 142, 284
Feynman, R. P., 6, 30
Field theory, 46, 48, 55, 70, 116, 129, 132, 133, 137, 142, 143, 146, 169, 206, 245, 254, 259, 260, 278, 283, 288
Finitism, 7, 75, 78, 84, 86–88, 133, 138, 146, 147
Finkelstein, D., xvi, xxv, xxvi, 126, 143, 146, 290
Form, xxx, xxxi, 4, 77, 82, 83, 213, 214, 218, 220–222, 227, 235, 238, 239, 298, 299, 306, 308–310, 331, 338, 340, 341
Franck, J., 52
Franz, H. R., 31
Frege, G., XV, 17
Frequency, 3, 4, 51, 60, 61, 64–68, 98, 186, 188, 191, 192, 213, 214, 218, 245, 246, 255, 263, 269, 270
Friedmann, A. A., 145
Future, 3, 9, 192, 262, 263, 278, 289, 304, 313, 316, 335

Gödel, K., 208
Görnitz, T., xvii, xxiv, 120, 141, 145
Gadamer, H.-G., 314
Galilean transformation, 26, 43
Galilei, G., 23, 26, 38, 40, 47, 265
Gauge group, 8, 113, 129, 135, 136, 145
Gauge theory, xxviii, 136
Gauss, C. F., 36, 46, 141
Geometry, Euclidean, 14, 15, 41, 83
Geometry, non-Euclidean, 6, 15, 36, 83

Gestalt, 213–215, 298, 341
Gibbs, H. W., 34, 187, 191, 204, 205
Glansdorff, P., 218, 220, 221, 223, 227
Goethe, J. W. v., 341
Golden Copenhagen rule, 322, 326
Graudenz, D., xxiv
Gravitation, 15, 22, 131, 145, 156, 245, 291
Green, H. S., 112–114
Gregor-Dellin, M., 93
Grosse, R., xvii
Gupta, S. N., 48, 108, 146, 259

H-theorem, 190–193, 201
Hügel, K., xvii
Habicht, C., 51
Haken, H., xvii, 215
Hamilton, W., 6, 29, 30
Hawking, S. W., xxix, 167, 169, 209
Hegel, G. W. F., 237, 328
Heidegger, M., xv, 287, 304
Heidenreich, W., xvii, 112, 135
Heisenberg, M., 238
Heisenberg, W., xiii–xv, xix, xxxi, 1, 2, 13, 30, 31, 36, 43, 48, 53, 54, 70, 72, 83, 84, 132–135, 137, 244, 246–248, 250–256, 259, 264, 265, 269, 282–284, 288, 293, 311–313, 325, 335, 336
Helmholtz, H. v., 34, 41
Heraclitus, 309
Hertz, G., 51, 52
Heyn, E., xvii
Hidden variable, 253, 291–293, 316
Hoffmann, B., 288
Holographic principle, xxix
Hooft, G. 't, xxix
Hume, D., 2, 67, 86, 207, 335
Husserl, E., 323
Hypothesis, 2, 8, 21, 23, 24, 33, 38, 51, 89, 94, 122, 125–127, 130, 146, 152, 201, 202, 204–206, 219, 244, 245, 248, 249, 274, 275, 316, 318, 339

Idealization, 18, 20, 37, 41, 83, 86
Identity, 6, 21, 42, 77, 90, 124, 125, 133, 134, 137, 212, 254, 299, 300, 327, 328

Indeterminism, 7, 9, 55, 72, 77, 78, 247, 249, 305, 313, 316
Inertia, 6, 23, 38, 44, 45, 84, 96, 141, 182, 339, 341
Infinity, 151, 154, 263, 309
Information, xi, xvi, xxi, xxv, xxviii–xxx, 4, 5, 177, 197, 211–239, 260, 262, 300–306
Information theory, 230, 262
Initial condition, 19–23, 29, 50, 185, 186, 218, 248
Interaction, xxviii, 85, 90, 266, 272, 324, 329
Interpretation of physics, xi, xv, xxxi, 2, 4, 5, 8–10, 15, 17, 18, 32, 34, 38, 42–44, 48, 55–57, 59, 60, 114, 133, 203, 204, 211, 212, 241, 243, 244, 247–249, 253, 255, 259–262, 264, 265, 268, 269, 275–278, 283, 284, 291–294, 297, 313–315, 321–323
Invariance, xxv, 31, 42, 47–49, 84, 98, 107, 121, 126, 136, 144, 145, 206
Irreversibility, xvi, 4, 9, 15, 28, 34, 50, 86, 182, 185–187, 190, 191, 197, 201, 208, 211, 215, 218, 266–268, 278, 279, 294, 304, 310, 313, 316, 320–323, 327, 328

Jacob, P., xvii, 112, 310
James, W., 234, 325
Jammer, M., 244, 247–249, 256, 259, 260, 282, 283, 286, 291–293
Joos, H., 135
Jordan, P., xiv, 147, 156, 254, 259, 344
Jung, C. G., 32

Künemund, T., xvii, 112
Küppers, B. O., xvii, 220
Kaluza, T., 146, 151
Kant, I., xv, xxi, 2, 3, 27, 29, 33, 41, 42, 50, 72, 82, 86, 93, 108, 212, 228, 237, 256, 258, 271, 300, 304, 307, 308, 334–337
Kepler, J., 29, 30, 38, 39
Kinematics, 55, 79, 83, 84, 251, 330
Kirchhoff, G. R., 51
Klein, F., 30, 83, 141, 146, 151
Kornwachs, K., xvii

Kramers, H. A., 245, 246, 248, 263
Kreisgang, xxii, 5, 10, 71, 82, 94, 212, 234, 235, 308, 331, 333, 346
Kuhn, T. S., 1, 13, 313, 335, 336
Kunsemüller, H., xvi

Lakatos, I., 59
Lambert, J. H., 36
Landau, L., 203
Lange, L., 41, 42
Laplace, P. S. de, 50, 66, 67, 212, 228
Lattice, xxii, 3, 6, 7, 64, 65, 67, 73–75, 78
Law of inertia, 29, 43, 44, 95, 96, 182, 184
Laws of nature, form of, 3, 14, 28, 29, 37, 50, 54, 300, 340
Lehmeier, T., xvii
Leibniz, G. W., 26, 38–42, 44–46, 143, 146, 339
Lenard, P., 51
Lepton, 123, 125, 126, 128–132, 135, 136
Leucippus, 25
Lie, S., 30, 141
Life, 1, 211, 215, 219, 232, 234, 236, 299, 303, 308, 310, 333, 339–341
Lindemann, F. v., 132
Lobachevsky, N. I., 141
Logic, xiii, xiv, xvi, xx, xxii, 3, 4, 6, 10, 16, 17, 39, 55, 63, 212, 234, 236, 239, 308, 311, 314, 327, 331, 333, 334, 345
Lorentz group, xxiii, 42, 82, 98, 102, 126, 127, 171
Lorentz invariance, 6, 42, 43, 49, 120, 121, 126, 132, 206
Lorentz, H. A., 206
Lorenz, K., 82, 234, 335, 336
Ludwig, G., 41, 156, 265

Mach's principle, 45, 48
Mach, E., 26, 30, 38, 39, 41, 42, 44–46, 48, 143, 251
Macrostate, 189, 190, 198–200, 202, 204, 205, 214, 216, 217, 222, 224, 268, 322
Magellan, 94
Maxwell field, 127–129, 131, 136, 247

Maxwell, J. C., 35, 45, 51, 52, 132, 265, 287
McTaggart, J. M. E., xxiii
Measurement, xxiii, xxviii, 6, 22, 24, 43, 49, 51, 61, 62, 69, 83, 86, 95, 102, 104, 138, 139, 146, 164, 249, 252, 257, 258, 260–268, 270–276, 281, 282, 284–286, 289, 291, 292, 294, 304–306, 314, 317, 321, 322, 324, 326, 336
Mehra, J., 46
Meré, A. G. de, 59
Metaphysics, 10, 243, 287–289, 315, 330, 331, 334, 340–346
Metric field, 6, 45, 48, 141, 143, 145
Meyer-Abich, K. M., xvi, 245, 256
Microstate, 189, 190, 199, 200, 203–205, 214, 216, 217, 222, 224, 227, 268, 322
Mind, 5, 213, 275, 299, 306, 307, 331
Mind–body problem, 269, 299
Minkowski space, xxiv, 8, 43, 48, 81, 82, 101, 103, 108, 109, 118–122, 125–127, 137, 141, 142, 145, 146, 154, 169, 171, 173, 175, 207
Minkowski, H., 44, 206
Mittelstaedt, P., 39, 48, 207
Modality, 3, 63, 212, 316, 320, 322, 327, 328, 331
Myth, 265, 314, 339

Napoleon, 39
Negentropy, 216, 217
Neumann, J. v., 6, 244, 258, 259, 266
Newton, I., 25–27, 29, 35, 38–41, 43–45, 50, 141, 145, 221, 265, 287
Nietzsche, F., 237
Nostradamus, 317–319
Novelty, 229, 230, 232, 233

Object, 7, 18, 19, 74–75, 269, 289, 328
Observer, 43, 66, 67, 86, 87, 102, 103, 120, 181, 202, 227, 232, 233, 238, 260–265, 267–271, 274, 275, 279, 281, 283–286, 290, 294, 299, 304, 307, 309, 310, 314, 322, 326
Ontology, xxx, xxxi, 44, 280, 325, 333
Ousia, 298

Pais, A., 259
Para-Bose operator, 116, 172
Paradoxes, 276–278
Parmenides, 25, 287, 288, 315, 342, 343
Pascal, B., 306, 310
Past, 2, 9, 192, 262, 263, 289, 311, 316, 317
Pauli, W., xiii, 135, 247, 248, 252, 259, 282, 284, 344
Penrose, R., xxv, xxvi
Perception, xxii, 4, 27, 39, 48, 72, 82, 83, 90, 93, 238, 248, 250, 256, 271, 280, 299, 303, 308, 314, 318, 319, 321, 327, 331, 335, 341, 344, 345
Petersen, A., 257
Phase space, 18, 20, 29, 47, 183, 305
Phenomenon, 256, 269, 290, 299, 309, 312, 339
Philosophy, xi, xii, xv, xix, xx, xxii, 1, 5, 25, 30–33, 39, 46, 71, 106, 149, 213, 234, 255, 257, 277, 286, 287, 298, 299, 306–309, 311, 315, 325, 331, 333–336, 338–342, 344, 345
Philosophy of science, xxx, xxxi, 219, 334
Picht, G., 25, 88, 288, 300, 304, 315, 343
Planck length, xxix, 157, 158
Planck, M., xiii, 51, 52
Plato, XV, 26, 30, 236, 237, 250, 287, 288, 298, 299, 307, 309, 335, 342, 343
Podolsky, B., xiv, 142, 244, 256, 277, 282, 283, 286, 289, 326
Poincaré group, xxiv, 8, 83, 85, 101, 102, 107, 108, 114, 127, 132, 134, 169–171, 173, 174
Poincaré, H., 205
Point mass, 6, 16, 18–22, 24, 25, 27, 36, 40, 73, 85, 131, 132, 141, 187, 254, 288, 289, 299–301, 305
Popper, K. R., 46, 234, 235, 289, 293, 335
Position space, xxi, xxvii, 6, 8, 9, 81, 84, 95, 98, 141, 151–153, 156, 160, 171, 305
Positivism, 251, 256, 277, 284
Possibility, 4, 19, 26, 28, 41, 54, 57, 67, 85, 194, 196, 200, 237, 262, 289,

292–294, 316, 320–322, 324, 327, 329, 330, 337, 340, 343, 346
Present, 3, 85–87, 182, 191, 193, 194, 202, 206, 207, 279, 309, 311, 315, 317, 327, 329, 343
Prigogine, I., 218, 220, 221, 223, 227
Probability, xiii, 2–4, 6, 34, 54, 56, 57, 59, 79, 213, 215–218, 246–248, 261
Property, xxx, 5, 19, 74, 77, 89, 172, 184, 213, 214, 238, 272, 301, 328
Proposition, xiii, 3, 6, 7, 16, 28, 40, 46, 55, 63, 64, 73–78, 85, 86, 207, 257, 275, 314, 327, 337
Prout, W., 33
Psyche, 299, 344

Quantization, xxvii, 6, 30, 46, 55, 70, 71, 92, 98, 116, 123–125, 127, 128, 142, 143, 171, 253–255
  multiple, xx, xxvii, xxviii, 6, 116, 132, 170
Quantum electrodynamics, xxvi, 8, 108, 123, 129, 130, 140, 282
Quantum field theory, xxiv, xxvii, 84, 153, 156, 171
Quantum gravity, xxvi, xxxi, 145
Quantum information, xxv, 163, 167, 176, 177
Quantum information theory, xxi, xxiii, xxiv
Quantum logic, xxii, xxiii, 7
Quantum theory, 1, 2, 4–8, 14, 51–53, 78, 243–295
Quark, 132, 133, 135, 136, 152
Quasiparticle, 116, 117, 122, 129, 134, 144
Qubit, xxi, xxiii, xxviii, 153, 176

Ray representation, 113
Realism, 82, 251, 256, 277, 284, 287, 289, 293, 334, 335
  structural, xxxi
Reality, 248, 260, 277, 286–289, 295, 334
Reconstruction, 4, 54, 71, 78, 254, 259
Reduction of wave packet, 249, 252, 261, 267, 268, 275, 281, 293, 295, 321–323
Refutation, 22
Reichenbach, H., 187

Relativity, general, xvii, 6, 8, 15, 27, 39, 40, 42–47, 50, 54, 55, 57, 82, 84, 95, 102, 106–108, 120–122, 141–145, 149–151, 159, 168, 201, 207, 208, 246, 260, 288, 289, 291
Relativity, special, xiii, xxiv, 2, 6, 8, 14, 15, 24, 28, 34, 35, 40–45, 47–49, 52, 54, 55, 57, 75, 81, 83, 84, 86, 94, 96, 98, 99, 101–103, 105, 107, 126, 133, 134, 141, 146, 154, 155, 206, 207, 252, 266, 278, 288, 300, 304, 305
Religion, xix, 315, 342
Rest mass, 8, 102, 108, 116, 118, 122, 127, 133–135, 137, 138, 140, 144, 145, 147, 167, 172, 176, 177, 303
Retrodiction, 263, 281, 317
Reversal of arguments, 55, 83, 84, 106, 121, 142, 144, 150
Reversibility, 182, 184, 185, 194, 205
Riemann, B., 46, 141
Roman, P., xvii
Rosen, N., xiv, 142, 244, 256, 277, 282, 283, 286, 289, 326
Rovelli, C., xxvi
Rubens, H., 51
Rudolph, E., 342
Russell, B., 39, 40
Rutherford, E., xiii, 52, 137

Süßmann, G., xvi, xx, xxvii, 123, 324
Saccheri, G. G., 36
Sartre, J.-P., 325
Scheibe, E., xvi, xx, xxvii, 123, 256
Schiller, F., 297
Schmutzer, E., 146
Scholem, G., 237
Schopenhauer, A., 93
Schrödinger's cat, 278, 279
Schrödinger, E., xiv, 53, 142, 246, 247, 254, 278, 279
Schwarzschild radius, 168, 169
Scientific revolution, 1, 13, 33, 35, 106, 313, 331, 335, 336, 339
Segal, J. E., 119
Semantic consistency, 14, 18, 35–37, 41–43, 107, 234, 258, 260–262,

264, 268, 269, 276, 284, 294, 312, 313, 337, 346
Semantics, 16, 18, 20, 22, 36, 37, 200, 229, 230, 252, 302, 312
Separability, 2, 3, 7, 10, 83, 87, 129, 130, 266, 308, 309, 321, 322, 328, 330, 341
Shannon, C. E., 4, 212, 214, 216, 222, 230, 232, 233
Simplicity, 34, 47, 48, 312, 336
Slater, J. G., 245, 246, 248, 263
Smolin, L., xxvi
Snyder, H. S., 146
Socrates, xiv, 346
Solar eclipse, 196, 317, 318
Sommerfeld, A., xiv, 132, 250
Spacetime continuum, xvii, xxviii, 8, 43–45, 47–49, 55, 79, 81, 82, 94, 95, 109, 117, 122, 133, 134, 141, 146, 201, 207, 288, 289
Spatiality, 81–83
Spencer, H., 219, 228
Spinorism, xxi, xxv, xxvi
Spinoza, B., 288, 315
Strawson, P. F., 238
Subject, 27, 39, 143, 261, 269, 271, 289, 307, 308
Substance, xxx, 5, 31–33, 39, 40, 42, 45, 46, 48, 52, 90, 143, 172, 297–301, 303, 304, 306–308, 338, 340
Supernova Ia, 164
Superposition, 77, 102, 103, 107, 153, 247, 282, 294, 304
Supersymmetry, 92, 111, 132
Symmetry, xi, xxi, xxiii–xxv, xxviii, 3, 8, 10, 28, 48, 67, 69, 81, 83, 84, 91, 93–95, 98, 113–115, 117, 118, 129, 133, 134, 142, 143, 151, 152, 154, 191, 202, 239, 247, 250, 266, 290, 291, 300, 304, 328, 340

Tataru-Mihaj, P., xvii
Teller, E., 257, 265, 266
Temporality, xxiii
Tensor space of urs, 91, 92, 97, 100, 104, 108, 109, 111, 113–116, 121, 127, 133, 136, 139, 142, 143, 145, 146

Tetrad, xxvii
Thales, 196, 317, 318
Thermodynamics, xiv, xxix, xxxi, 4, 5, 14, 15, 33, 34, 51, 54–56, 106, 158, 184, 186, 188, 193, 204, 211, 218–220, 266, 268, 291, 300, 301, 317, 326
  second law, xiv, 4, 15, 34, 186, 193–195, 197–206, 208, 209, 212, 218–223, 228, 300
Thirring, W., 48, 108, 146
Thought experiment, xiv, 142, 256, 260, 264, 274, 277–282, 315, 326
Time, xi, xxiii, xxxi, 3, 4, 13, 24, 27, 28, 34, 56, 57, 86, 88, 93, 96, 99, 102, 121, 141, 147, 158, 193–195, 200, 201, 204, 206, 208, 221, 222, 266, 279, 288, 289, 297, 298, 304, 306, 309, 312, 315, 316, 329, 334, 337, 339, 343
  direction of, 190, 194, 201, 202, 204, 206
  modes of, xxiii, 6, 134, 194, 289
Trieste theory, 323, 326
Truth, 1–3, 10, 38, 63, 195, 214, 234–239, 289, 298, 310, 315, 323, 342


Uncertainty relation, xiv, 9, 43, 78, 79, 248, 251–253, 265, 292, 322, 328, 330, *see* Uncertainty relation
Unity of Nature, xxxii
Unity of nature, xvi, 2, 236, 338, 341
Unity of physics, xi, xiii, xv, 2, 6, 11, 13, 27, 105, 243
Universe, 106
Ur hypothesis, 91
Ur theory, xi, xxiii, 98, 140
Ur alternative, xxiii
Ur hypothesis, xxiii, 8, 9, 56, 91, 93, 94, 98, 99, 103, 108, 111, 130, 133, 136, 144, 151, 177, 255, 259, 266, 278, 290
Ur theory, xi, xvii, xix–xxi, xxiii–xxxi, 30, 101, 108, 112, 120, 121, 133–137, 140, 144, 146, 150–152, 161, 171

Vacuum, xxiv, 26, 48, 49, 92, 96, 112–115, 161, 168, 170, 173–176, 305
Void, 25, 26, 342

Waerden, B. L. v. d., 37, 251
Wagner, C., 93
Wagner, R., 93
Wave mechanics, 6, 15, 30, 31, 254, 258
Wave picture, 54, 253, 254
Wave–particle duality, 254, 255
Weizsäcker, C. v., xvi, 212, 229, 230, 232
Weizsäcker, E. v., xvi, 212, 229, 230, 232
Weizsäcker, V. v., 325
Wesendonck, M., 93
Weyl equation, 123, 125
Weyl, H., 45, 46
Wheeler, J. A., xxv, 281, 290
Wigner's friend, 279
Wigner, E. P., 29, 75, 127, 132, 254, 267, 279
WMAP, 166
Wolff, M., 341
World, 49–51

Zucker, F. J., xvi

# Carl Friedrich von Weizsäcker Foundation

## Preamble

In Carl Friedrich von Weizsäcker the nowadays more than ever necessary intercultural and interdisciplinary dialog has found one of its most important proponents. He is one of the few great thinkers who combine the perspectives of science, philosophy, religion, and politics with a view toward the challenges but also the responsibilities of our times.

Two quotes characterize the intention and manner of his endeavor:
"*I am ready to criticize a position if I could equally defend it*" — "*Our ethics must not stay behind the development of our technology, our observing reason not behind our analytical intellect, our love not behind our power.*"

What will we do? What do we need to know? What do we know already?
To develop effective longterm strategies which contribute to recognizing, planning, and implementing the necessary approaches in the tension of challenge and responsibility—this is the aim of the Carl Friedrich von Weizsäcker-Foundation along the lines of the concerns of Carl Friedrich von Weizsäcker.

If you want to know more about the Carl Friedrich von Weizsäcker Society "Wissen und Verantwortung e.V." and the Carl Friedrich von Weizsäcker-Foundation, please write to:

Carl Friedrich von Weizsäcker-Stiftung
Bielefelder Straße 8
D-32130 Enger
fax: 05224/977 898
e-mail: stiftung@cfvw.de

| *Executive Directors* | *Board* |
|---|---|
| Dr. Bruno Redeker | Dr. Walter Kroy |
| Bernhard Winzinger | Prof. Dr. Thomas Görnitz |
|  | Bishop Dr. Reinhard Marx |

# Fundamental Theories of Physics

*Series Editor: Alwyn van der Merwe, University of Denver, USA*

1. M. Sachs: *General Relativity and Matter.* A Spinor Field Theory from Fermis to Light-Years. With a Foreword by C. Kilmister. 1982  ISBN 90-277-1381-2
2. G.H. Duffey: *A Development of Quantum Mechanics.* Based on Symmetry Considerations. 1985  ISBN 90-277-1587-4
3. S. Diner, D. Fargue, G. Lochak and F. Selleri (eds.): *The Wave-Particle Dualism.* A Tribute to Louis de Broglie on his 90th Birthday. 1984  ISBN 90-277-1664-1
4. E. Prugovečki: *Stochastic Quantum Mechanics and Quantum Spacetime.* A Consistent Unification of Relativity and Quantum Theory based on Stochastic Spaces. 1984; 2nd printing 1986  ISBN 90-277-1617-X
5. D. Hestenes and G. Sobczyk: *Clifford Algebra to Geometric Calculus.* A Unified Language for Mathematics and Physics. 1984  ISBN 90-277-1673-0; Pb (1987) 90-277-2561-6
6. P. Exner: *Open Quantum Systems and Feynman Integrals.* 1985  ISBN 90-277-1678-1
7. L. Mayants: *The Enigma of Probability and Physics.* 1984  ISBN 90-277-1674-9
8. E. Tocaci: *Relativistic Mechanics, Time and Inertia.* Translated from Romanian. Edited and with a Foreword by C.W. Kilmister. 1985  ISBN 90-277-1769-9
9. B. Bertotti, F. de Felice and A. Pascolini (eds.): *General Relativity and Gravitation.* Proceedings of the 10th International Conference (Padova, Italy, 1983). 1984  ISBN 90-277-1819-9
10. G. Tarozzi and A. van der Merwe (eds.): *Open Questions in Quantum Physics.* 1985  ISBN 90-277-1853-9
11. J.V. Narlikar and T. Padmanabhan: *Gravity, Gauge Theories and Quantum Cosmology.* 1986  ISBN 90-277-1948-9
12. G.S. Asanov: *Finsler Geometry, Relativity and Gauge Theories.* 1985  ISBN 90-277-1960-8
13. K. Namsrai: *Nonlocal Quantum Field Theory and Stochastic Quantum Mechanics.* 1986  ISBN 90-277-2001-0
14. C. Ray Smith and W.T. Grandy, Jr. (eds.): *Maximum-Entropy and Bayesian Methods in Inverse Problems.* Proceedings of the 1st and 2nd International Workshop (Laramie, Wyoming, USA). 1985  ISBN 90-277-2074-6
15. D. Hestenes: *New Foundations for Classical Mechanics.* 1986  ISBN 90-277-2090-8; Pb (1987) 90-277-2526-8
16. S.J. Prokhovnik: *Light in Einstein's Universe.* The Role of Energy in Cosmology and Relativity. 1985  ISBN 90-277-2093-2
17. Y.S. Kim and M.E. Noz: *Theory and Applications of the Poincaré Group.* 1986  ISBN 90-277-2141-6
18. M. Sachs: *Quantum Mechanics from General Relativity.* An Approximation for a Theory of Inertia. 1986  ISBN 90-277-2247-1
19. W.T. Grandy, Jr.: *Foundations of Statistical Mechanics.* Vol. I: *Equilibrium Theory.* 1987  ISBN 90-277-2489-X
20. H.-H von Borzeszkowski and H.-J. Treder: *The Meaning of Quantum Gravity.* 1988  ISBN 90-277-2518-7
21. C. Ray Smith and G.J. Erickson (eds.): *Maximum-Entropy and Bayesian Spectral Analysis and Estimation Problems.* Proceedings of the 3rd International Workshop (Laramie, Wyoming, USA, 1983). 1987  ISBN 90-277-2579-9
22. A.O. Barut and A. van der Merwe (eds.): *Selected Scientific Papers of Alfred Landé.* [*1888-1975*]. 1988  ISBN 90-277-2594-2

# Fundamental Theories of Physics

23. W.T. Grandy, Jr.: *Foundations of Statistical Mechanics.* Vol. II: *Nonequilibrium Phenomena.* 1988 ISBN 90-277-2649-3
24. E.I. Bitsakis and C.A. Nicolaides (eds.): *The Concept of Probability.* Proceedings of the Delphi Conference (Delphi, Greece, 1987). 1989 ISBN 90-277-2679-5
25. A. van der Merwe, F. Selleri and G. Tarozzi (eds.): *Microphysical Reality and Quantum Formalism, Vol. 1.* Proceedings of the International Conference (Urbino, Italy, 1985). 1988 ISBN 90-277-2683-3
26. A. van der Merwe, F. Selleri and G. Tarozzi (eds.): *Microphysical Reality and Quantum Formalism, Vol. 2.* Proceedings of the International Conference (Urbino, Italy, 1985). 1988 ISBN 90-277-2684-1
27. I.D. Novikov and V.P. Frolov: *Physics of Black Holes.* 1989 ISBN 90-277-2685-X
28. G. Tarozzi and A. van der Merwe (eds.): *The Nature of Quantum Paradoxes.* Italian Studies in the Foundations and Philosophy of Modern Physics. 1988 ISBN 90-277-2703-1
29. B.R. Iyer, N. Mukunda and C.V. Vishveshwara (eds.): *Gravitation, Gauge Theories and the Early Universe.* 1989 ISBN 90-277-2710-4
30. H. Mark and L. Wood (eds.): *Energy in Physics, War and Peace.* A Festschrift celebrating Edward Teller's 80th Birthday. 1988 ISBN 90-277-2775-9
31. G.J. Erickson and C.R. Smith (eds.): *Maximum-Entropy and Bayesian Methods in Science and Engineering.* Vol. I: *Foundations.* 1988 ISBN 90-277-2793-7
32. G.J. Erickson and C.R. Smith (eds.): *Maximum-Entropy and Bayesian Methods in Science and Engineering.* Vol. II: *Applications.* 1988 ISBN 90-277-2794-5
33. M.E. Noz and Y.S. Kim (eds.): *Special Relativity and Quantum Theory.* A Collection of Papers on the Poincaré Group. 1988 ISBN 90-277-2799-6
34. I.Yu. Kobzarev and Yu.I. Manin: *Elementary Particles. Mathematics, Physics and Philosophy.* 1989 ISBN 0-7923-0098-X
35. F. Selleri: *Quantum Paradoxes and Physical Reality.* 1990 ISBN 0-7923-0253-2
36. J. Skilling (ed.): *Maximum-Entropy and Bayesian Methods.* Proceedings of the 8th International Workshop (Cambridge, UK, 1988). 1989 ISBN 0-7923-0224-9
37. M. Kafatos (ed.): *Bell's Theorem, Quantum Theory and Conceptions of the Universe.* 1989 ISBN 0-7923-0496-9
38. Yu.A. Izyumov and V.N. Syromyatnikov: *Phase Transitions and Crystal Symmetry.* 1990 ISBN 0-7923-0542-6
39. P.F. Fougère (ed.): *Maximum-Entropy and Bayesian Methods.* Proceedings of the 9th International Workshop (Dartmouth, Massachusetts, USA, 1989). 1990 ISBN 0-7923-0928-6
40. L. de Broglie: *Heisenberg's Uncertainties and the Probabilistic Interpretation of Wave Mechanics.* With Critical Notes of the Author. 1990 ISBN 0-7923-0929-4
41. W.T. Grandy, Jr.: *Relativistic Quantum Mechanics of Leptons and Fields.* 1991 ISBN 0-7923-1049-7
42. Yu.L. Klimontovich: *Turbulent Motion and the Structure of Chaos.* A New Approach to the Statistical Theory of Open Systems. 1991 ISBN 0-7923-1114-0
43. W.T. Grandy, Jr. and L.H. Schick (eds.): *Maximum-Entropy and Bayesian Methods.* Proceedings of the 10th International Workshop (Laramie, Wyoming, USA, 1990). 1991 ISBN 0-7923-1140-X
44. P. Pták and S. Pulmannová: *Orthomodular Structures as Quantum Logics.* Intrinsic Properties, State Space and Probabilistic Topics. 1991 ISBN 0-7923-1207-4
45. D. Hestenes and A. Weingartshofer (eds.): *The Electron.* New Theory and Experiment. 1991 ISBN 0-7923-1356-9

# Fundamental Theories of Physics

46. P.P.J.M. Schram: *Kinetic Theory of Gases and Plasmas.* 1991   ISBN 0-7923-1392-5
47. A. Micali, R. Boudet and J. Helmstetter (eds.): *Clifford Algebras and their Applications in Mathematical Physics.* 1992   ISBN 0-7923-1623-1
48. E. Prugovečki: *Quantum Geometry.* A Framework for Quantum General Relativity. 1992
    ISBN 0-7923-1640-1
49. M.H. Mac Gregor: *The Enigmatic Electron.* 1992   ISBN 0-7923-1982-6
50. C.R. Smith, G.J. Erickson and P.O. Neudorfer (eds.): *Maximum Entropy and Bayesian Methods.* Proceedings of the 11th International Workshop (Seattle, 1991). 1993   ISBN 0-7923-2031-X
51. D.J. Hoekzema: *The Quantum Labyrinth.* 1993   ISBN 0-7923-2066-2
52. Z. Oziewicz, B. Jancewicz and A. Borowiec (eds.): *Spinors, Twistors, Clifford Algebras and Quantum Deformations.* Proceedings of the Second Max Born Symposium (Wrocław, Poland, 1992). 1993   ISBN 0-7923-2251-7
53. A. Mohammad-Djafari and G. Demoment (eds.): *Maximum Entropy and Bayesian Methods.* Proceedings of the 12th International Workshop (Paris, France, 1992). 1993
    ISBN 0-7923-2280-0
54. M. Riesz: *Clifford Numbers and Spinors* with Riesz' Private Lectures to E. Folke Bolinder and a Historical Review by Pertti Lounesto. E.F. Bolinder and P. Lounesto (eds.). 1993
    ISBN 0-7923-2299-1
55. F. Brackx, R. Delanghe and H. Serras (eds.): *Clifford Algebras and their Applications in Mathematical Physics.* Proceedings of the Third Conference (Deinze, 1993) 1993
    ISBN 0-7923-2347-5
56. J.R. Fanchi: *Parametrized Relativistic Quantum Theory.* 1993   ISBN 0-7923-2376-9
57. A. Peres: *Quantum Theory: Concepts and Methods.* 1993   ISBN 0-7923-2549-4
58. P.L. Antonelli, R.S. Ingarden and M. Matsumoto: *The Theory of Sprays and Finsler Spaces with Applications in Physics and Biology.* 1993   ISBN 0-7923-2577-X
59. R. Miron and M. Anastasiei: *The Geometry of Lagrange Spaces: Theory and Applications.* 1994   ISBN 0-7923-2591-5
60. G. Adomian: *Solving Frontier Problems of Physics: The Decomposition Method.* 1994
    ISBN 0-7923-2644-X
61. B.S. Kerner and V.V. Osipov: *Autosolitons.* A New Approach to Problems of Self-Organization and Turbulence. 1994   ISBN 0-7923-2816-7
62. G.R. Heidbreder (ed.): *Maximum Entropy and Bayesian Methods.* Proceedings of the 13th International Workshop (Santa Barbara, USA, 1993) 1996   ISBN 0-7923-2851-5
63. J. Peřina, Z. Hradil and B. Jurčo: *Quantum Optics and Fundamentals of Physics.* 1994
    ISBN 0-7923-3000-5
64. M. Evans and J.-P. Vigier: *The Enigmatic Photon.* Volume 1: The Field $B^{(3)}$. 1994
    ISBN 0-7923-3049-8
65. C.K. Raju: *Time: Towards a Constistent Theory.* 1994   ISBN 0-7923-3103-6
66. A.K.T. Assis: *Weber's Electrodynamics.* 1994   ISBN 0-7923-3137-0
67. Yu. L. Klimontovich: *Statistical Theory of Open Systems.* Volume 1: A Unified Approach to Kinetic Description of Processes in Active Systems. 1995   ISBN 0-7923-3199-0;
    Pb: ISBN 0-7923-3242-3
68. M. Evans and J.-P. Vigier: *The Enigmatic Photon.* Volume 2: Non-Abelian Electrodynamics. 1995   ISBN 0-7923-3288-1
69. G. Esposito: *Complex General Relativity.* 1995   ISBN 0-7923-3340-3

# Fundamental Theories of Physics

70. J. Skilling and S. Sibisi (eds.): *Maximum Entropy and Bayesian Methods.* Proceedings of the Fourteenth International Workshop on Maximum Entropy and Bayesian Methods. 1996
ISBN 0-7923-3452-3
71. C. Garola and A. Rossi (eds.): *The Foundations of Quantum Mechanics Historical Analysis and Open Questions.* 1995   ISBN 0-7923-3480-9
72. A. Peres: *Quantum Theory: Concepts and Methods.* 1995 (see for hardback edition, Vol. 57)
ISBN Pb 0-7923-3632-1
73. M. Ferrero and A. van der Merwe (eds.): *Fundamental Problems in Quantum Physics.* 1995
ISBN 0-7923-3670-4
74. F.E. Schroeck, Jr.: *Quantum Mechanics on Phase Space.* 1996   ISBN 0-7923-3794-8
75. L. de la Peña and A.M. Cetto: *The Quantum Dice.* An Introduction to Stochastic Electrodynamics. 1996   ISBN 0-7923-3818-9
76. P.L. Antonelli and R. Miron (eds.): *Lagrange and Finsler Geometry.* Applications to Physics and Biology. 1996   ISBN 0-7923-3873-1
77. M.W. Evans, J.-P. Vigier, S. Roy and S. Jeffers: *The Enigmatic Photon.* Volume 3: Theory and Practice of the $B^{(3)}$ Field. 1996   ISBN 0-7923-4044-2
78. W.G.V. Rosser: *Interpretation of Classical Electromagnetism.* 1996   ISBN 0-7923-4187-2
79. K.M. Hanson and R.N. Silver (eds.): *Maximum Entropy and Bayesian Methods.* 1996
ISBN 0-7923-4311-5
80. S. Jeffers, S. Roy, J.-P. Vigier and G. Hunter (eds.): *The Present Status of the Quantum Theory of Light.* Proceedings of a Symposium in Honour of Jean-Pierre Vigier. 1997
ISBN 0-7923-4337-9
81. M. Ferrero and A. van der Merwe (eds.): *New Developments on Fundamental Problems in Quantum Physics.* 1997   ISBN 0-7923-4374-3
82. R. Miron: *The Geometry of Higher-Order Lagrange Spaces.* Applications to Mechanics and Physics. 1997   ISBN 0-7923-4393-X
83. T. Hakioğlu and A.S. Shumovsky (eds.): *Quantum Optics and the Spectroscopy of Solids.* Concepts and Advances. 1997   ISBN 0-7923-4414-6
84. A. Sitenko and V. Tartakovskii: *Theory of Nucleus.* Nuclear Structure and Nuclear Interaction. 1997   ISBN 0-7923-4423-5
85. G. Esposito, A.Yu. Kamenshchik and G. Pollifrone: *Euclidean Quantum Gravity on Manifolds with Boundary.* 1997   ISBN 0-7923-4472-3
86. R.S. Ingarden, A. Kossakowski and M. Ohya: *Information Dynamics and Open Systems.* Classical and Quantum Approach. 1997   ISBN 0-7923-4473-1
87. K. Nakamura: *Quantum versus Chaos.* Questions Emerging from Mesoscopic Cosmos. 1997
ISBN 0-7923-4557-6
88. B.R. Iyer and C.V. Vishveshwara (eds.): *Geometry, Fields and Cosmology.* Techniques and Applications. 1997   ISBN 0-7923-4725-0
89. G.A. Martynov: *Classical Statistical Mechanics.* 1997   ISBN 0-7923-4774-9
90. M.W. Evans, J.-P. Vigier, S. Roy and G. Hunter (eds.): *The Enigmatic Photon.* Volume 4: New Directions. 1998   ISBN 0-7923-4826-5
91. M. Rédei: *Quantum Logic in Algebraic Approach.* 1998   ISBN 0-7923-4903-2
92. S. Roy: *Statistical Geometry and Applications to Microphysics and Cosmology.* 1998
ISBN 0-7923-4907-5
93. B.C. Eu: *Nonequilibrium Statistical Mechanics.* Ensembled Method. 1998
ISBN 0-7923-4980-6

# Fundamental Theories of Physics

94. V. Dietrich, K. Habetha and G. Jank (eds.): *Clifford Algebras and Their Application in Mathematical Physics.* Aachen 1996. 1998 ISBN 0-7923-5037-5
95. J.P. Blaizot, X. Campi and M. Ploszajczak (eds.): *Nuclear Matter in Different Phases and Transitions.* 1999 ISBN 0-7923-5660-8
96. V.P. Frolov and I.D. Novikov: *Black Hole Physics.* Basic Concepts and New Developments. 1998 ISBN 0-7923-5145-2; Pb 0-7923-5146
97. G. Hunter, S. Jeffers and J-P. Vigier (eds.): *Causality and Locality in Modern Physics.* 1998 ISBN 0-7923-5227-0
98. G.J. Erickson, J.T. Rychert and C.R. Smith (eds.): *Maximum Entropy and Bayesian Methods.* 1998 ISBN 0-7923-5047-2
99. D. Hestenes: *New Foundations for Classical Mechanics (Second Edition).* 1999 ISBN 0-7923-5302-1; Pb ISBN 0-7923-5514-8
100. B.R. Iyer and B. Bhawal (eds.): *Black Holes, Gravitational Radiation and the Universe.* Essays in Honor of C. V. Vishveshwara. 1999 ISBN 0-7923-5308-0
101. P.L. Antonelli and T.J. Zastawniak: *Fundamentals of Finslerian Diffusion with Applications.* 1998 ISBN 0-7923-5511-3
102. H. Atmanspacher, A. Amann and U. Müller-Herold: *On Quanta, Mind and Matter Hans Primas in Context.* 1999 ISBN 0-7923-5696-9
103. M.A. Trump and W.C. Schieve: *Classical Relativistic Many-Body Dynamics.* 1999 ISBN 0-7923-5737-X
104. A.I. Maimistov and A.M. Basharov: *Nonlinear Optical Waves.* 1999 ISBN 0-7923-5752-3
105. W. von der Linden, V. Dose, R. Fischer and R. Preuss (eds.): *Maximum Entropy and Bayesian Methods Garching, Germany 1998.* 1999 ISBN 0-7923-5766-3
106. M.W. Evans: *The Enigmatic Photon Volume 5: O(3) Electrodynamics.* 1999 ISBN 0-7923-5792-2
107. G.N. Afanasiev: *Topological Effects in Quantum Mecvhanics.* 1999 ISBN 0-7923-5800-7
108. V. Devanathan: *Angular Momentum Techniques in Quantum Mechanics.* 1999 ISBN 0-7923-5866-X
109. P.L. Antonelli (ed.): *Finslerian Geometries A Meeting of Minds.* 1999 ISBN 0-7923-6115-6
110. M.B. Mensky: *Quantum Measurements and Decoherence Models and Phenomenology.* 2000 ISBN 0-7923-6227-6
111. B. Coecke, D. Moore and A. Wilce (eds.): *Current Research in Operation Quantum Logic.* Algebras, Categories, Languages. 2000 ISBN 0-7923-6258-6
112. G. Jumarie: *Maximum Entropy, Information Without Probability and Complex Fractals.* Classical and Quantum Approach. 2000 ISBN 0-7923-6330-2
113. B. Fain: *Irreversibilities in Quantum Mechanics.* 2000 ISBN 0-7923-6581-X
114. T. Borne, G. Lochak and H. Stumpf: *Nonperturbative Quantum Field Theory and the Structure of Matter.* 2001 ISBN 0-7923-6803-7
115. J. Keller: *Theory of the Electron.* A Theory of Matter from START. 2001 ISBN 0-7923-6819-3
116. M. Rivas: *Kinematical Theory of Spinning Particles.* Classical and Quantum Mechanical Formalism of Elementary Particles. 2001 ISBN 0-7923-6824-X
117. A.A. Ungar: *Beyond the Einstein Addition Law and its Gyroscopic Thomas Precession.* The Theory of Gyrogroups and Gyrovector Spaces. 2001 ISBN 0-7923-6909-2
118. R. Miron, D. Hrimiuc, H. Shimada and S.V. Sabau: *The Geometry of Hamilton and Lagrange Spaces.* 2001 ISBN 0-7923-6926-2

# Fundamental Theories of Physics

119.  M. Pavšič: *The Landscape of Theoretical Physics: A Global View*. From Point Particles to the Brane World and Beyond in Search of a Unifying Principle. 2001    ISBN 0-7923-7006-6
120.  R.M. Santilli: *Foundations of Hadronic Chemistry*. With Applications to New Clean Energies and Fuels. 2001    ISBN 1-4020-0087-1
121.  S. Fujita and S. Godoy: *Theory of High Temperature Superconductivity*. 2001
    ISBN 1-4020-0149-5
122.  R. Luzzi, A.R. Vasconcellos and J. Galvão Ramos: *Predictive Statitical Mechanics*. A Nonequilibrium Ensemble Formalism. 2002    ISBN 1-4020-0482-6
123.  V.V. Kulish: *Hierarchical Methods*. Hierarchy and Hierarchical Asymptotic Methods in Electrodynamics, Volume 1. 2002    ISBN 1-4020-0757-4; Set: 1-4020-0758-2
124.  B.C. Eu: *Generalized Thermodynamics*. Thermodynamics of Irreversible Processes and Generalized Hydrodynamics. 2002    ISBN 1-4020-0788-4
125.  A. Mourachkine: *High-Temperature Superconductivity in Cuprates*. The Nonlinear Mechanism and Tunneling Measurements. 2002    ISBN 1-4020-0810-4
126.  R.L. Amoroso, G. Hunter, M. Kafatos and J.-P. Vigier (eds.): *Gravitation and Cosmology: From the Hubble Radius to the Planck Scale*. Proceedings of a Symposium in Honour of the 80th Birthday of Jean-Pierre Vigier. 2002    ISBN 1-4020-0885-6
127.  W.M. de Muynck: *Foundations of Quantum Mechanics, an Empiricist Approach*. 2002
    ISBN 1-4020-0932-1
128.  V.V. Kulish: *Hierarchical Methods*. Undulative Electrodynamical Systems, Volume 2. 2002
    ISBN 1-4020-0968-2; Set: 1-4020-0758-2
129.  M. Mugur-Schächter and A. van der Merwe (eds.): *Quantum Mechanics, Mathematics, Cognition and Action*. Proposals for a Formalized Epistemology. 2002    ISBN 1-4020-1120-2
130.  P. Bandyopadhyay: *Geometry, Topology and Quantum Field Theory*. 2003
    ISBN 1-4020-1414-7
131.  V. Garzó and A. Santos: *Kinetic Theory of Gases in Shear Flows*. Nonlinear Transport. 2003
    ISBN 1-4020-1436-8
132.  R. Miron: *The Geometry of Higher-Order Hamilton Spaces*. Applications to Hamiltonian Mechanics. 2003    ISBN 1-4020-1574-7
133.  S. Esposito, E. Majorana Jr., A. van der Merwe and E. Recami (eds.): *Ettore Majorana: Notes on Theoretical Physics*. 2003    ISBN 1-4020-1649-2
134.  J. Hamhalter. *Quantum Measure Theory*. 2003    ISBN 1-4020-1714-6
135.  G. Rizzi and M.L. Ruggiero: *Relativity in Rotating Frames*. Relativistic Physics in Rotating Reference Frames. 2004    ISBN 1-4020-1805-3
136.  L. Kantorovich: *Quantum Theory of the Solid State: an Introduction*. 2004
    ISBN 1-4020-1821-5
137.  A. Ghatak and S. Lokanathan: *Quantum Mechanics: Theory and Applications*. 2004
    ISBN 1-4020-1850-9
138.  A. Khrennikov: *Information Dynamics in Cognitive, Psychological, Social, and Anomalous Phenomena*. 2004    ISBN 1-4020-1868-1
139.  V. Faraoni: *Cosmology in Scalar-Tensor Gravity*. 2004    ISBN 1-4020-1988-2
140.  P.P. Teodorescu and N.-A. P. Nicorovici: *Applications of the Theory of Groups in Mechanics and Physics*. 2004    ISBN 1-4020-2046-5
141.  G. Munteanu: *Complex Spaces in Finsler, Lagrange and Hamilton Geometries*. 2004
    ISBN 1-4020-2205-0

# Fundamental Theories of Physics

142. G.N. Afanasiev: *Vavilov-Cherenkov and Synchrotron Radiation*. Foundations and Applications. 2004  ISBN 1-4020-2410-X
143. L. Munteanu and S. Donescu: *Introduction to Soliton Theory: Applications to Mechanics*. 2004  ISBN 1-4020-2576-9
144. M.Yu. Khlopov and S.G. Rubin: *Cosmological Pattern of Microphysics in the Inflationary Universe*. 2004  ISBN 1-4020-2649-8
145. J. Vanderlinde: *Classical Electromagnetic Theory*. 2004  ISBN 1-4020-2699-4
146. V. Čápek and D.P. Sheehan: *Challenges to the Second Law of Thermodynamics*. Theory and Experiment. 2005  ISBN 1-4020-3015-0
147. B.G. Sidharth: *The Universe of Fluctuations*. The Architecture of Spacetime and the Universe. 2005  ISBN 1-4020-3785-6
148. R.W. Carroll: *Fluctuations, Information, Gravity and the Quantum Potential*. 2005  ISBN 1-4020-4003-2
149. B.G. Sidharth: *A Century of Ideas*. Personal Perspectives from a Selection of the Greatest Minds of the Twentieth Century. Planned 2006.  ISBN 1-4020-4359-7
150. S.H. Dong: *Factorization Method in Quantum Mechanics*. Planned 2007.  ISBN to be announced
151. R.M. Santilli: *Isodual Theory of Antimatter with applications to Antigravity, Grand Unification and Cosmology*. 2006  ISBN 1-4020-4517-4
152. A. Plotnitsky: *Reading Bohr: Physics and Philosophy*. 2006  ISBN 1-4020-5253-7
153. V. Petkov: *Relativity and the Dimensionality of the World*. Planned 2007.  ISBN to be announced
154. H.O. Cordes: *Precisely Predictable Dirac Observables*. 2006  ISBN 1-4020-5168-9
155. C.F. von Weizsäcker: *The Structure of Physics*. Edited, revised and enlarged by Thomas Görnitz and Holger Lyre. 2006  ISBN 1-4020-5234-0